Fundamentals of Mechanics of Robotic Manipulation

International Series on

MICROPROCESSOR-BASED AND INTELLIGENT SYSTEMS ENGINEERING

VOLUME 27

Fundamentals of Mechanics of Robotic Manipulation

by

MARCO CECCARELLI
University of Cassino,
Cassino (FR), Italy

SPRINGER-SCIENCE+BUSINESS MEDIA, B.V.

A C.I.P. Catalogue record for this book is available from the Library of Congress.

DOI 10.1007/978-1-4020-2110-7

Printed on acid-free paper

Table of contents

Preface

This book has evolved from a course on Mechanics of Robots that the author has thought for over a dozen years at the University of Cassino at Cassino, Italy.
It is addressed mainly to graduate students in mechanical engineering although the course has also attracted students in electrical engineering. The purpose of the book consists of presenting robots and robotized systems in such a way that they can be used and designed for industrial and innovative non-industrial applications with no great efforts.
The content of the book has been kept at a fairly practical level with the aim to teach how to model, simulate, and operate robotic mechanical systems. The chapters have been written and organized in a way that they can be red even separately, so that they can be used separately for different courses and readers. However, many advanced concepts are briefly explained and their use is empathized with illustrative examples. Therefore, the book is directed not only to students but also to robot users both from practical and theoretical viewpoints. In fact, topics that are treated in the book have been selected as of current interest in the field of Robotics. Some of the material presented is based upon the author's own research in the field since the late 1980's.
In Chapter 1 an introductory overview of robots and Robotics is given by presenting basic characteristics and motivation for robot uses and by outlining a historical development. Chapter 1 outlines briefly the history and development that led to robots and robotized systems. The arguments are presented as a motivation for the interest on the applications of robots as well as for an in-depth study of the theoretical aspects. The technical evolution of automatic systems to robots is outlined by finally presenting the technical characteristics of a robot as a special automatic component. Indeed robots can be considered the most advanced automatic systems. With the aim to justify extensive use of robots an economic evaluation of using robots is presented and discussed. In addition, many other problems and aspects are outlined with the aim to give a wide view of the concerns related to the use of the robots not only in an industrial plant, but also in the human society. Thus, arguments from economy, psychology, and sociology are listed to give an account of all the problems that should be considered when using robots. However, the book concerns with technical aspects that are still addressing great attention even from research viewpoint. Thus, at the end of the chapter there is a list of forum where the technical aspects of Robotics are usually discussed. This information can be considered important even for further study of readers on the topics of this book and related arguments.
Chapter 2 deals with analysis and synthesis of manipulation by using men, automatic systems, or robots. The problem is attached in a way that any of the above-mentioned systems can apply the results of the study of a manipulation to give a rational flexible plan for a manipulative task. In particular, the aim of a flexible programming and design of a manipulation sequence is achieved by decomposing a manipulation into operations, phases, and elementary actions that have a decreasing content of

manipulative actions. A general procedure is outlined for analyzing manipulations as based on the above-mentioned decomposition. The procedure is formulated with no specific reference to robots, although the rest of the chapter describes its application with robots. Thus, the programming issue is discussed by referring to industrial robots and particularly to VAL-II language, which is used to describe peculiarities of grammar and syntax of a robot language. Several examples are illustrated and discussed by using teaching experiences and practical applications of industrial robots. The aim of describing many examples consists of presenting how a theoretical study of manipulation can be of great help for practical applications, even for a flexible programming of robotized solutions.

Chapter 3 is devoted to explain the behavior and operation of robots from Mechanics viewpoint with the aim to deduce formulations that are useful for analysis and design purposes. The arguments are attached in the frame of Mechanics of Robots, but the treatment is synthetic with the aim to deduce useful formulation even for simulation purpose on PC. Kinematics is approached by defining a model through HD (Hartenberg -Denavit) parameters. The position analysis is formulated by using Transformation Matrix (in homogenous coordinates), which is described from several viewpoints. The joint space and actuator spaces are illustrated and a simple example shows numerical details. A specific attention is addressed to workspace analysis, which is formulated in several algorithms. The position analysis is inverted to define the manipulator design problem. General formulations are outlined by using techniques deduced from the synthesis of mechanisms, but also an optimization problem is formulated and discussed. The path planning is approached starting from formulation of Inverse Kinematics. The two subjects are described as strongly related to each other, although some specific formulations are deduced for each of them. The singularity problem is illustrated by using the Jacobian analysis. The Jacobian matrix of a manipulator is defined and algorithms are outlined for its evaluation. Addressing to velocity and acceleration analysis with recursive algorithms completes the Kinematics, which makes use of the Transformation Matrix. The static behavior of a manipulator is studied through a suitable model, which is described with specific attention and comments. Then, the static equilibrium is formulated by using the traditional approach with D'Alembert Principle. An example is included to show closed-form expressions as well as to discuss numerical issues. The static model is further completed with inertia characteristics in order to study the Dynamics of a manipulator. The two fundamental approaches of Newton-Euler Equations and Lagrange Formulation are treated to give the equations of motion and constraints forces. Examples are used to illustrate and discuss details of the algorithms and numerical problems of simulation. Specific sections are devoted to discuss other main characteristics of a manipulator: repeatability, precision, and stiffness. Formulation and comments are deduced with the aim to help the understanding of their effect on the efficiency of a robot. The end of the chapter is devoted to introduce the Mechanics of parallel manipulators, mainly referring to a prototype CaPaMan (Cassino Parallel Manipulator) that has been designed and built at Laboratory of Robotics and Mechatronics in Cassino. Synthetically the fundamentals of

the Mechanics of parallel manipulators is outlined similarly to the previous sections for serial manipulators with the aim to show differences but analogies in the formulation and behavior.
Chapter 4 deals with problems of grasping as an important aspect for a manipulator component and manipulative task. Gripping devices are reviewed together with their fundamental characteristics in order to give a view of existent variety. However, the attention is focused on two-finger grippers, which are justified and motivated by considering the model of grasp and relevant Statistics for two-finger grasp applications. Thus, the two-finger grasp is studied in depth by means of models and formulations, which are also useful for design algorithms and force control purposes. The design issue is treated by looking at the mechanisms that are used in grippers and formulating the design problem, even as an optimization problem when suitable performance indices are defined. The grasp force control is described by referring to the Mechatronics operation and design of a gripper. In particular, different control schemes are described by using electropneumatic systems, since they are the most used in industrial applications. Some laboratory experiences are described and discussed as example of designing, operating, practicing with a gripping system.
The Index consists of an alphabetically ordered list of subjects treated in the book with the indication of the corresponding pages.
Bibliography consists of a list of main books in the field of Robotics, and specifically on Mechanics of Robots with the aim to give further sources of reading and references of different approaches. In the book there is not any reference to Bibliography. Bibliography has been limited to few significant books, although the literature on Robotics and mechanical aspects of Robots is very rich. However, main sources have been indicated in Chapter 1 as publications in Journals and Conference Proceedings in which readers can also find new material.
The author would like to acknowledge the collaboration with his former professor Adalberto Vinciguerra at University "La Sapienza" of Rome. The author is thankful also to his pupils and now colleagues Erika Ottaviano, Chiara Lanni, Giuseppe Carbone, and to professor Giorgio Figliolini for the collaboration at LARM: Laboratory of Robotics and Mechatronics. The author will thank also the former students who taught to the author how to teach and transmit his theoretical and practical experience on robotic systems to them within the limited period of a semester. Also acknowledged is the professional assistance by the staff of Kluwer Academic Publishers, Dordrecht (The Netherlands), especially by Miss Nathalie Jacobs.
The author is grateful to his wife Brunella, daughters Elisa and Sofia, and son Raffaele for their patience and comprehension over the many years during which this book was developed and written.

Marco Ceccarelli
Cassino, Italy
December 2003

Chapter 1:
Introduction to Automation and Robotics

1.1 Automatic systems and robots

Robots can be considered as the most advanced automatic systems and robotics, as a technique and scientific discipline, can be considered as the evolution of automation with interdisciplinary integration with other technological fields.
An automatic system can be defined as a system which is able to repeat specific operations generally with a low degree of intellectual and manipulative levels, but that can be easily programmed in agreement with demands of productivity.
It is worthy of note that an automatic system is generally able to perform one operation for which its mechanical structure has been designed. Aspects of flexibility depend on the possibility of reprogramming the control unit, which is generally able only to modify the time sequence of the designed operations. Therefore, an automatic system of industrial type can be thought of as composed of two parts:
- hardware with mechanical, electrical, pneumatic, and hydraulic components that provide the mechanical capability to perform an *a priori*-determined operation of movement and/or manipulation;
- control and operation counterpart with electronic components and software that provide the capability of autonomy and flexibility to the working of the system.
The two parts are essential in an automatic system and are integrated in the sense that their design and operation must be considered as a unique goal in order to obtain and operate an automatic system with the best performances. Therefore, an automatic system in which the hardware part is preponderant cannot have a suitable flexibility for a flexible production in agreement with the demands of productivity and market. In some cases these limited solutions are still required from productivity , when the product of a massive production is absorbed from the market with certain regularity during a period of time longer than the amortization time of the automatic system. Those systems with low flexibility are generally denominated as 'rigid automatic systems'. In Fig. 1.1 the variety of production systems is represented as a function of the productivity level.
Automatic systems are often designed with the aim of having the possibility of an easy updating of their structure and operation with the purpose of being adjusted more quickly to the mutable demands of production and design of products. These very flexible systems are denominated as 'flexible automatic systems'.
In addition, it is worthy of note that a system with only the mechanical counterpart, although versatile, cannot be considered an automatic system since the updating of its operation is not obtained by means of control units. Such a system can be properly reprogrammed, but it requires the manual action of one or more human operators to change some components of the machinery or their running. Therefore, these systems can be properly named as machinery or mechanical systems.

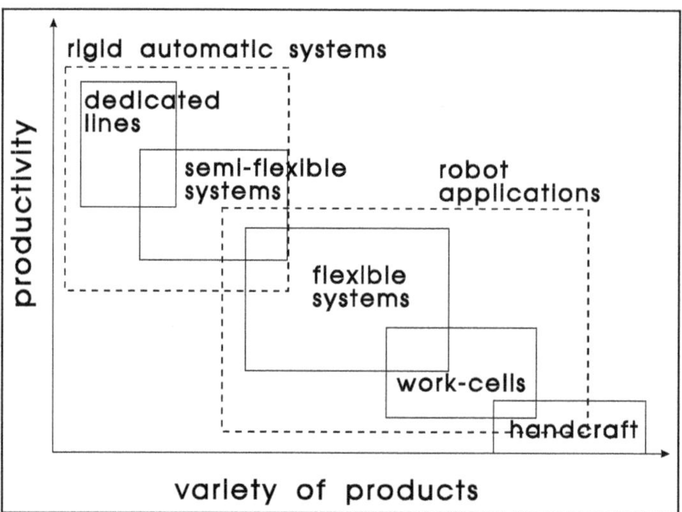

Fig. 1.1: Variety of automatic and robotized systems as functions of productivity level and product demand.

Similarly, control systems cannot be considered as automatic systems, since they are not able to perform mechanical tasks, although they can be provided with a high level of flexibility in terms of re-programmability and memory capability. Therefore these systems can be properly named as control units or electrical–electronic–informatics systems.

However, an automatic system is designed and built by using a suitable combination and integration of mechanical systems and control units. Indeed, the success and engineering application of automatic systems strongly depends on the practical integration of the above-mentioned counterparts.

A robot can be defined as a system which is able to perform several manipulative tasks with objects, tools, and even its extremity (end-effector) with the capability of being re-programmed for several types of operations. There is an integration of mechanical and control counterparts, but it even includes additional equipment and components, concerned with sensorial capabilities and artificial intelligence. Therefore, the simultaneous operation and design integration of all the above-mentioned systems will provide a robotic system, as illustrated in Fig. 1.2.

In fact, more than in automatic systems, robots can be characterized as having simultaneously mechanical and re-programming capabilities. The mechanical capability is concerned with versatile characteristics in manipulative tasks due to the mechanical counterparts, and re-programming capabilities concerned with flexible characteristics in control abilities due to the electric-electronics-informatics counterparts. Therefore, a robot can be considered as a complex system that is composed of several systems and devices to give:

- mechanical capabilities (motion and force);
- sensorial capabilities (similar to human beings and/or specific others);
- intellectual capabilities (for control, decision, and memory).

Initially, industrial robots were developed in order to facilitate industrial processes by substituting human operators in dangerous and repetitive operations, and in unhealthy environments. Today, additional needs motivate further use of robots, even from pure technical viewpoints, such as productivity increase and product quality improvements. Thus, the first robots have been evolved to complex systems with additional capabilities.

Nevertheless, referring to Fig. 1.2, an industrial robot can be thought of as composed of:

- a mechanical system or manipulator arm (mechanical structure), whose purpose consists of performing manipulative operation and/or interactions with the environment;
- sensorial equipment (internal and external sensors) that is inside or outside the mechanical system, and whose aim is to obtain information on the robot state and scenario, which is in the robot area;
- a control unit (controller), which provides elaboration of the information from the sensorial equipment for the regulation of the overall systems and gives the actuation signals for the robot operation and execution of desired tasks;
- a power unit, which provides the required energy for the system and its suitable transformation in nature and magnitude as required for the robot components;
- computer facilities, which are required to enlarge the computation capability of the control unit and even to provide the capability of artificial intelligence.

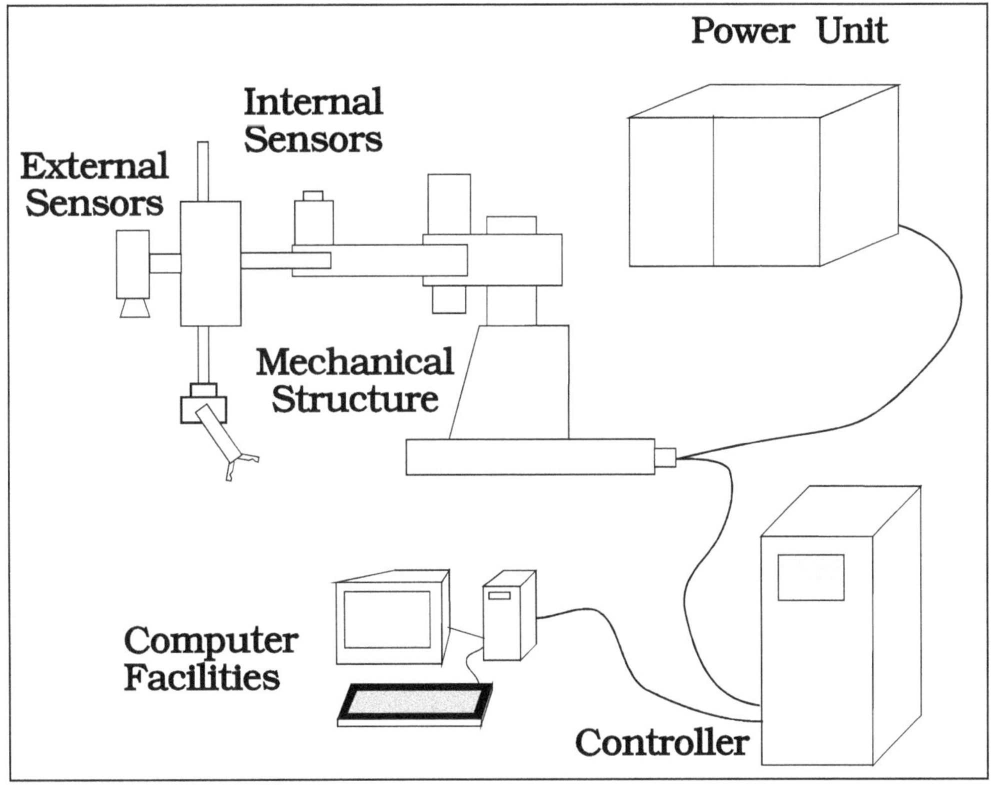

Fig. 1.2: Components of an industrial robot.

Thus, the above-mentioned combination of sub-systems gives the three fundamental simultaneous attitudes to a robot, i.e. mechanical action, data elaboration, and re-programmability.
Consequently, the fundamental capability of robotic systems can be recognized in:
- mechanical versatility;
- re-programmability.

Mechanical versatility of a robot can be understood as the capability to perform a variety of tasks because of the kinematic and mechanical design of its manipulator arm.
Re-programmability of a robot can be understood as the flexibility to perform a variety of task operations because of the capability of its controller and computer facilities.
These basic performances give a relevant flexibility for the execution of several different tasks in a similar or better way than human arms. In fact, nowadays robots are well-established equipment in industrial automation since they substitute human operators in operations and situations, which are:
- dangerous – for manipulative tasks and/or unhealthy environments;
- repetitive – with low-level cultural and technical content;
- tiresome – for manipulative tasks requiring energy greater than that provided by a human operator;
- difficult – for human operators.

In addition, the use of robots is well-established in industrial production since robots are:
- versatile, as they have the ability to operate in different situations;
- useful for unhealthy or limited environments, which can be dangerous or unfeasible for human operators.

Besides the above-mentioned motivations of technical nature, robots and robotic systems are currently used with the aim of:
- decreasing production costs, which can be related to a better use of the machinery;
- increasing productivity, with an increase of the operation velocity;
-enhancing product quality, in terms of constant characteristics and improved manufacturing;
- achieving flexible production as a capability of rapid adjustment of production to required changes in manufacturing.

The mechanical capability of a robot is due to the mechanical sub-system that generally is identified and denominated as the 'manipulator', since its aim is the manipulative task.
The term manipulation refers to several operations, which include:
- grasping and releasing of objects;
- interaction with the environment and/or with objects not related with the robot;
- movement and transportation of objects and/or robot extremity.

Consequently, the mechanical sub-system gives mechanical versatility to a robot through kinematic and dynamic capability during its operation. Manipulators can be classified according to the kinematic chain of their architectures as:
- serial manipulators, when they can be modeled as open kinematic chains in which the links are jointed successively by binary joints;

- parallel manipulators, when they can be modeled as closed kinematic chains in which the links are jointed to each other so that polygonal loops can be determined.

In addition, the kinematic chains of manipulators can be planar or spatial depending on which space they operate. Most industrial robotic manipulators are of the serial type, although recently parallel manipulators have aroused great interest and are even applied in industrial applications.

In general, in order to perform similar manipulative tasks as human operators, a manipulator is composed of the following mechanical sub-systems:

- an arm, which is devoted to performing large movements, mainly as translations;
- a wrist, whose aim is to orientate the extremity;
- an end-effector, which is the manipulator extremity that interacts with the environment.

Several different architectures have been designed for each of the above-mentioned manipulator sub-systems as a function of required specific capabilities and characteristics of specific mechanical designs. It is worthy of note that although the mechanical design of a manipulator is based on common mechanical components, such as all kinds of transmissions, the peculiarity of a robot design and operation requires advanced design of those components in terms of materials, dimensions, and designs because of the need for extreme lightness, compactness, and reliability.

The sensing capability of a robot is obtained by using sensors suitable for knowing the status of the robot itself and surrounding environment. The sensors for robot status are of fundamental importance since they allow the regulation of the operation of the manipulator. Therefore, they are usually installed on the manipulator itself with the aim of monitoring basic characteristics of manipulations, such as position, velocity, and force. Additionally, an industrial robot can be equipped with specific and/or advanced sensors, which give human-like or better sensing capability. Therefore, a great variety of sensors can be used, to which the reader is suggested to refer to in specific literature.

The control unit is of fundamental importance since it gives capability for autonomous and intelligent operation to the robot and it performs the following aims:

- regulation of the manipulator motion as a function of current and desired values of main kinematic and dynamic variables by means of suitable computations and programming;
- acquisition and elaboration of sensor signals from the manipulator and surrounding environment;
- capability of computation and memory, which is needed for the above-mentioned purposes and robot re-programmability.

In particular, an intelligence capability has been added to some robotic systems concerned mainly with decision capability and memory of past experiences by using the means and techniques of expert systems and artificial intelligence. Nevertheless, most of the current industrial robots have no intelligent capability since the control unit properly operates for the given tasks within industrial environments. The control systems that are used in robots can be classified as:

- electro-mechanical sequencer units with end-stroke stops; units with pneumatic logic; electronic units with logic; PLC (Programmable Logic Controller) units;
- micro-processors;

- minicomputers;
- computers.

Indeed, nowadays industrial robots are usually equipped with minicomputers, since the evolution of low-cost PCs has determined the wide use of PCs in robotics so that sequencers, which are going to be restricted to PLC units only, will be used mainly in rigid automation or low-flexible systems.

Since robots can be considered complex systems both in design and operation even within work-cells, it is of fundamental importance and convenience to have the possibility to simulate robot operation and even the overall robotic systems. Using numerical methods and/or experimental tests even with laboratory prototypes can simulate robots. Therefore, an expert in automation and robotics should be aware of and be able to work with techniques and methodologies which are useful in designing and running the above-mentioned simulations.

1.2 Evolution and applications of robots

It is well known that the word 'robot' was coined by Karel Capek in 1921 for a theater play dealing with cybernetic workers, who/which replace humans in heavy work.

Indeed, even today the word 'robot' has a wide meaning that includes any system that can operate autonomously for a given class of tasks. Sometimes intelligent capability is included as a fundamental property of a robot, as shown in many works of fiction and movies, although many current robots, mostly in industrial applications, are far from being intelligent machines.

From a technical viewpoint a unique definition of a robot has taken time to be universally accepted. In 1988 the International Standard Organization (ISO) states: "An industrial robot is an automatic, servo-controlled, freely programmable, multipurpose manipulator, with several axes, for the handling of work pieces, tools or special devices. Variably programmed operations make possible the execution of a multiplicity of tasks".

However, different definitions are still used. In 1991 the International Federation for the Promotion of Mechanisms and Machine Science (formerly the International Federation for the Theory of Machines and Mechanisms) (IFToMM)gives its own definitions. Robot is defined as "Mechanical system under automatic control that performs operations such as handling and locomotion"; and manipulator is defined as "Device for gripping and controlled movements of objects".

Even roboticists still use their own definition for robots to emphasize some peculiarities. For example in 2001 from the IEEE (the Institute of Electrical and Electronics Engineers) community one can find the statement: "a robot is a machine constructed as an assemblage of joined links so that they can be articulated into desired positions by a programmable controller and precision actuators to perform a variety of tasks".

Different meanings for robots are still persistent from nation to nation, from technical field to technical field, and from application to application.

Nevertheless, a robot or robotic system can be recognized when it has three main characteristics: mechanical versatility, re-programming capacity, and intelligent capability.

However, the word 'automaton' can still be conveniently used in many cases, when the semantic meaning is fully understood from the originating word. The word means "something having the power of spontaneous motion" and it comes from the Greek word 'automaton', which is composed of the terms 'auto' and 'matos' that have the meaning of 'self' and 'moving/acting', respectively.
Thus, the term 'robot' can be addressed to systems, whose meaning and characteristics are those reported in the above-mentioned modern definitions for robots. The automaton/automata term can still be used for systems that do not correspond fully to the current robot concept, but can still show autonomous operation capability.
Therefore, one should properly name as robots only those systems that have been designed in the last five decades. Nevertheless, those concepts can be recognized in many systems of the past and one can still use the word 'robot' in a broad sense when she/he refers to historical solutions.
The idea of a substitute/help for men in heavy/unpleasant work can be found since the beginning of humanity. Thus, tools and machines were conceived, built, and used as intermediate solutions with increasing performance over time. Substitutes for manpower were also considered in Antiquity. Slaves were the first efficient 'intelligent machinery' solutions! but artificial solutions were also considered, mainly from theoretical viewpoints.
In VIIIth century B.C., in the 18th book of The Iliad, Homer describes maidservants that are built by Vulcan for the service of the Gods: they are mobile by using wheels, are nicely human shaped, are able to speak and with some intelligence. Homer described several other intriguing machines provided with automatic operation and even artificial intelligence. They show that in Homeric times the idea of a robot (properly 'automaton') for doing practical work, even for manufacturing purposes, was not considered altogether impossible, but within human reach.
However, in Homer's poetry the robots are mainly used to astonish the reader. This aim, indeed, was quite common in the Greek Theater that made use regularly of automatons and automatic machines to some extent. Emblematic are those that are mentioned in the plays by Aeschylus, and later by Aristophan.
The theatric machines were basically large mechanisms consisting of booms, wheels, and ropes that could raise weights/persons and even in some cases move back and forth quickly to simulate space motions. Unfortunately, none of the Theater machines still exist since they were made of perishable materials. However, one can find several references to these theatric machines in the Greek plays and even in vase paintings.
The engineering of Theater machines was persistent in Antiquity in Greece and in the Roman Empire. The Greeks reached heights in knowledge even in technical fields. The emblematic example is the School of Alexandria in Egypt where since the IIIrd century B.C. there was intense activity in teaching and also research on automatic devices. A brilliant example is the work by Heron of Alexandria who, in particular, treated in depth pneumatics and automatic mechanisms for several applications. The *Treatise on Pneumatics* by Heron was a masterpiece for designing automatic machines not only in Antiquity, but it was rediscovered and also used during the Renaissance.
Greek culture evolved and circulated as combined with Roman technology. The

Romans also further developed a technical culture in other fields, such as civil structures (roads, buildings, bridges, etc.) and in military applications (war machines, defense structures, etc.). They conceived and built several machines that could help humans in the work, although they made extensive use of slaves as perfect robots, since sometimes slaves were used as machinery without any moral attention. However, Roman engineers enhanced the mechanical design and automation.
A brilliant example is Vitruvius, who lived in 1st century B. C. and wrote an encyclopedic treatise 'De Architectura' that was rediscovered and used during the Renaissance. In the liber X Vitruvius treats in detail mechanisms and mechanical design of machines.
But technical evolution in machine design was important not only in the western world. For example, in China the culture reached heights that also needed mature technology. In the field of automation Chinese designers developed brilliant solutions.
A very significant example is the Wooden Cow that was used for transportation purposes of heavy loads by using a one degree of freedom (d.o.f.) walking machine, as shown in Fig. 1.3.
Another very significant example can be selected from the Arabic world which achieved great technical designs during the Middle Ages.
An example is reported in Fig. 1.4, where an intriguing mechanisms is shown to operate as an automaton that provides water for washing hands and then a towel for drying hands, and finally it goes back to a start configuration. The mechanical design of the

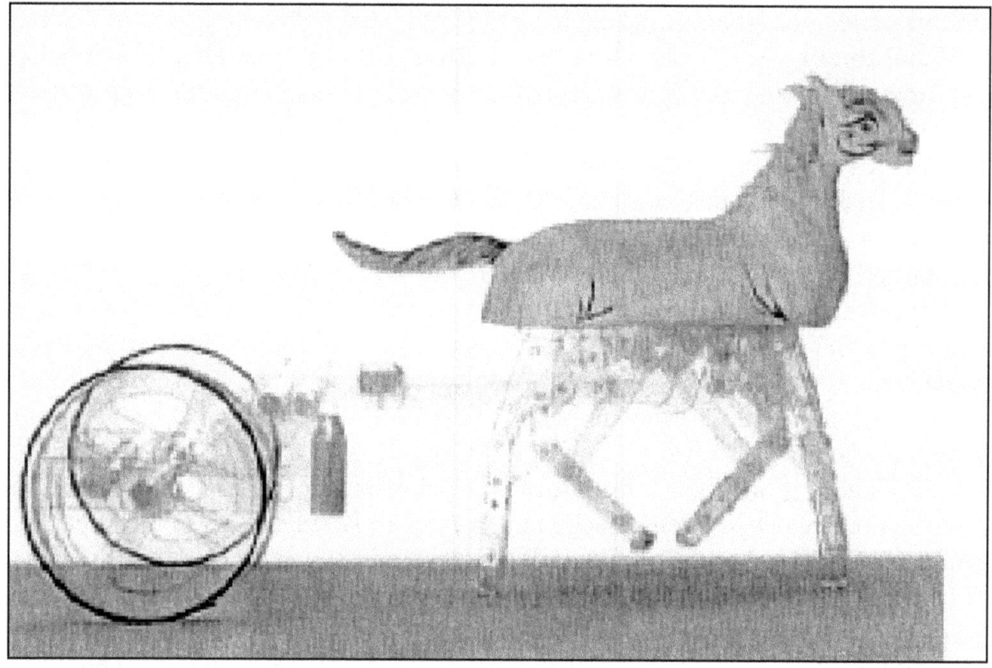

Fig. 1.3: A modern reconstruction of the Chinese Wooden Cow that was designed and built in Vth century B.C. (from the Ancient Chinese Machines Foundation at National Cheng Kung University in Tainan).

automaton is based on a great knowledge of mechanics and hydraulics, that was only partially a heritage from the Greek culture, but it was evolved in practical applications to consciously interact with the environment.
Since the XIIIth century early automata that can recall robotic systems and indeed promoted evolution in modern robots, can be recognized in early systems related to clocks, theater machines, looms, automatic toys, and mechanical calculators.
Since the early Renaissance automation was attempted consciously and obtained relevant results with practical applications in work activity mainly to increase production and/or product quality.
Mechanisms, mechanical design, and automation evolved considerably during the Renaissance because of the increased needs of technical means. At the time of the Renaissance there was the establishment of a mature technical culture whose expression is the work of many personalities, Leonardo da Vinci being the most famous, because of his great encyclopedic knowledge and skill.

Fig. 1.4: An Arabic design of an automaton by Al-Jazari in the XIIth century.

A specific important field of automation that gave considerable stimulus towards robotic systems can be considered as textile manufacturing. In particular, looms evolved from manually operated systems at the beginning of the XVIIIth century to fully automatic machines at the end of the XIXth century. The mechanisms used in the looms were improved with suitable designs and complex architectures in order to obtain several kinds of manufacturing and high-speed productions. At the same time they were made as automatic devices by using steam power and later, electric motors. Programming was achieved by using suitable mechanisms that were able to perform the work that was coded in suitable cards. Other suitable mechanical devices regulated the operation speed.
Much of the robot evolution can be considered as based on computer evolution. But the computer is not a conception of the XXth century, only its electronic construction. In fact, calculators were built in the past by using mechanical devices to perform several kinds of computations. The most famous are the computing machines by Des Cartes, Pascal, and Leibniz. Some computing machines were very successful and were produced for the market, mainly in the XIXth century.
In the 1820s Charles Babbage designed a computing machine based on systems of gears that addressed great attention. He obtained considerable funds to build the final design of a 'universal calculator'. But Babbage never completed the project, although his early prototype and even today partial reconstruction have proved the efficiency of the systems of the Babbage design. These calculator machines can also be considered a result of the improvements in automation, and specifically in the design of complex mechanisms.
In addition, early robots can be considered as the many automatic toys that were developed for the pleasure/astonishment of aristocratic people since the Renaissance. They reached high levels of automation and autonomous operation so that they were considered as automata during the XVIIIth century. An emblematic example is the writer automaton shown in Fig. 1.5, where one can recognize the robot characteristics even if the system is driven by mechanisms only. Many of these automata could be easily re-programmed by changing only some mechanical parts or by using punched cards.
Figure 1.6 shows a Japanese automaton, which is similar to that of Fig. 1.5. The tea maidservant is a design of the XVIIth century and is capable of moving and offers a cup of tea. Its design is based on suitable mechanisms, partially shown in Fig.1.6b), that have been developed in several versions until the end of the XVIIIth century. Like the automaton of Fig. 1.5, the Japanese automaton of Fig. 1.6 was used mainly for the amusement of the aristocracy . However it gives a clear example of the level of expertise of Japanese culture in the technical field of automata.
The advent of the Industrial Revolution increasingly promoted the design and construction of automatic devices. Thus, at the beginning of the XIXth century automatic devices were regularly designed and used. Such a high level of expertise and knowledge was achieved for the mechanical design of automatic devices in the second half of the XIXth century.

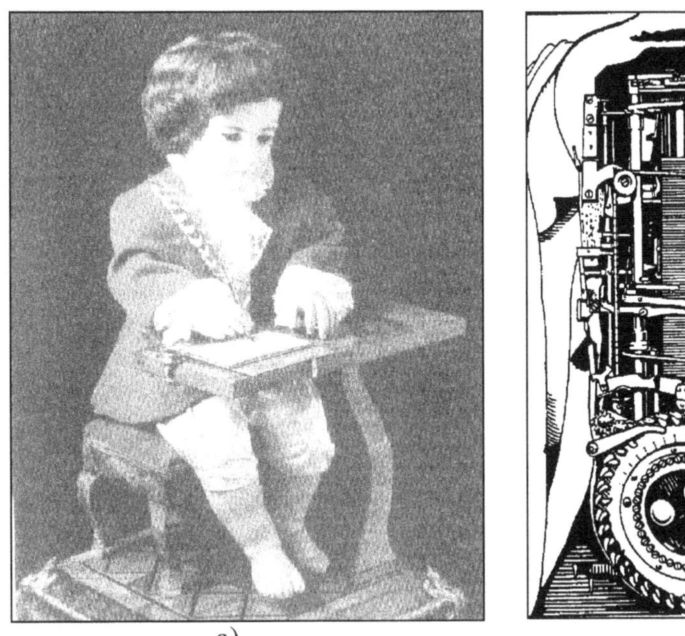

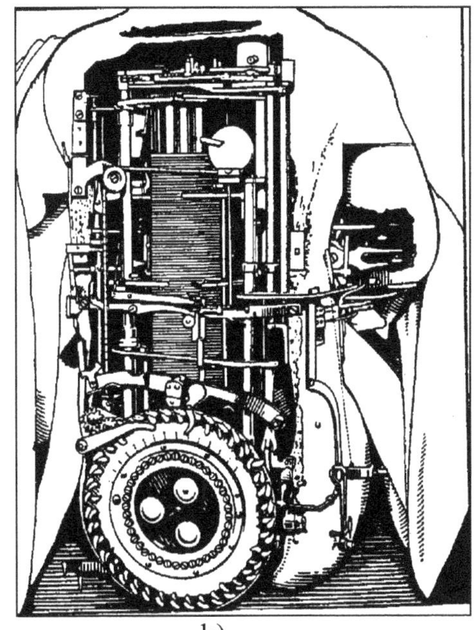

a) b)

Fig. 1.5: The writer automata by Jaquet-Droz in 1760: a) the robot shape; b) the mechanisms.

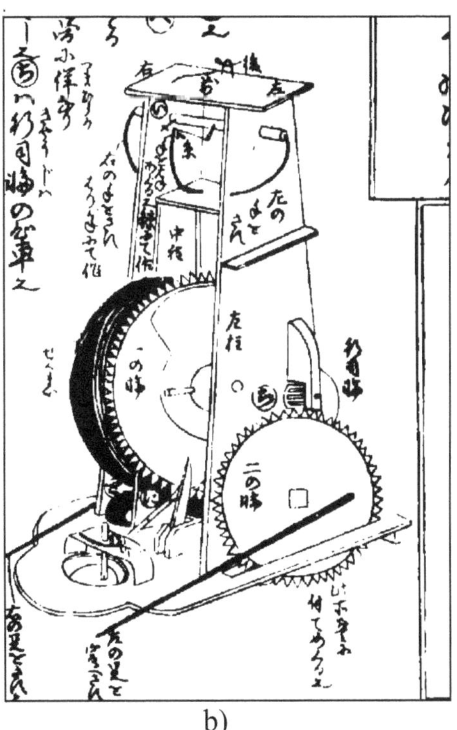

a) b)

Fig. 1.6: The Japanese tea maidservant automaton whose design was developed in the XVIIth century: a) a modern reconstruction; b) an original design scheme.

Still in the XXth century many systems are based on designs or even constructions of mechanisms conceived at that time. These automatic machines also show how a complex manipulative task can be performed successfully by mechanisms and not by robots.
A very particular field can be considered as the so-called war machines, as the field of military applications was named until the XVIIIth century. Since early Renaissance the most clever designers were employed for conception and construction of new and more powerful weapons, devices for attack and defense activities against castles and fortresses, and new and more resistant defensive civil structures. Many new systems were conceived, designed, built and in some cases even used, although many others were re-designed from ancient designs mainly from the rediscovered Roman literature. Now famous devices are the military tank, the helicopter, and even the first humanoid robot that were designed by Leonardo da Vinci. In addition many other engineers designed weapons and new mechanisms to improve existing military machinery, for example the feeding mechanisms for automatic guns. But very often this design and construction activity was not recorded since secrecy was an important requirement that did not allow documents or publications on the designed systems, although the benefit of acquired knowledge and expertise also somehow passed to the non-military applications, as still happens today.
The advent of electronics gave a further stimulus and jump in technology, mainly because it allows regulation and control of several motors contemporaneously. Indeed, this is probably the main technical motivation that has made possible the development of modern robots which today we mean for multi-degrees of freedom systems.
The following dates are fundamental and well known in the recent history of modern robotics:

- 1947: Raymond Goertz built and used a servo electric-powered teleoperator;
- 1954: George C. Devol obtained the first patent for a manipulator with memory;
- 1956: Joseph Engelberger and George C. Devol started the first robot builder company 'Unimation';
- 1968: Victor D. Scheinman designed the Stanford Arm;
- 1971: the Japan Industrial Robot Association JIRA is founded;
- 1975: the U.S. Robot Industries Association RIA is founded;
- 1975: robots are implemented in a large-scale production at the Ford plant in Detroit
- 1981: robots are implemented in a large-scale production at the FIAT plant in Cassino;
- 1984: the IEEE Journal of Robotics and Automation was established.

The 1981 event is a significant example to indicate how fast was the diffusion of robots in the industrialized world, when one also considers that the robot builder company COMAU of the FIAT group very quickly reached great prominence as robot builders in the world market.
Of course, the above-mentioned list of events is not exhaustive and is reported with the aim of emphasizing how fast the evolution process was in modern robotics. Many other significant events occurred and many other important enhancements were achieved.

One can remark that modern robotics started by copying and mimicking the human arm, but soon attention was addressed to other solutions that gained inspiration from the living world in nature and/or were conceived with artificial architectures to obtain the nature solutions.

Finally, at the end of the 1990s people began learning not only how to design robots, but mainly how to make them in order to operate efficiently in many industrial applications.

A general report on the recent status of robotics in the world is almost impossible, although some international projects help to have information, since the fragmentation of application fields is considerable and the secrecy of new or successful solutions is still a persistent behavior in the commercial world.

Aspects of high-speed evolution in modern robotics are illustrated by means of the examples of Figs 1.7 and Table 1.1.

Figure 1.7 together with Table 1.1 shows an illustrative example of how industrial robots evolved in 10 years: the shape of the arm changed from a very anthropomorphic design to a well-balanced architecture with better dynamic performances; the performances have been improved considerably mostly because of the components enhancement. Nevertheless, one can observe that architectures of industrial robots did not evolve as much as other robot components and characteristics. Indeed, great evolution took place and is still ongoing, mostly in the field of electronics. This gives great enhancements in robotic systems that sometimes are not fully exploited because of a lack of analogous improvements in the mechanical counterparts.

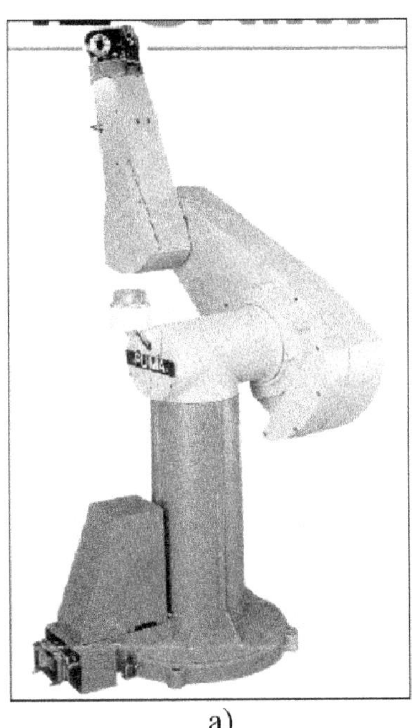

a)

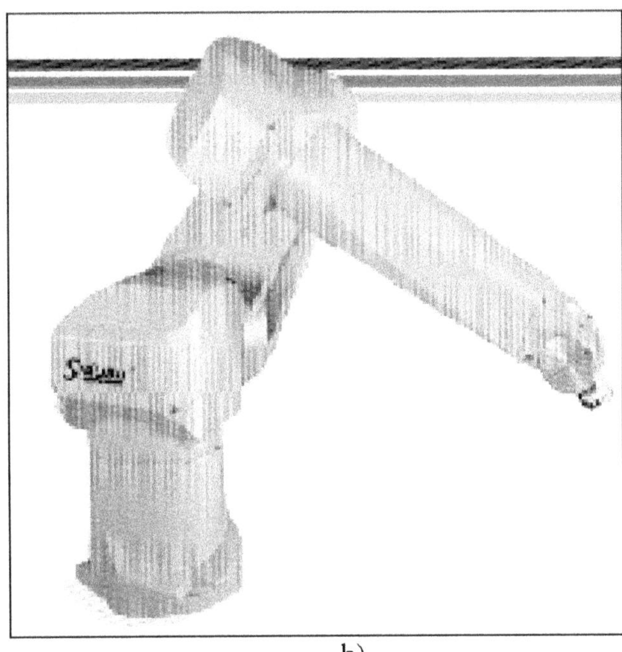

b)

Fig. 1.7: The design of the PUMA-type robot:
a) the PUMA562 at the beginning of the 1980s; b) the RX90L at the end of the 1990s.
(Photos are taken from sheets for commercial advertisement)

Table 1.1: Performance characteristics of PUMA-type robots in Fig.1.7. (Data are taken from sheets for commercial advertisement)

	PUMA562 (Fig. 1.7a)	RX90L (Fig. 1.7b)
d.o.f.s	6	6
Payload (N)	40	120
Maximum reach (mm)	878	985
Repeatability (mm)	± 0.1	± 0.02

Optimism in the future of robotics can also be justified by considerations based on the evolution that is outlined in the diagrams of Figs. 1.8 and 1.9. These diagrams have been traced by taking into account past evolution.
In particular, Fig. 1.8 indicates that the number of installed robots can suddenly increase when new areas for robot applications are discovered as suitable, although saturation can occur in other fields of application. In addition, Fig. 1.9 stresses the fact that the cost of robots will decrease more than the cost of automatic devices and even more than the cost of manual operation, so that the range of economic profit by using robots will enlarge the range of the fields of applications.
Modern robots for industrial applications have been involved from both historical and technical viewpoints according to the following so-called generations:
- first-generation robots with low flexibility and limited versatility for specific manipulative tasks in structured environments, as developed in the 1970s;
- second-generation robots with medium flexibility with a considerable re-programming capability, feedback control, and sensing, as developed in the 1980s;
- third-generation robots with high flexibility, with considerable sensorial capability, and some characteristics of Artificial Intelligence(AI) (decision capability and memory), even with light and miniaturized designs, as developed at the end of the 1980s;

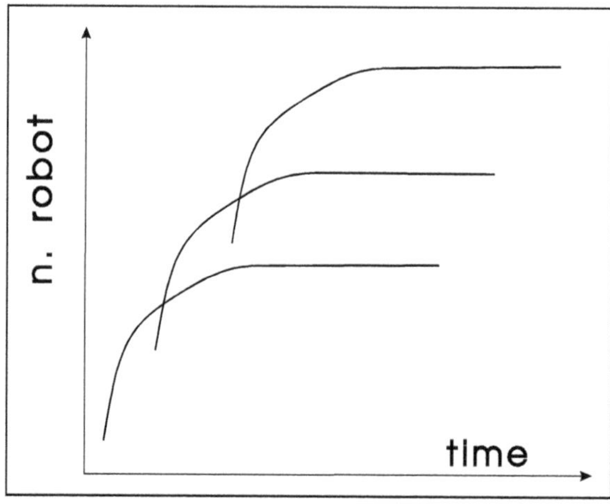

Fig. 1.8: Trend for new applications of robots.

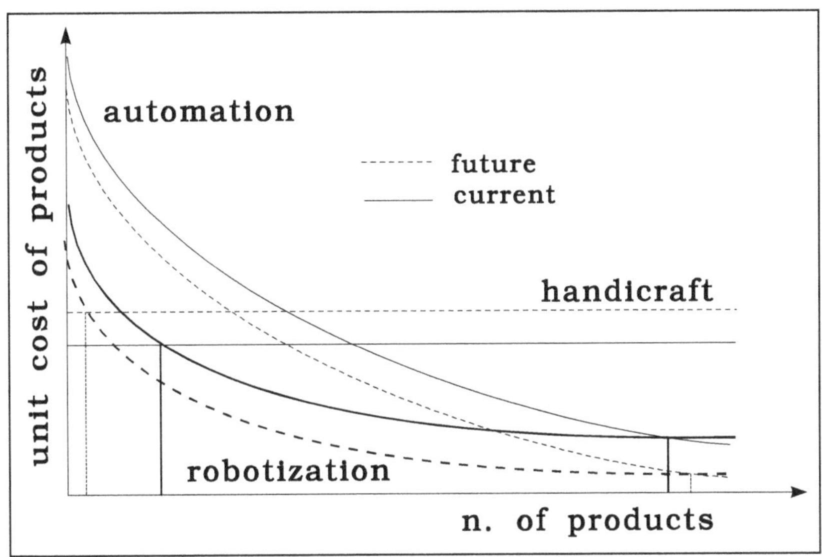

Fig. 1.9: An estimation of cost variation for production systems as function of product variety.

- fourth-generation robots with innovative mechanical architectures, considerable sensing and AI, vision systems, and advanced control systems for very accurate operation, as developed in the 1990s;
- fifth-generation robots in the form of integrated systems for manipulation and walking, with light and miniaturized design, advanced sensing and AI, as developed in laboratories since the beginning of the 1990s.

It is worthy of note that each robot generation was used and is still used in certain fields that can be considered as traditional for the application of these robots so that they are still built and operate successfully.

In the beginning modern robots were used to substitute human operators in dangerous actions and in unhealthy environments. But soon their peculiar characteristics of versatility and flexibility were exploited in industrial applications for repetitive operations with low-level technological and intellectual contents. Therefore, these are the areas of industrial activity in which a 'traditional' considerable use of robots and robotized systems has been widely accepted and is still in progress. Typical applications are the handling of dangerous objects, painting, welding, and assembly. In these processes great experience has been developed at such a high level that robots are often considered the only solution in those areas.

Consequently robots have also been used in others areas not only because of acceptance from designers and operators, and improvement need of productivity and product quality, but mainly because of a natural evolution of automation to robotics once the robots have been evolved and considered as industrial machinery. In addition, significant has been and is the need by the industrial management to show that production is obtained also by using the latest and most advanced equipment for financial and publicity purposes.

Robots are applied for technical reasons in non-industrial fields, such as:
- spatial and submarine explorations
- bio-mechanics

but industrial applications of robots are experienced in areas, such as:
- measure and control of production quality
- precise handling
- sealing and gluing
- loading and unloading of tools and machinery
- cutting

with an extension in all areas of industry.

Significant robot applications can be considered as those for servicing and maintenance or during the assembly and installation of industrial plants. Robots have been used in the control and maintenance of nuclear plants for energy production, and recently even in procedures for inactivated plants. In addition, a considerable number of robots are used in research centers and laboratories working in robotics but also in other fields where they use the robots only as assistant machinery.

The International Federation of Robotics (IFR) lists the number of robots as in Table 1.2 referring to the most industrialized countries by using published data.

However, it is worthy of note that small companies or users themselves obtain a significant production of robots. Therefore, the data of Table 1.3 for the largest robot builders are listed only to give the size of robot builders and robot production.

Table 1.2: Existing robots in the world (data from IFR publications).

	USA	Germany	Italy	France	UK	Japan	World
Robots in use in 1995	66,000	51,000	23,000	13,000	8,000	387,000	650,000
New robots in 1999	15,063	10,548	5,224	3,092	1,392	35,609	81,508

Great emphasis and attention is already addressed to innovative solutions for future applications, mostly at universities and research institutions.

Innovation is expected not only in terms of new robotic systems but also related to low-cost and easy operated systems that can be dedicated to friendly use of robots.

In particular, regarding short-term evolution, new robotic systems will be developed for:
- service robots
- space and submarine exploration and work
- humanoid robots
- robots for medical applications
- robots for agriculture

without yet mentioning military applications.

Table 1.3: Production of main robot constructors (data from IFR publications).

Robot constructor	Nationality	New robots in 1990	In service in 1999
YASKAWA	Japan	4,400	60,000
GMF	Japan–USA	3,300	70,000
ABB	Switzerland–Sweden	3,000	90,000
HITACH	Japan	2,600	70,500
KAWASAKI	Japan	2,000	30,000
NACHI	Japan	1,700	37,000
KUKA	Germany	800	30,000
COMAU	Italy	500	9,500
REANULT Automation	France	420	7,500
STAUBLIUNIMATION	Switzerland–France	400	8,000
CLOOS	Germany	400	8,000

Service robots will be increasingly developed with specific designs but mostly for friendly use and co-operation with humans, who may not yet be trained in using robots. This means that great efforts should be spent to make the robot acceptable as a friend or company in many service applications, by evolving in terms of shape and esthetic appearance. In Japan they are already working on letting a robot express some feeling by means of body posture and/or facial expressions, and this they call the field with the multi-meaning name 'Kansei'.

Dangerous applications, such as space and submarine exploration, even work and colonization, will require more and more robots, but with a robust design and very advanced features to survive in a hard environment for a long duration with a very reliable operation capability.

Specific attention has been addressed recently to humanoid robots and great results are expected to achieve the reality of the utopia of having a robot that can substitute a man in a work activity without any supervision. This seems to be a real need for the future when low-level jobs will be performed only by machinery since humans will not accept such kind of jobs, because they are not able to give intellectual satisfaction to educated people.

The use of robots in the medical field is a typical example of new applications of robots with great potentiality that only now is being accepted. Many important achievements are expected mostly to help the surgery, assistance, rehabilitation and other activities with patients. The hard task, besides the complex technical content, consists of stimulating a successful collaboration of two worlds that are antithetic: medicine and (robot) engineering. Indeed, many other collaborations that are not possible today will be asked of roboticists.

Innovative solutions are expected mainly for spreading robots more and more by simplifying the execution and making robotic systems cheaper . This trend is already a current issue for robotics but in the future it could become critical for the success and

even survival of robots.
Thus, a challenging problem can be considered of how to obtain sophisticated, advanced solutions that are required by technical needs but with no great operational complexity.
Another very important challenge can be considered the use of robots in agriculture. In the 1980s there were first attempts at robotic harvesting of horticultural products and fruits from trees, mainly citrus. But the economic motivation disappeared in the 1990s when a great influx of immigrants made available low-cost human operators. However, in the future it should again be of interest, even to promote robots in agriculture at large. The use of robots in agriculture is a serious goal since it has all the characteristics that still need to be further investigated and particularly made practical since they are related to outdoor applications, but with additional constraints due to the avoidance of damage in cultivation.
Finally, the most important challenge for roboticists in the future can be considered as how to attach problems and develop solutions in less time in newer fields of applications. New problems and fields will be defined more suddenly, probably without the chance that we have today to study and practice with prototypes in research centers.

1.3 Examples and technical characteristics of industrial robots

The most important parameters are summarized in Table 1.4 to give an overview of the characteristics that can be used for a selection of a robot.
Figures 1.10, 1.11, and 1.12 are illustrative examples of technical advertisement sheets that are usually available for a first choice and evaluation of robots since they give the most important technical characteristics of the robots.
The models of manipulator and control unit identify the system in the robot production. The sizes and weights of the manipulator, controller, power unit, and other sub-systems give data which are useful for proper installation and location of the robot.
The usual application given by the constructor generally refers to previous successful experiences, but it does not exclude other applications.
The number of degrees of freedom gives a measure of the kinematic capability of manipulation and generally also indicates the number of independent actuators that operate in the manipulator architecture.
The architecture is classified by using a classification of its kinematic chain. The manipulator architecture can be anthropomorphic when it has mobility, which is similar to that of the human arm. Among those non-anthropomorphic architectures, one of the most interesting and most applied is the SCARA architecture (Selective Compliant Arm for Robot Assembly) due to its high performance in object handling and assembling.
The workspace is defined as the set of reachable points by a reference point on the robot extremity. It is of great interest since it gives and illustrates the manipulation capability of a robot and when it is known one can design and operate a work-cell in a suitable way so that the robot can manipulate properly the flux of objects/products.
A robot can be programmed according to several procedures. Generally, the procedures used are concerned with the capability of feasible trajectories that can be performed by means of point to point (PTP) or continuous control.

Table 1.4: Basic technical characteristics in advertisement sheets for industrial robots

General data:	model, size, and weight of the manipulator model, size, and weight of the controller model, size, and weight of the power unit model, size, and weight of the other sub-systems usual application declared by the constructor
Kinematic characteristics:	degrees of freedom architecture type workspace shape and dimensions path types mobility ranges of joints and extremity
Dynamic characteristics:	payload (mass and/or moments of inertia) maximum velocity maximum acceleration type of balancing systems
Actuators:	actuator type mechanical power operation range
Control unit:	hardware type programming language input/output capability input/output characteristics
Sensors:	type of sensors sensing characteristics operation range
Manipulation:	precision error repeatability error

The mobility range is usually given in terms of the ranges for the joints. The mobility in Cartesian space is given as workspace.
The payload gives the capability of a robot in carrying weights, but it can also be specified with moments of inertia.
The dynamic characteristics are usually expressed by means of maximum values for velocity and acceleration of the robot extremity, that are evaluated by taking into account the manipulator dynamics and joint constraints.
Using suitable balancing designs can enhance the static and dynamic characteristics. But the type and characteristics of a balancing design can give additional limits to the robot operation and a proper knowledge can help for optimum use of the robot.
The type of actuation (which can be electric, pneumatic, or hydraulic) is given in order to provide the necessary power in the suitable form and magnitude that are usually indicated as general data. The characteristics of the controller are given in terms of the used hardware, memory capability, interfaces, and software performances.

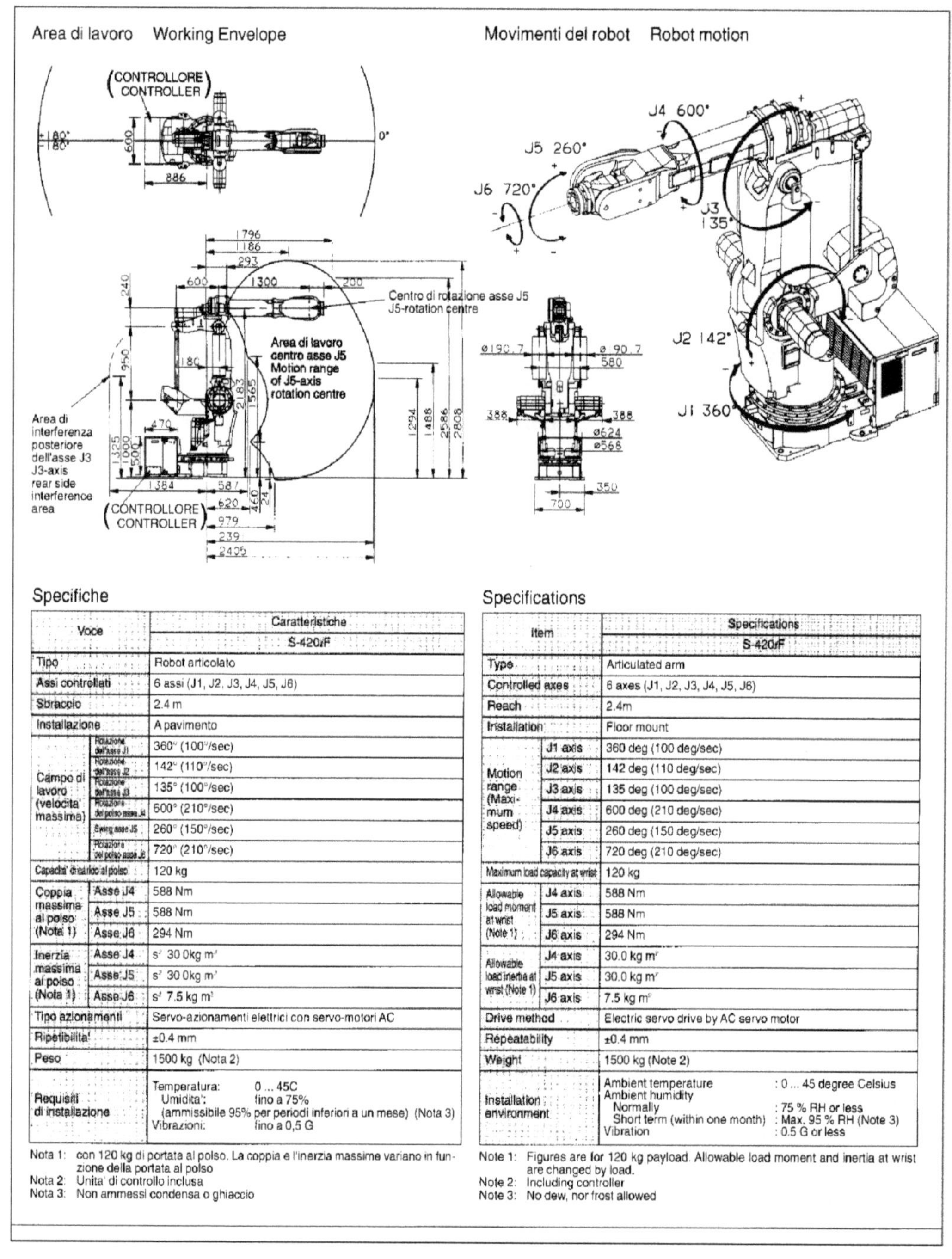

Specifiche

Voce		Caratteristiche S-420iF
Tipo		Robot articolato
Assi controllati		6 assi (J1, J2, J3, J4, J5, J6)
Sbraccio		2.4 m
Installazione		A pavimento
Campo di lavoro (velocita' massima)	Rotazione dell'asse J1	360° (100°/sec)
	Rotazione dell'asse J2	142° (110°/sec)
	Rotazione dell'asse J3	135° (100°/sec)
	Rotazione del polso asse J4	600° (210°/sec)
	Swing asse J5	260° (150°/sec)
	Rotazione del polso asse J6	720° (210°/sec)
Capacita' di carico al polso		120 kg
Coppia massima al polso (Nota 1)	Asse J4	588 Nm
	Asse J5	588 Nm
	Asse J6	294 Nm
Inerzia massima al polso (Nota 1)	Asse J4	s² 30 0kg m²
	Asse J5	s² 30 0kg m²
	Asse J6	s² 7.5 kg m²
Tipo azionamenti		Servo-azionamenti elettrici con servo-motori AC
Ripetibilita'		±0.4 mm
Peso		1500 kg (Nota 2)
Requisiti di installazione		Temperatura: 0 ... 45C Umidita': fino a 75% (ammissibile 95% per periodi inferiori a un mese) (Nota 3) Vibrazioni: fino a 0,5 G

Nota 1: con 120 kg di portata al polso. La coppia e l'inerzia massime variano in funzione della portata al polso
Nota 2: Unita' di controllo inclusa
Nota 3: Non ammessi condensa o ghiaccio

Specifications

Item		Specifications S-420iF
Type		Articulated arm
Controlled axes		6 axes (J1, J2, J3, J4, J5, J6)
Reach		2.4m
Installation		Floor mount
Motion range (Maximum speed)	J1 axis	360 deg (100 deg/sec)
	J2 axis	142 deg (110 deg/sec)
	J3 axis	135 deg (100 deg/sec)
	J4 axis	600 deg (210 deg/sec)
	J5 axis	260 deg (150 deg/sec)
	J6 axis	720 deg (210 deg/sec)
Maximum load capacity at wrist		120 kg
Allowable load moment at wrist (Note 1)	J4 axis	588 Nm
	J5 axis	588 Nm
	J6 axis	294 Nm
Allowable load inertia at wrist (Note 1)	J4 axis	30.0 kg m²
	J5 axis	30.0 kg m²
	J6 axis	7.5 kg m²
Drive method		Electric servo drive by AC servo motor
Repeatability		±0.4 mm
Weight		1500 kg (Note 2)
Installation environment		Ambient temperature : 0 ... 45 degree Celsius Ambient humidity Normally : 75 % RH or less Short term (within one month) : Max. 95 % RH (Note 3) Vibration : 0.5 G or less

Note 1: Figures are for 120 kg payload. Allowable load moment and inertia at wrist are changed by load.
Note 2: Including controller
Note 3: No dew, nor frost allowed

Fig. 1.10: Technical advertisement sheet for a FANUC S-420iF robot.

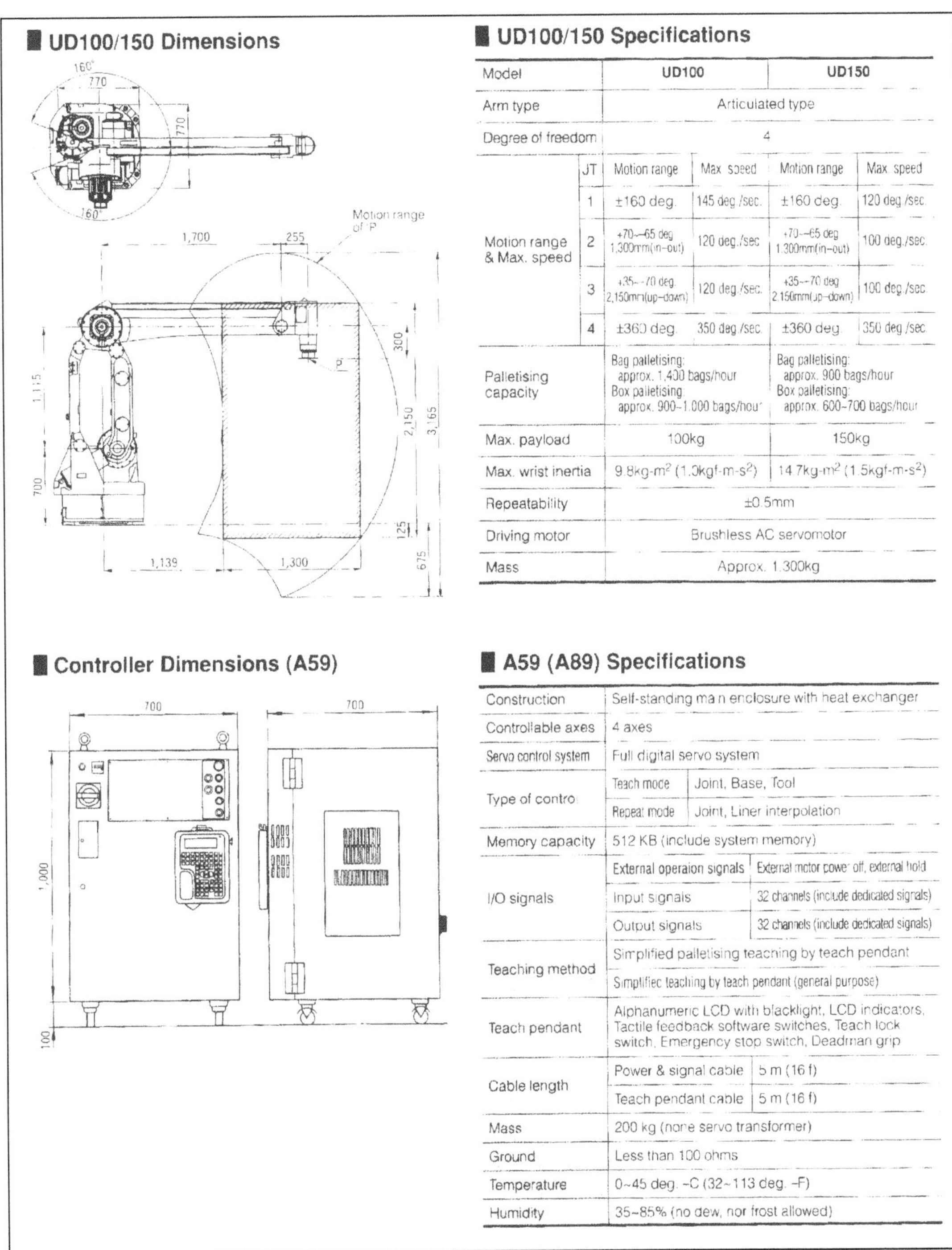

UD100/150 Dimensions

UD100/150 Specifications

Model		UD100		UD150	
Arm type		Articulated type			
Degree of freedom		4			
Motion range & Max. speed	JT	Motion range	Max. speed	Motion range	Max. speed
	1	±160 deg.	145 deg./sec.	±160 deg.	120 deg./sec.
	2	+70~-65 deg 1,300mm(in-out)	120 deg./sec.	+70~-65 deg 1,300mm(in-out)	100 deg./sec.
	3	+35~-70 deg 2,150mm(up-down)	120 deg./sec.	+35~-70 deg 2,150mm(up-down)	100 deg./sec.
	4	±360 deg.	350 deg./sec.	±360 deg.	350 deg./sec.
Palletising capacity		Bag palletising: approx. 1,400 bags/hour Box palletising: approx. 900~1,000 bags/hour		Bag palletising: approx. 900 bags/hour Box palletising: approx. 600~700 bags/hour	
Max. payload		100kg		150kg	
Max. wrist inertia		9.8kg-m² (1.0kgf-m-s²)		14.7kg-m² (1.5kgf-m-s²)	
Repeatability		±0.5mm			
Driving motor		Brushless AC servomotor			
Mass		Approx. 1,300kg			

Controller Dimensions (A59)

A59 (A89) Specifications

Construction	Self-standing main enclosure with heat exchanger	
Controllable axes	4 axes	
Servo control system	Full digital servo system	
Type of control	Teach mode	Joint, Base, Tool
	Repeat mode	Joint, Liner interpolation
Memory capacity	512 KB (include system memory)	
I/O signals	External operaion signals	External motor power off, external hold
	Input signals	32 channels (include dedicated signals)
	Output signals	32 channels (include dedicated signals)
Teaching method	Simplified palletising teaching by teach pendant	
	Simplified teaching by teach pendant (general purpose)	
Teach pendant	Alphanumeric LCD with blacklight, LCD indicators, Tactile feedback software switches, Teach lock switch, Emergency stop switch, Deadman grip	
Cable length	Power & signal cable	5 m (16 f)
	Teach pendant cable	5 m (16 f)
Mass	200 kg (none servo transformer)	
Ground	Less than 100 ohms	
Temperature	0~45 deg. -C (32~113 deg. -F)	
Humidity	35~85% (no dew, nor frost allowed)	

Fig. 1.11: Technical advertisement sheet for a Kawasaki UD100/150 robot and its control unit.

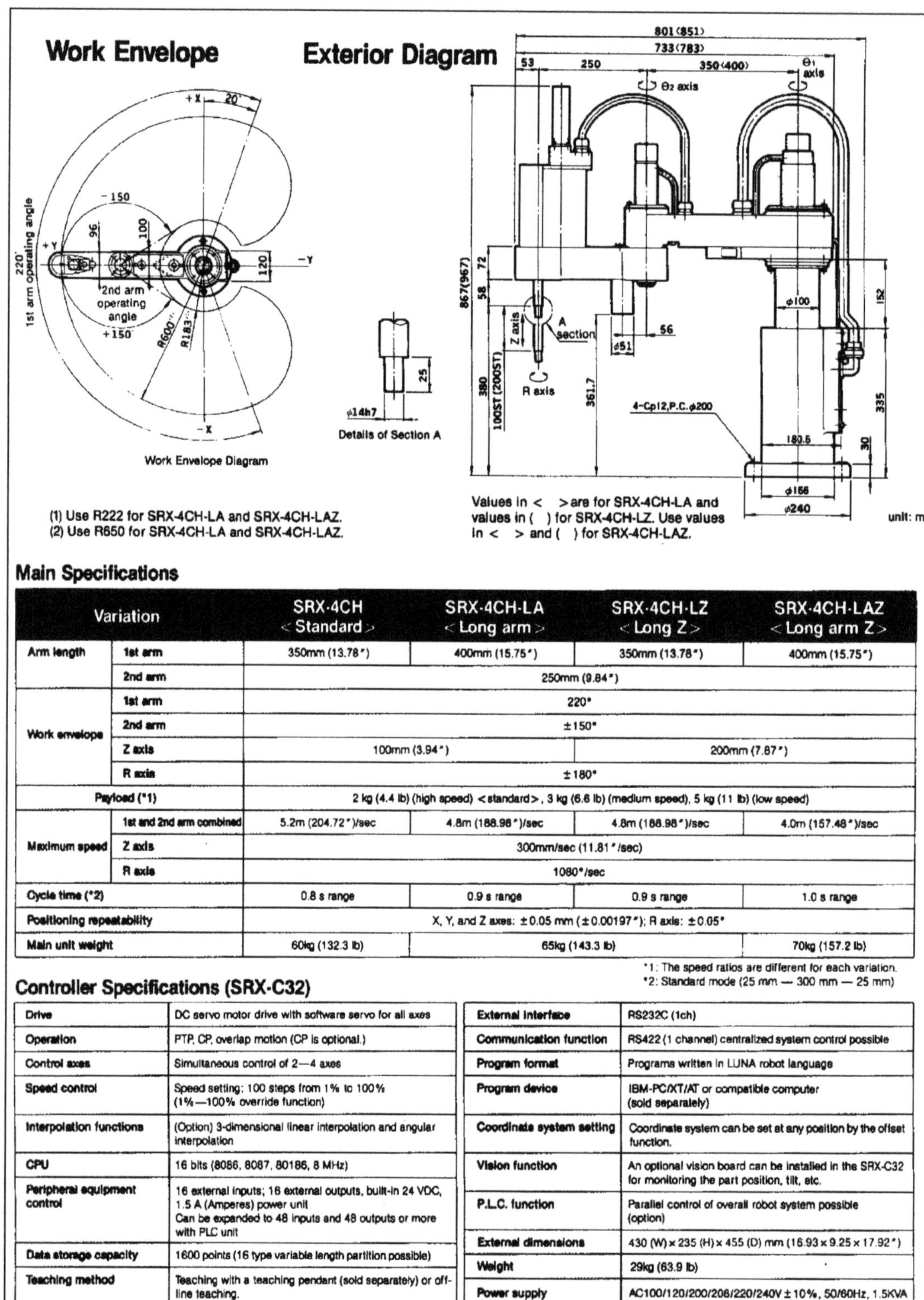

Main Specifications

Variation		SRX-4CH < Standard >	SRX-4CH-LA < Long arm >	SRX-4CH-LZ < Long Z >	SRX-4CH-LAZ < Long arm Z >
Arm length	1st arm	350mm (13.78")	400mm (15.75")	350mm (13.78")	400mm (15.75")
	2nd arm	250mm (9.84")			
Work envelope	1st arm	220°			
	2nd arm	±150°			
	Z axis	100mm (3.94")		200mm (7.87")	
	R axis	±180°			
Payload (*1)		2 kg (4.4 lb) (high speed) <standard>, 3 kg (6.6 lb) (medium speed), 5 kg (11 lb) (low speed)			
Maximum speed	1st and 2nd arm combined	5.2m (204.72")/sec	4.8m (188.98")/sec	4.8m (188.98")/sec	4.0m (157.48")/sec
	Z axis	300mm/sec (11.81"/sec)			
	R axis	1080°/sec			
Cycle time (*2)		0.8 s range	0.9 s range	0.9 s range	1.0 s range
Positioning repeatability		X, Y, and Z axes: ±0.05 mm (±0.00197"); R axis: ±0.05°			
Main unit weight		60kg (132.3 lb)	65kg (143.3 lb)		70kg (157.2 lb)

*1: The speed ratios are different for each variation.
*2: Standard mode (25 mm — 300 mm — 25 mm)

Controller Specifications (SRX-C32)

Drive	DC servo motor drive with software servo for all axes
Operation	PTP, CP, overlap motion (CP is optional.)
Control axes	Simultaneous control of 2—4 axes
Speed control	Speed setting: 100 steps from 1% to 100% (1%—100% override function)
Interpolation functions	(Option) 3-dimensional linear interpolation and angular interpolation
CPU	16 bits (8086, 8087, 80186, 8 MHz)
Peripheral equipment control	16 external inputs; 16 external outputs, built-in 24 VDC, 1.5 A (Amperes) power unit Can be expanded to 48 inputs and 48 outputs or more with PLC unit
Data storage capacity	1600 points (16 type variable length partition possible)
Teaching method	Teaching with a teaching pendant (sold separately) or off-line teaching.
External interface	RS232C (1ch)
Communication function	RS422 (1 channel) centralized system control possible
Program format	Programs written in LUNA robot language
Program device	IBM-PC/XT/AT or compatible computer (sold separately)
Coordinate system setting	Coordinate system can be set at any position by the offset function.
Vision function	An optional vision board can be installed in the SRX-C32 for monitoring the part position, tilt, etc.
P.L.C. function	Parallel control of overall robot system possible (option)
External dimensions	430 (W) × 235 (H) × 455 (D) mm (16.93 × 9.25 × 17.92")
Weight	29kg (63.9 lb)
Power supply	AC100/120/200/208/220/240V ± 10%, 50/60Hz, 1.5KVA

Fig. 1.12: Technical advertisement sheet for a SONY SCARA SRX robot.

The programming language is indicated since it can be specific for the robot so that it requires specific training. The input/output capability is given with the characteristics of the signals that the controller can receive and send so that the robot can be integrated in a robotic work-cell with other machinery under the supervision of a unique control system.
The type and characteristics of sensors are useful to indicate the autonomy level and necessary control signals. Generally position sensors and force sensors are indicated to give information on controlled manipulations.
Of great importance are the characteristics of the robot manipulation that are generally expressed in terms of precision and repeatability errors. The precision error gives a measure of the uncertainty of extremity positioning for the robot extremity in one programmed operation. The repeatability error gives a measure of the uncertainty of extremity positioning for the robot extremity in repetitive operations.
In addition, there are norms that give reference characteristics and procedures for characterization of robots.
International norms are defined and continuously updated by the ISO (International Standard Organization) even referring specifically to robotics.
The choice of a robot can be achieved by using criteria which mainly concern: feasibility of the robotic manipulation, technical data of available robots, and financial convenience of a robotization.
Indeed, a robot can be selected not only by using data in the technical advertisement sheet, but even on the basis of personal considerations and mainly previous experiences with specific robots. These considerations can be a constraint for fully exploiting the potentiality of a robotization and consequently for an optimum choice of a robot, although they can be of great help in achieving the acceptance of a robotization and selected robot.
Therefore, it is convenient to refer to previous experiences in similar productions and robotizations, but also to study new solutions by designing and operating a robotization with the help of experts in design and management of the productive process.
In particular, the design approach of mimicking human labor with specific robots, although very attractive, sometimes does not give the suitable desired results in terms of productivity and product quality. Indeed, when a robot is used one should consider the peculiarity of the alternative manual manufacturing, but an efficient use of robots requires a certain abstraction of the operations in order to design even the production for the robots from the beginning.

1.4 Evaluation of a robotization

A successful use of robots in industrial applications depends on many aspects from several fields that nevertheless should be properly considered both singularly and in a panoramic view.
In this section only a few considerations are discussed to give a general view of those relations, which are analyzed in specific literature to which the reader can refer.
Fundamental relations with the use of robots can be recognized in economical consequences, social impacts, and cultural effects.

Regarding financial aspects, the convenience can be evaluated by analyzing the costs and earnings when a robot will substitute traditional machinery with human operators. But problems may arise for cost analysis since the market evaluation for labor and mainly for robots is very unpredictable.
In addition, in novel robot applications there are not available data from previous experiences so the cost can only be considered a desired estimation.
Financial convenience can also be considered as strongly related to the number of human operators and quality products.
It is often said that robot use causes unemployment, but this is difficult to prove since, although it is true that robots substitute human operators with a low level of education, one should also consider that automation and mainly robotics have motivated the formation and employment of many high-level professionals. Moreover, in general robot use has improved the education level of operators working with robots.
Regarding the production results, it is worthy of note that quality is not always improved but high productivity and mainly regular and uniform production can be achieved with robot use.
Social impact can be considered as related to the evolution of the employee's education. In fact, the use of robots requires technicians and mainly engineers with a high level of education so that the ratio of the social classes will change both in terms of numerical and social components.
It is recognized that the decrease of human labor and increase of high-level technicians changes the composition of the traditional community in production plants with so-called inverted pyramids, which means that there is not one plant director and many operators but many directors and few operators . This social change in a production plant changes the relations and generally the scenario not only in the industrial community but in society at large.
Regarding the cultural aspects, it is evident that generally the use of robots requires a high level of education. This is achieved with a longer formation and even with an evolution of the teaching. Nevertheless, the cultural impact can be recognized also in a better education and consequently quality of life, since the working environment will be improved as a need for well-educated people.
The above-mentioned few considerations can be further discussed both with positive and negative remarks since in general the use of advanced automation and mainly robots do not yet give clear answers to the many questions and aspects. This is because there is not yet enough experience, tradition, and finally history.
Nevertheless, it has been considered meaningful to give a short overview in order to give a certain panorama of the aspects that are involved when using robots. Of course, those aspects are discussed in depth in different specific communities and sources, to which the reader can refer.
Nevertheless, one of the most important issues for using and selecting a robot can be considered the financial convenience, even from an engineering viewpoint.
However, in some cases a robotization can be convenient even when the financial evolution gives a negative judgement, since security aspects or social considerations make it necessary. This is the case, for example, of the handling of radioactive

dangerous material, or when the risk of breaking-down the process can cause environmental damage. In the last case a robot can be conveniently used because of its sensitivity to the changes of the process that can be used as an alarm system.
In addition, the psychological impact or esthetic reasons for publicity can justify the use of robots, regardless of the direct economic convenience.

1.4.1 An economic estimation

In this section a simple procedure has been outlined for an economic evaluation of a robot or robotization.
A critical aspect is the analysis of costs for a robotization that should be compared with the costs for an industrial process with traditional machinery and human operators.
The aim of the robot use can be either a new design or a modification of the industrial process under estimation.
The analysis of costs is a very complicated problem since the use of robots has direct consequences but it can also be related to many other indirect facts. For example, the use of robots can affect the change of strategy and equipment for product design and production, the need of a different technical environment and expertise, only to cite some of the technical viewpoints.
Therefore, an analysis of costs can be carried out at different levels of accuracy, depending on the depth of financial investigation, procedures, and available data.
Using the following items the analysis of costs for a robot evaluation can be carried out:

- cost of human labor, which consists of the costs of substituted operators but also of extra technicians; indeed, this cost will include the cost for training and periodic re-training of personnel, and the cost of temporary employees who are needed for the operation of the process;
- cost of the robot system, which will include all the devices (tools, complementary equipment, etc.) to achieve the desired task;
- cost of installation of the robot and robotization, in terms of the work-cell with the proper structures, energy flux, production connections, and security systems;
- cost of the robotization design, which may require the study of new production strategies with new design of the production process and machinery;
- cost of robot operation, which includes the cost of used energy and materials, but also the maintenance costs;
- cost of temporary change of activity, which is related to programmed or non-programmed stops for updating/adjusting the production.

In addition, the use of robots can require an increase of indirect work, even of irregular type, such as periodic consultation for re-programming or updating of robots, additional maintenance for frequent change of the production or continuous non-programmed operations.
Economic evaluations, that can be considered usual in robot applications, make use of a determination of the pay back time (T) and return of investment (ROI). In both cases, the evaluation is based on main characteristic parameters of the industrial application, such as:

- P, which indicates the cost of the robot that can also be considered with items for

robot main system cost (as 50 to 70% of P), tools cost (as 10 to 15% of P), and installation cost (as 10 to 20% of P);
- F, which indicates discount of taxes that can be available for encouraging industrialization;
- L, which is the hourly cost of human labor;
- M, which is the manufacturing cost of the variation of used material for the production;
- R, which is the running cost of operation and maintenance of the system;
- D, which is the annual depreciation of the robot/robotization;
- Q, that is a coefficient by means of which it is possible to consider the variation of the product quality and productivity.

The payback time indicates the time of the robotization operation, which is needed to pay the investment cost through income.
Usually, this time is assumed as five years for industrial applications. Indeed, this value depends on the frequency and duration of the running of the process. In the case of flexible automation using robots this time is usually fixed at three years. However, in some cases it can be limited to one year only (as often occurs in Japan), but it can also be increased up to ten years, as for example in large industrial plants.
A first evaluation can be obtained through a simplified expression for the payback time in the form

$$T_S = \frac{P}{kH\,L} \tag{1.4.1}$$

However, a correct expression for the pay back time (T) can be given as

$$T = \frac{P - F}{kH\,(L + M - R) - D + Q(kHL - D)} \tag{1.4.2}$$

which consider the parameters of the above-mentioned list.
However, it is worthy of note that the depth of an analysis of costs and even the use of different expressions with more terms can give results which can be very different and even opposite to the judgement based on the value given by Eq. (1.1).
Another fundamental parameter for an economic evaluation is considered the return of investment (ROI), which is defined as the ratio between the earning with the new solution and needed investment. By using the above-mentioned list for the analysis costs, ROI can be evaluated as

$$ROI\% = \frac{kH\,(L + M - R) - D}{P - F}100 \tag{1.4.3}$$

Equation (1.3) does not consider that the value of earnings will decrease over time. Positive values of ROI are considered as those that are at least comparable with the

return that a bank will give for a deposit similar to the investment.
Equations (1.1) to (1.3) are illustrative examples of a numerical financial evaluation of a robotization or robot use/selection, but industrial applications may require deeper analysis and consideration of several other aspects as outlined in this section.

1.5 Forum for discussions on robotics

Intense activity is carried out in the scientific community in order to enhance robot capability and enlarge robot applications. These efforts and consequent continuous evolution of robots give great possibility to robot users who, nevertheless, are urged to be continuously aware of new results and innovations, even by participating in discussion and investigation.
The literature is very rich and often becomes obsolete very quickly. Therefore it has been considered convenient for the purposes of this book not to include a long list of references, but in the following is given a list of main sources in which technical aspects of robots and robotics are treated:

- Conferences

ACRA (Asian Conference on Robotics and Its Application)
AIM (IEEE/ASME International Conference on Advanced Intelligent Mechatronics)
ARK (International Symposium on Advances in Robot Kinematics)
ASME Biennial Conference on Mechanisms and Robotics
ICRA (IEEE International Conference on Robotics and Automation)
ISR (International Symposium on Robotics of IFR)
ICAR (International Conference on Advances in Robotics)
IFToMM World Congress
IROS (IEEE/RSJ International Conference on Intelligent Robots and Systems)
ROMANSY (CISM-IFToMM Symposium on Theory and Practice of Robots and Manipulators)
RAAD (International Workshop on Robotics in Alpe-Adria-Danube Region)
Congresses of the National Associations of Robotics

- Journals

Advanced Robotics (VSP)
Journal of Mechanical Design (ASME Transactions)
Journal of Dynamic Systems, Measurement and Control (ASME Transactions)
Journal of Robotics and Automation (IEEE Transactions)
Journal of Robotics and Mechatronics (Fuji)
Journal of Robotic Systems (Wiley)
Mechanism and Machine Theory (Pergamon Press)
Mechatronics (Pergamon Press)
Robotica (Cambridge University Press)
Systems, Man and Cybernetics (IEEE Transactions)
The Industrial Robot (IFS)
The International Journal of Robotics Research (MIT Press)
National Journals on Automation and Robotics
In addition, several exhibitions and fairs are organized on national and international

frames for automation and robotics in which the constructors show their robots in standard applications and new solutions.

Chapter 2:
Analysis of manipulations

2.1 Decomposition of manipulative actions

A manipulation is generally thought of as a complex set of actions, which are used to perform a transportation of an object in the space with a movement similarly to that of a human arm when its hand holds an object. Therefore, a manipulation can be considered as related to movements of objects with a specific purpose. For example both the transportation of components for assembly and movements for painting can be considered as manipulations, although the first operations are preliminary actions and the latter perform completely a task.

Indeed, similar to a human manipulation, a manipulative task can be considered as composed of:

- grasping actions
- movements (which are manipulations in a broad sense)
- releasing actions

Therefore, a manipulation can be efficient and successful when all the above-mentioned phases are properly considered. In this chapter manipulative activity is discussed as referring to automatic systems and particularly robots.

It is fundamental to consider a manipulative operation as related to the capability of automatic and/or robotic systems. Consequently, it is convenient to study a decomposition of the manipulative operations corresponding to the capability of automatic and robotic systems.

For this purpose, the concept of elementary action is fundamental , which can be defined as the smallest manipulative entity that can be performed by the simplest action of the actuation in an automatic or robotic system through one or few programming instructions. An elementary action can be considered as due to a small manipulative operation but in conjunction with a simple programming capability. The practical usefulness of elementary actions is clear when it is thought as fundamental for the operation of an automatic system and particularly a robot.

But the decomposition of a manipulation requires the identification and analysis of entities with different manipulative content as those that one can recognize in so-called operations, phases, and elementary actions.

An operation can be defined as a set of manipulative activities that achieves a completed manipulative goal. It can be composed of two or more phases that can also be repeated within it.

A phase can be defined as a set of manipulative activities for intermediate goals, although it refers to a completed manipulation. It can be composed of two or more elementary actions that can be grouped and repeated within it.

For example, the manipulation related to the writing can be decomposed of operations referring to the writing of words, phases referring to the writing of letters, and elementary actions referring to single movements for tracing segments of letters.

2.2 A procedure for analyzing manipulation tasks

A key issue for an optimum use of robots lies in the fact that a robot should be introduced to a user by stressing that versatility and flexibility are equally important. Too much attention on one of the two aspects may give a wrong impression of relative significance and lead a user to think incorrectly that one aspect predominates and may solve any deficiency due to the other one. Therefore, from an educational perspective, the best approach consists in imparting knowledge separately on the manipulative process to be performed and on the robot to be used.

Thus, it is convenient to study manipulative tasks without any reference to any particular robot in order to determine manipulation requirements independently from the limits of a specific robot. By the same token, a robot's capabilities can be conveniently learned in terms of programming facilities and mechanical capabilities with no reference to specific applications, which may reduce the interest for overall capabilities.

Then, the problem of matching the results of learning activities and analysis arises as consisting of finding a feasible way to make the robot perform a given manipulative task with the required flexibility and versatility. Thus, a user can gain an insight into how to rationally use a given robot for a given manipulation, and particularly how to further improve with minor adjustments the robot's operation or industrial process in terms of productivity, efficiency, or changes in manufacturing production.

In Fig. 2.1 the aforementioned considerations have been translated into a practical procedure used by the author for students' practice with robots. Figure 2.1 indicates the steps that are required to analyze a manipulative task in order to identify a specific robot and operate it through suitable programming, which can be verified for flexibility and versatility optimum performances.

In particular, according to item one of the flowchart in Fig. 2.1 a given process has to be analyzed in terms of manipulative operations, even mimicking human actions in order to establish rigorously what manipulations are required.

Then, the environment should be studied or designed to identify obstacles and constraints for the manipulation. Usually in industrial applications the environment is named as 'structured' when its scenario is completely known and fixed together with objects that are known mainly in terms of shape, size, weight, and location. Alternatively an environment can be considered as 'unstructured' when the scenario and objects in it are not known and even change over time.

This environment analysis may yield to geometric and kinematic models of the environment, which the robot will work in. At this stage students–users can be encouraged to think in a more abstract way in the sense that the process should be studied with no reference to a predetermined automated or robotized solution.

According to the next item in the flowchart, process operations have to be identified as proper manipulative actions, i.e. movements of objects including tools and subsidiary machinery. This can also be done by drawing sketches for the robotized lay-out to identify manipulative tasks. This modeling activity, even through drawing sketches, will help to check both manipulation and robot performances.

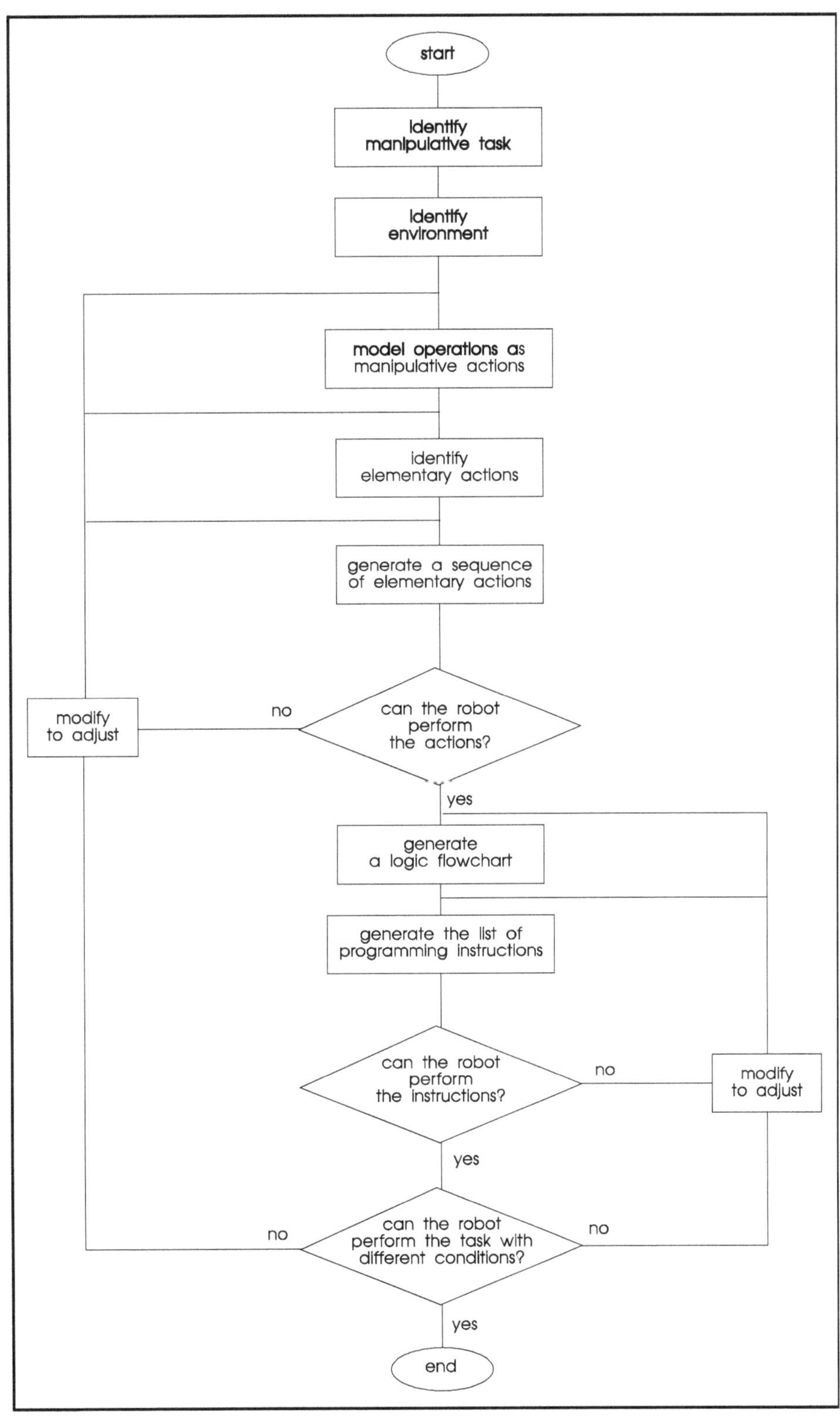

Fig. 2.1: A procedure for manipulation analysis and programming.

The next two steps deal with a more detailed manipulation analysis with the aim of identifying elementary actions and qualifying their nature and timing.
Once the manipulation entities have been identified and partitioned, it has to be established whether the robot's capabilities are suitable for the operation's purpose. In this item robot performances are verified with respect to the given manipulation.
This step can be useful either to check operation feasibility of a given robot or to select a better-robotized solution. In Fig. 2.1 a first adjustment of previous steps can be required following an unsatisfactory response, i.e. when the given robot or the chosen robot cannot properly perform the previously obtained manipulation sequence. Adjustments may consist of rethinking the modeling of the manipulative tasks by also changing elementary actions and/or the design of the elementary action sequence.
The item 'modify to adjust' can be developed at different levels of detail and it may describe and include specific algorithms to obtain optimum solutions for each specific problem.
Indeed, a manipulation analysis can be a complex activity and it includes several difficult problems that can be solved separately with suitable techniques within a common frame, so that constraints on one aspect can be the result of another one and vice versa. In this sense each main block in the procedure of Fig. 2.1 is completed by the item 'modify to adjust' which is connected to. In addition this item specifies how to obtain the modifications, which can be usually evaluated by means of sophisticated algorithms for any specific problem. These analyses can be carried out by using CAD/CAM software, simulation graphics, or multi-body dynamics depending on the level of sophistication and details of the required study. In this contest we do not make use of these algorithms since we want to focus on manipulation analysis for teaching purposes only. Therefore, at a first level of analysis this approach can be conveniently thought of as empirical nature only.
However, the item 'modify to adjust' encompasses a series of elements, such as considerations, computations, and evaluations, which are useful to provide better solutions for a problem that has not been considered or solved in a feasible or acceptable way.
However, it is worth pointing out that in order to solve difficult problems suitable techniques and procedures are available in the literature, such as robotized work-cell lay-out design, choice and optimal placement of an appropriate robot, optimal planning for robot trajectory, and object assembly. Therefore, the proposed manipulation analysis can also be considered as a common frame capable of grouping these different problems and consequently connecting their partial solutions and constraints within a unique problem solving approach.
Once the robot is considered feasible because its versatility matches the manipulation requirements, a programming design can be started for a flexible solution by developing a logic flowchart for programming instructions. A flowchart can be very useful since students–users have a better response (quicker and easier) in adjusting the programming when later changes in the manipulative process are required. This is true, indeed, for any software design: in a software program a flowchart may be of great help for future updating of instructions, and also helps subsequent users understand. Furthermore, in

this way a user obtains a programming solution, which is not yet directed to a given robot, because it is not oriented to a given programming language. Indeed, the logic flowchart can be used to program manipulations in several different robot languages. This obviously means that the flowchart is independent of the robot choice.
The next step consists of generating a list of programming instructions for a given robot with the instructions of a given programming language, which requires knowledge of instruction syntax, as well as the limits of robot movements and performance.
According to the item 'can the robot perform the instructions?' a first check on the feasibility of the sequence for programming instructions may require adjusting the previous programming solutions.
Finally, the flexibility of the robotized solution can be verified by introducing some changes in the manipulative process. This will show whether the robot's versatility allows it to operate properly with some further change. Although this last item may not be necessary for a given application, it is however of primary importance for an optimum use of robots because it reveals how versatile and flexible a designed robotized solution is. The above-mentioned checks can be performed by a debugging process, as well as by means of online practical tests or starting-up calibration.
Furthermore, for an optimum (versatile and flexible) use of a robot, the validation items can also suggest adjustments in the robot structure or particularly in the robot end-effector, so that gripper design can be included in the analysis of the manipulation.
A manipulation is a process performing 'controlled movements of object'. In order to achieve maximum efficiency in robot operation the manipulation process, which can be very complex, must be analyzed in depth by breaking down the manipulation into manipulative elementary actions to be performed for object and robot movements. The object can be a grasped object or the robot end-effector itself since the manipulative goal may require intermediate and subsidiary actions on the robot. Basically, in the first stage of a manipulation analysis only the movements of the robot end-effector or object have to be taken into account, with no reference to robot capabilities nor to human mimic, although this last comparison may help to understand the complexity or simplicity of the manipulative task. In practice this can be achieved by looking at the movements of a rigid body representing either a grasped object or the robot end-effector.
The identification of the elementary actions performed for the rigid body of a reference object to be manipulated provides a useful tool to break down the manipulative process.
A first result may consist in an easy evaluation of the number of degrees of freedom that is required to perform the process. This constitutes a basic criterion for choosing the appropriate robot or a class of robots for a given manipulation.
At this stage a schematic drawing of the object's fundamental positions and configurations is particularly useful. This determines the working area and mechanical versatility of a robot, and thus it helps in rationally choosing the appropriate robot.
Once the process has been analyzed in terms of elementary actions, the robotized portion can be selected and a further analysis may be of help in carrying out flexible programming, as Fig. 2.1 illustrates.
The elementary actions to be performed by a robot can be conveniently identified and

classified as transport, movement, passive pause, and active pause.
A transport elementary action T can be defined as a robot motion performing a movement of the object, which is moved from one position to another. A movement elementary action M can be defined as a robot motion performing a change of the manipulator configuration to obtain a movement of the robot end-effector, when this is not considered the primary object to be manipulated in the process. A pause can be considered an elementary action of the robotic system during which no movements are performed but some other activities are being undergone. The robot is in a state of active pause A when the manipulator is not moving but the robot is operating through its equipment (sensors, control, programming unit, and power supply). A passive pause P can be identified when no activity is performed and the robot is waiting to start again.
Once all elementary actions have been identified and classified, they can also be organized or re-organized for a suitable manipulation sequence by grouping them in some set of actions, which may perform a well identified sub-task. This can be useful if parts of this sub-task are repeated within the process.
Subsequently, a user can generate the programming list by considering that each elementary action may correspond to one or few instructions. The logic flowchart can be of great help to generate a flexible programming and for future adjustment/improvement of the robot operation also with respect to changes in the programmed manipulation, since it allows to easily identify the instructions or the set of instructions to be modified or changed.
For an optimum use of a robot a numerical evaluation of the performed manipulation can be obtained by recording the operation time for each elementary action. This gives an assessment of the robot operation and a tool to identify possible optimization with respect to productivity and manipulation changes.
Indeed, several other criteria have been proposed in the literature and are also currently used to optimize, for example, energy consumption, actuator torques, velocity and acceleration values, and operation cost. Only cycle time has been used in this book since it is of evident interpretation and interest for a practical synthetic evaluation.
Some examples have been reported in following sections to illustrate in detail the application of the proposed ideas and procedure of a manipulation analysis.

2.3 Programming for robots

A manipulative task of a robot is performed through the execution of a program with instructions for a sequence of elementary actions by the robot. Such a program can be obtained through the following activities:
- online programming, which consists of identifying and storing the sequence of elementary actions, while the robot operates, through techniques of guided teaching or by means of the teach pendant commands;
- offline programming, which consists of listing the instructions for elementary actions in the robot control unit while the robot does not operate in the manipulation under programming.

In general a suitable flexibility is obtained by using offline programming since it allows to determine instructions and programming variables that can be specified for a specific

application of the robot and therefore can be easily updated. In addition, during the offline programming the robot can still work on other manipulative tasks without the need of operation breaks.
The online programming is usually used during the set-up, calibration, and start of a programmed manipulation, mainly in the activity concerned with the identification of precision points for a specific application of the offline programming. Precision points can be considered as positions of robot extremity that are asked to be reached exactly during the robot operations. They can be determined by position and orientation of robot extremity or by the robot configuration through the values of joint coordinates.
The online programming with guided teaching consists of a program construction that is obtained by a human operator who accompanies, directly or by telemanipulation, the robot through the manipulative task. During the teaching mode the operator provides the storage of the suitable positions and path segments of the robot, and elementary actions of the other robot components into the unit control as a sequence of instructions. Updating and correction of stored positions can be easily obtained even by an offline programming. But if one needs to change the manipulative task or path, by adding or removing precision points, a new online activity is required.
The online programming with teach pendant consists of a program construction that is obtained by a human operator who guides the robot along a manipulative path through commands via the keyboard of the teach pendant. The commands are stored in the control unit together with the precision points, characteristic segments, and other elementary actions of the manipulative path. Usually, the teach pendant programming gives a better ability of robot guiding as compared with the guided teaching and additionally it gives a better identification of the manipulative characteristics with a greater agility and safety due to the cable length of the teach pendant.
The offline programming of a robot consists of developing a list of instructions for elementary actions in the programming language used by the robot. Generally, in industrial robots, a specific operating system for the software can be used in the microprocessors of the control units so that programming can be obtained only in the same hardware systems. However, in offline programming it is possible to check the feasibility of the program in terms of instruction sequence and syntax by its debugging within the same software environment. In addition, running the program in suitable CAD/CAM environments with 3D visualization can simulate the manipulation efficiency. Indeed the CAD/CAM simulation can also be useful in the design process of a robotization. Today such a simulation can be used to both check and optimize the robot use and manipulation process. However, these simulations can be performed at different levels of performance evaluation depending on the used models that are based on kinematic behavior only and/or even on dynamics of multi-body schemes.
Offline programming can be obtained by using several programming languages that can be classified as:
- specific robot languages, such as VAL for PUMA-Unimation robots or ACL for Scorbot–Eshed;
- general purpose languages for automation systems, such as AML for IBM robots;
- scientific oriented languages, such as AR-Basic, which is based on Basic.

The programming languages are different from each other mainly by the dictionary of instruction terms, which are related to the elementary actions that can be executed by the robot. However, those dictionaries are similar to the current human dictionary in the sense that terms are used to program elementary actions that are similar to human actions. Nevertheless, the syntax of the instructions can be very different from one language to another depending on the variables for the control unit and signals that are needed to perform the programmed elementary actions.
The specific robot languages are developed by the robot builders and/or controller builders as an evolution of programming systems for automatic devices and they are usually based on specific hardware for the control and programming of robot actuation. Therefore, they can be very different from each other both in the dictionary and syntax.
The general-purpose languages are usually developed to control and program not only robots but also a complete robotic work-cell with its components that can be any other machine or device.
The scientific oriented languages have been developed from scientific languages with the aim to enhance their capability and even to have computational capability for production processes within the same robotic system.
Unfortunately, there is not yet a standard view for programming languages, although the need of standardization is required to facilitate the use and programming of robots within a common frame. Therefore, a suitable programming of industrial robots, that exploits properly the characteristics of flexibility and versatility of a robot, requires a specific in-depth knowledge of programming languages for specific robots.
Instructions for robot programming can be classified as depending on the purpose of:
- program execution;
- program editing;
- executing computations and logical evaluations;
- executing elementary actions.

The instructions for program execution are useful to run and manipulate programs (with actions like storage and copying files, debugging of program lines , ...) that have been elaborated for operating a robot.
The instructions for program editing are those instructions that are used to build, update, or change a program. Generally, a specific dictionary and syntax, depending on the robotic system, for which they serve, characterize them.
The instructions for executing computations and logical evaluations are those instructions that are used to perform those type of computations that can be useful to obtain a required level of flexibility, re-programming, and simplicity. Although the mathematics, generally a specific dictionary and syntax, depending on the robotic system, for which they serve, characterize them.
The instructions for executing elementary actions are those instructions that are specifically devoted to operate, regulate, and control the movements of a robotic system by means of simple actions that are strongly related to elementary actions.
Each programming language shows peculiarities although a certain common frame can be recognized, so that a common designing of the programming can be carried out as explained above. Nevertheless, those small differences that can be in the dictionary or in

the syntax are still so significant that a certain specific expertise is required for each programming language.
In the next sections two programming languages, namely VAL-II and ACL, are illustrated briefly with the aim of showing the similarity but the small differences that can make the same programming capability impossible to use. The reader can carry out an interesting comparison between VAL-II and ACL in order to appreciate the similarity but those small differences in the dictionary and syntax that make the same instruction list of a program impossible to use. At the end of the chapter examples will show practical application of the analysis and programming of manipulative tasks by using those programming languages.

2.3.1 A programming language for robots: VAL-II[1]

In this section basic instructions of the programming language VAL-II for industrial robots are illustrated in order to show the characteristics and requirements for a suitable programming of these robots. The Unimation VAL-II language is presented briefly since it can be considered significant from historical and technical viewpoints to give examples of dictionary and syntax of instructions that are used to program the operation of industrial robots.
The syntax of each instruction is usually indicated through the necessary data for the instruction execution within bracket symbols < > and through additional data within bracket symbols [ ] when they are of complementary information.
As examples of the above-mentioned groups of instructions one can refer to the basic and most used instructions as in the following.
Among the instructions for program execution let us consider: LOAD, STORE, PLIST, EXECUTE.
The instruction LOAD <file name> loads the contents of the specified file into the memory of the controller.
The instruction STORE <file name> [program-1,..., program-n] stores the following programs and data in the specified file name: the programs that are named within the square brackets as program-1 to program-n, all the subroutines that are linked to them, all the location variables and real-valued variables that are used in the storing programs and subroutines.
PLIST [<file name 1>, ..., <file name-n>] displays all the steps of the named programs or all users programs in the system memory.
The instruction EXECUTE [<file name>], [<loops>], [<step>] runs the specified file program as many times as the number <loops> indicates by starting at the specified program step. If no file program is specified, the system executes the last executed program.
Among the instructions for program editing let us consider: EDIT, D (delete), I (insert), M(modify), R(replace), T(teach) , TS(teach straight line), E(exit).
The editor command EDIT [<file name>], [<step>] is used to start an editing session

[1] The information on VAL-II herein reported is based on the Programming Manual – User's Guide to VAL-II version 2.0, Unimation, 1986.

with the specified file program at the indicated program step. If no options are included, editing proceeds from the first step of the last edited program.

In the editing session, command D <steps> deletes the specified number of program steps starting with the current step. Only the current step is deleted if no number <steps> is specified.

The command I moves instructions at the current step down one step and inserts the program instruction that is typed next.

The command M <old string> <new string> replaces the characters that are specified in the old string with those specified in the new string.

The command R <string> replaces characters in the program step that is displayed immediately above the line containing the cursor.

The command T <location variable> initiates a teaching session for recording robot locations and program instructions for joint-interpolated motions to the taught locations. In the teaching mode, each time RECORD pushbutton on the teach pendant is pressed, a MOVET instruction is inserted into the program being edited or under construction. The generated instruction contains as its arguments a location variable and current hand opening. The value of the location variable is set equal to the position and orientation of the robot tool at the time the RECORD pushbutton is pressed. Pressing RETURN on the terminal keyboard terminates the teaching session.

The command TS <location variable> initiates a teaching session similar to the command T, but it generates straight-line interpolated motion between recorded robot locations.

The command E closes the editing session and returns to display mode.

Among the instructions for executing computations and logical evaluations let us consider DO, FOR-END, GO TO, IF, SET, ABS, COS, INT, SQR, and ATAN2.

The DO instruction is useful to control the execution of a group of instructions and can be used in the form of DO <group of instructions> UNTIL, or in the form of WHILE <logical expression> DO <group of instructions> END. The DO-UNTIL structure permits the execution of the group of instruction steps until the result of a logical expression in the group becomes TRUE changing from a FALSE value. The WHILE-DO-END structure permits the repeated execution of the group of instruction steps until when the result of the logical expression is TRUE. Otherwise the loop is ended and the program execution continues with instructions after END.

The structure FOR <real variable> = <initial> TO <final> [STEP <increment>] <group of steps> END is used to repeat the specified group of steps as many times as the counter variable is increased from its initial value to the specified final value by using the specified increment step. If increment is omitted STEP 1 is assumed.

The instruction IF <logical expression> GO TO <label> permits the execution of instructions at the step that is specified by the label number when the logical expression is TRUE. Otherwise the next step of the program is executed.

The structure IF <logical expression> THEN <first group of steps> [ELSE] [<second group of steps>] END is used for conditionally executing the first group of instructions or alternate second group when the logical expression is TRUE or FALSE, respectively. The program execution continues at the first step after END. If no ELSE option is

specified the IF structure is useful to execute or not the first group of steps only.
The instruction GO TO <label> performs an unconditional branch to the program step that is specified by the label number.
The instruction SET <variable> = <location value> assigns the value of the variable equal to the specified location value.
The function ABS <expression> returns the magnitude of the given expression.
The function SIN <expression> gives the trigonometric sine of the given argument in degrees. Similarly functions COS and TAN are trigonometric cosine and tangent, respectively.
The function ATAN2 <expression1>, <expression2> is the arctangent function giving the angle in degrees for the trigonometric tangent that is equal to the ratio between expression1 and expression2.
The function INT <expression> gives the integer part of the specified expression with its sign.
The function SQRT <expression> returns the square root of the specified expression when it is positive.
Among the instructions for executing elementary actions let us consider DO READY, MOVE, MOVES, MOVET, DRIVE, APPRO, APPROS, DEPART, DEPARTS, DELAY, SPEED, SIGNAL, CLOSE, OPEN, RELAX.
The READY instruction moves the robot to a standard starting configuration, which is at the top part of the workspace. The command DO READY can also be executed in display mode.
The instruction MOVE <location> moves the robot from the current configuration to the position and orientation for the extremity that is described by the six values in the location variable. The movement is obtained by interpolating between the initial and final joint positions.
The instruction MOVES <location> is similar to MOVE, but it moves the robot extremity along a straight-line path with a smooth rotation to the final orientation.
The instruction MOVET <location> is similar to MOVE, but additionally it opens the robot hand.
The instruction DRIVE <joint>, <change>, <speed> operates the single specified joint by changing its joint variable of the specified amount at given speed percentage value.
The instruction APPRO <location>, <distance> moves the robot extremity to the specified location but offsets along the tool Z-axis by the given distance.
The instruction APPROS is similar to APPRO, but the tool robot extremity is moved along a straight-line path and is smoothly rotated to its final orientation.
The instruction DEPART <distance> moves the tool at a given distance along the current Z-axis of the tool. DEPARTS is similar to DEPART, but the tool is moved along a straight-line.
The instruction DELAY <time> permits robot motion to stop for the specified period of time in seconds.
The instruction SPEED <value> ensures that subsequent robot motion be performed at the specified speed value as a percentage of the normal speed.
The instruction SIGNAL <signal1>, ..., <signal-n> turns the specified signal channels

on or off when positive or negative signal value are given, respectively.
The instructions OPEN and CLOSE cause the pneumatic control valves to open or close the robot hand, respectively.
The instruction RELAX turns off the electrovalves for the robot hand.
In addition, there are instructions for giving the configuration of the robot and position of the robot extremity, as specific action of pause elementary actions. They are, for example: WHERE, PPOINT, TRANS, HERE.
The instruction/command HERE <variable> sets the variable value of a transformation or a precision point equal to the current robot location.
The function TRANS [<expression1>, ..., <expression6>] gives a so-called transformation value from Cartesian position displacements and orientation rotations as the specified six expressions that are ordinate for the Cartesian coordinates and Euler angles. A zero value is assumed for any omitted expression.
The command WHERE displays the current location of the robot in term of Cartesian World coordinates and joint variables.
The function PPOINT [<expression1>, ..., <expression6>] returns a precision point value for each joint variable from the given expressions that are Cartesian coordinates and Euler angles of the robot extremity tool.
The above-mentioned instructions do not complete the programming capability of VAL-II language and the reader is kindly suggested to refer to the specific User's Manual for a complete analysis, even referring to any other specific programming language of industrial robots.

2.3.2 A programming language for robots: ACL[2]

In this section basic instructions of the programming language ACL (Advanced Control Language) for teaching robots are illustrated in order to show the characteristics and requirements for a suitable programming of those robots. The ACL language is presented briefly since it can be considered significant from technical viewpoints to give examples of dictionary and syntax of instructions that are used to program the operation of teaching robots.
The syntax of each instruction is usually indicated through the necessary data for the instruction execution within bracket symbols < > and through additional data within bracket symbols [ ] when they are of complementary information.
As examples of the above-mentioned groups of instructions one can refer to the basic and most used instructions as in the following.
Among the instructions for program execution let us consider RUN, LIST, REMOVE, RECEIVE, and APPEND.
The command RUN <file name> executes the program with the specified name.
The command LIST <file name> displays all instruction lines of the specified program.
The command REMOVE <file name> erases the specified program
The command RECEIVE <file name> loads the contents of the specified program from

---

[2] The information on ACL herein reported is based on the ACL Reference Guide 3rd Edition, Eshed Robotec, 1982.

a back-up file.
The command APPEND <file name> loads the specified external program.
The instructions for program editing are included in the command EDIT that starts an editing session. The command EDIT [<file name>] starts an editing session for the specified program.
The editing session can be quitted by the command EXIT.
The command S goes to the specified line of the program; P goes to the previous line; DEL erases the current line; L <label1> <label2> displays the instructions between the lines of label1 and label2.
Among the instructions for executing computations and logical evaluations let us consider: SET, GET, FOR-ENDFOR, IF-ENDIF, GO TO and basic mathematical functions.
The instruction SET <variable1> = <variable2> assigns the specified value of variable2 to the variable1.
The instruction GET <variable> permits the assignment of a value to the specified variable.
The mathematical function SIN <variable>, COS, TAN, and ATAN refer to trigonometric functions.
The instruction IF <variable1> <condition> <variable2> permits the check of the given condition (i.e. smaller than, equal to, and so on) between the specified variables. By using an additional line, instruction ANDIF or ORIF can combine other conditions with a first IF check. In the case of alternative conditions, if required the line instruction ELSE can be used. Then, the instruction ENDIF closes the IF loop.
The instruction FOR <variable> = <value1> TO <value2> <group of instructions> ENDFOR permits a repeat of the group of instructions within the loop until the specified variable is within the value1 and value2 being increased by one.
The instruction GO TO <label> continues the program execution at the instruction with the specified label.
Among the instructions for executing elementary actions let us consider: MOVE, MOVED, MOVEC, MOVEL, MOVES, SHIFT, OPEN, CLOSE, JAW, SPEED, HOME, and DELAY.
The instruction MOVE < position> [time] moves the robot to the specified position of robot extremity at the current speed or within the specified time.
The instruction MOVED is similar to MOVE, but it permits continuing to next instruction only when the robot has accurately reached the specified position. Indeed the suffix D is used to obtain this motion mode for the robot also for other types of motion, such as the following.
The instruction MOVEC <position1> <position1> moves the robot extremity to the specified position1 along a circular path passing through the specified position2.
The instruction MOVEL <position> moves the robot extremity along a straight-line path to the specified position at current speed.
The instruction MOVES <vector> <initial> <final> moves the robot smoothly through all positions in the specified vector from the specified initial to final positions.
The instruction SHIFT <position> BY <axis> <value> shifts robot location of a position

by the specified value for the specified joint axis.
The instruction SHIFTC <position> BY <value> <coordinates> shifts robot location of a position by the specified value for the specified Cartesian coordinate.
The instructions OPEN and CLOSE permit opening and closing of the gripper until the end of motion, respectively.
The instruction JAW <variable> [<time>] gives the gripper opening to the specified size within the specified time.
The instruction SPEED <variable> sets the speed for all joint axes at the specified value that can be a percentage of the normal speed.
The instruction HOME runs all robot axes to home starting position.
The instruction DELAY <variable> stops the program executions for the specified time in units of 10 milliseconds.
In addition, there are instructions for giving the configuration of the robot and position of the robot extremity, as specific action of pause elementary actions, such as for example HERE.
The command HERE <variable> records the robot configuration at a position through the joint angles.
The above-mentioned instructions do not complete the programming capability of ACL language and the reader is kindly suggested to refer to specific User's Manual.

2.4 Illustrative examples

In this section examples are illustrated from different viewpoints and for different applications with the aim of showing the practical feasibility of the above-mentioned concepts and procedures for analyzing and/or designing robotized manipulations.
Industrial robots are generally considered the most complicated systems in automation so that their use is usually restricted to applications requiring users with a high level of education and continuous training.
But the use of the basic concept of elementary actions can reduce the complexity of a robotized operation and deduce a flexible programming in terms of both clear manipulative items and program instructions.
Two sets of examples are illustrated concerned with teaching robot use and industrial applications, respectively, in order to show the difference but similarity for analysis and design of robotized manipulations in different cases of study.

2.4.1 Education practices

The teaching for robot use and programming can be carried out through practices with robots that can recall practical industrial applications or their basic characteristics. In the following, a few illustrative examples are described to show how to analyze or design a robotized manipulation for education purposes with laboratory equipment but significant characteristics.

2.4.1.1 Simulation of an industrial process

This example illustrates a practice simulating an industrial process for immersion of a ceramic plate into a recipient with crystalline fluid to end the finishing of a production.

The process has been examined in agreement with the procedure in Fig. 2.1. Manipulative tasks have been identified for a suitable robotization as shown in Figs 2.2a) and 2.3.

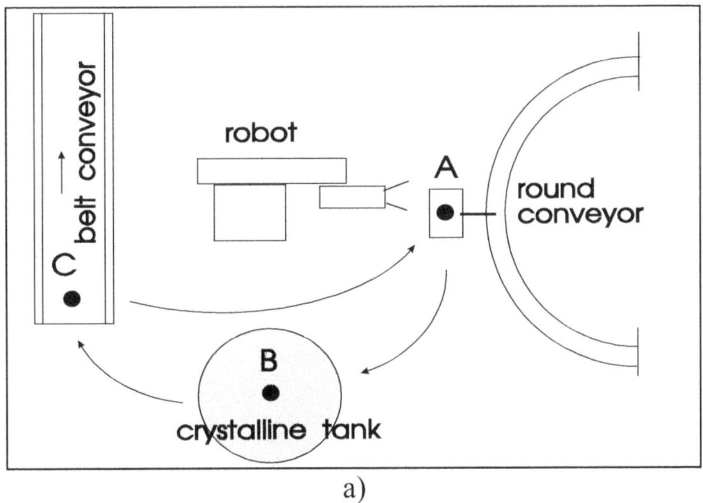

a)

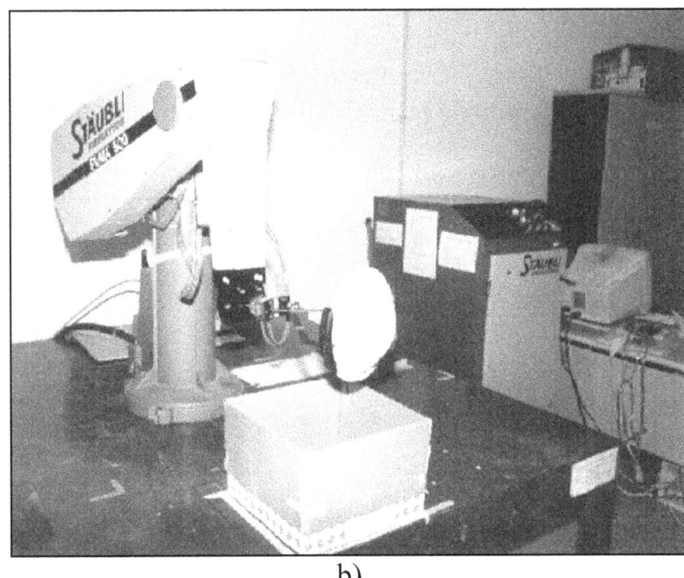

b)

Fig. 2.2: An illustrative example from teaching activity by using the industrial robot PUMA562 with the proposed procedure for robotized manipulation design: a) a scheme of the environment; b) the practising layout.

A robotized work-cell has been designed and its environment has been specified as is the case with the robotized work-cell of Fig. 2.2a), where a robot has been placed in a central position with respect to a round feeder, the fluid crystalline tank, and an unloading belt conveyor. The criterion for placing the robot in this position can be considered a result of the manipulation analysis, aiming at optimizing robot size and

manipulative trajectory through empirical adjustments during laboratory experiments.
In this student practice optimal trajectory planning has not been set on the basis of a specific analytical formulation but it has been estimated by only taking into account the robot movements in a general way by experimental set-up.
The practising layout is shown in Fig. 2.2b), in which a PUMA robot has been used because of its six d.o.f.s, which have been computed as needed to perform the prescribed actions. The required manipulation movements can be considered as three Cartesian translations and three rotations. The translations are necessary to move the robot extremity in the working space and rotations are necessary to conveniently orientate the gripper and plate as necessary at points A, B, and C.
The manipulation analysis has been carried out according to the procedure of Fig. 2.1 and the results have been reported in Fig. 2.3. In particular, Fig. 2.3 shows a model for the manipulative actions in which the reference points A, B, C refer to those of Fig. 2.2a) have been indicated with the manipulative trajectory between them. In particular, points A_1, B_1, and C_1 have been specified as those reference points from which accurate motion of the robot starts to final positions A, B, and C, respectively. Thus, fast motions can be programmed between two 1-points and slow accurate motions can be limited to the short path in each linear segment only.
Manipulations have been detailed in the form of elementary actions and Fig. 2.4 shows a sequence of the operations. Particularly the dripping motion at point B_1 in Fig. 2.3 has been prescribed in the form illustrated in the dripping action in Fig. 2.4 after some experiences that have been made following the item 'modify to adjust' of the procedure of Fig. 2.1.

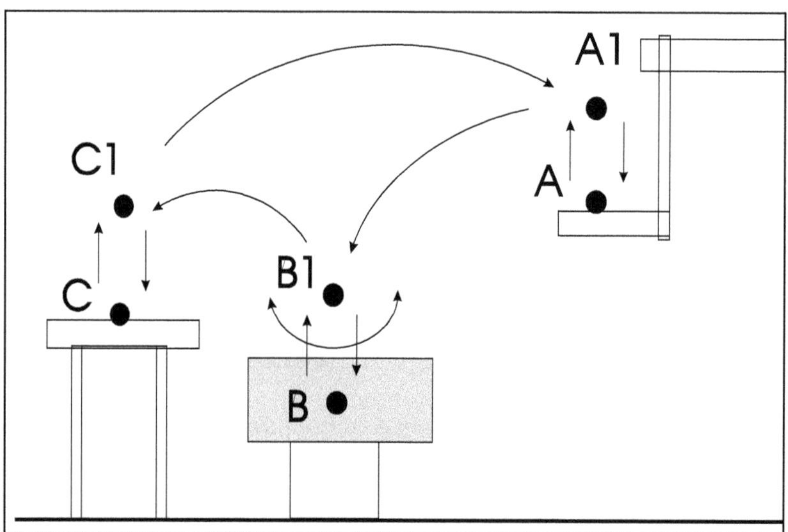

Fig. 2.3: A scheme for the analysis of manipulations in the example of Fig. 2.2

These experiences have been used to obtain a suitable motion sequence for the dripping to ensure appropriate draining and uniform distribution of the crystalline liquid on the

plate's surface. Number N in the flowchart of Fig. 2.4 refers to the number of plates in a production cycle and it can be prescribed for continuous functioning.
Two led checks have been designed as located in the work-cell to have the robot pause in positions suitable both for the manufacturing process and the robot. The first led check gives information on free place in the unloading conveyor and it controls a passive pause of the robot. The second led check sensor recognizes the presence of a place in the round feeder and gives the start signal to the robot.
Figure 2.4 lists elementary actions by grouping them in appropriate operations, which are functions of one reference point each so that programming flexibility depends on the reference point coordinates only.
In addition, one can recognize the elementary motions, named in such a way to recall as many feasible instructions as possible for robot programming. The flowchart of Fig. 2.4 is a clear example of how to achieve a high level of versatility and flexibility for a robot application by properly studying the basic characteristics of a manipulation in terms of sequence of elementary actions.
The nature of the elementary actions is identified through a manipulation analysis as shown in Fig. 2.4 and the corresponding list of instructions is reported in Fig. 2.5
The programming in Fig. 2.5 has been obtained by using the sequence of elementary actions as shown in operations and phases in Fig. 2.4.
In particular, once at the beginning of the program the precision points are listed and determined as functions of their coordinates in position and orientation through the VAL function TRANS, the robot starts at a high speed to reach the loading station in A. Then it approaches and grasps a plate at a lower speed 100 and successively goes to B where it operates the dripping with movements that are specified by means of precision points R1 and R2. Finally the robot goes to C to unload the plate and very rapidly (SPEED 200) goes to a wait position W or back to A1 if the counter has not reached the end value of a cycle.
The practical calibration and operation efficiency have been obtained by looking at several operation speeds as outlined at the end of the general flowchart of Fig. 2.1.
The elementary actions have been indicated in Table 2.1 according to the classification of the previous section as movement M, transport T, active pause A, and passive pause P in order to optimize both the manipulation and programming not just for cycle time.
The time evaluation can be very interesting since it gives a measurement of the length of each action within the overall cycle. Moreover, it can be used to identify what operation can be conveniently modified to improve productivity.
Thus, from Fig. 2.4 and Table 2.1 it is evident that the dripping operation can be easily speeded up, although only within a given range that ensures proper draining. The transport operations to B and C can also be speeded up but in a different way since the last transportation to C concerns a plate covered with crystalline, and excessive acceleration may alter the crystalline distribution on the plate.
Finally, Table 2.1 reports a final evaluation of the experienced practices also carried out in terms of operation timings in which some elementary actions have been grouped according to Fig. 2.4.

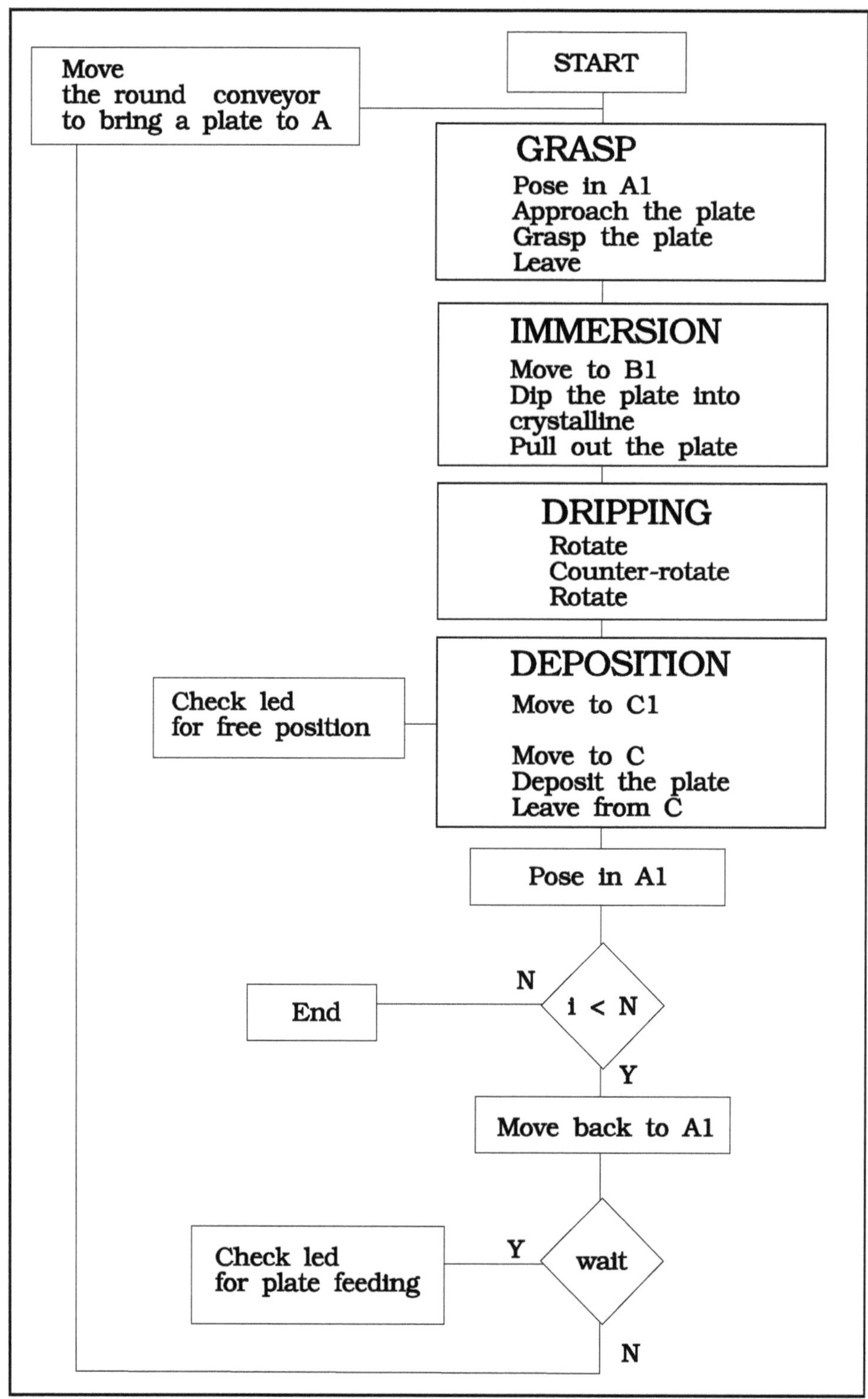

Fig. 2.4: A flowchart for the manipulation of Fig. 2.3 through elementary actions.

```
SET A= TRANS (xa,ya,za,oa,aa,ta)
SET A1= TRANS (xa1,ya1,za1,oa1,aa1,ta1)
SET B= TRANS (xb,yb,zb,ob,ab,tb)
SET B1= TRANS (xb1,yb1,zb1,ob1,ab1,tb1)
SET C= TRANS (xc,yc,zc,oc,ac,tc)
SET C1= TRANS (xc1,yc1,zc1,oc1,ac1,tc1)
SET F= TRANS (xf,yf,zf,of2,af,tf)
SET R1= TRANS (xr1,yr1,zr1,or1,ar1,tr1)
SET R2= TRANS (xr2,yr2,zr2,or2,ar2,tr2)
SET W= TRANS (xw,yw,zw,ow,aw,tw)
SPEED 110
COUNT = 0
OPEN
10 MOVE A1
SPEED 100
MOVE A
DELAY 1
CLOSE
MOVE A1
MOVE B1
MOVE B
DELAY 1
MOVE B1
DELAY 0.5
MOVE R1
DELAY 0.5
MOVE R2
DELAY 0.5
MOVE B1
DELAY 0.5
MOVE C1
MOVE C
DELAY 1
OPEN
MOVE C1
SPEED 200
COUNT = COUNT + 1
IF COUNT < 10 THEN
GO TO 10
ELSE
GO TO 20
20 MOVE C2
MOVE W
STOP
END
```

Fig. 2.5: The list of programming instructions for the flowchart in Fig. 2.4.

2.4.1.2 Writing with a robot

This example illustrates a practice for writing with a robot not only as an attractive application for a substitute of human activity, but also as manipulations representing precise manipulations for industrial applications, such as welding or assembling.

The writing process has been examined in conjunction with the manipulation of a sheet, which the robot will write on. In this case the environment has been considered as specific for a robotization and the manipulations have been conveniently thought for a 6 d.o.f. robot, but without mimicking the human operations. Thus, the working area has been designed as shown in Fig. 2.6 in order to adjust the manipulations to the available Scorbot robot ER-Vplus.

Two sets of manipulations have been examined as related to the feeding of a sheet and the writing, respectively.

Table 2.1: An evaluation of the robotized manipulative elementary actions of Fig. 2.3.

Description	**T**	**M**	**A**	**P**	**Time (sec)**
Robot goes to A1		•			0.2
goes to A grasps a plate goes to A1	 • •	• •	 •		 0.2–0.5
goes to B1	•				2.5
dips the plate to B	•			•	1.0
pulls out the plate to B1	•				0.5
moves to drip (up & down and rotate)	•		•		2.2–2.5
goes to C1	•				2.5
deposits the plate to C	•		•		1.0
goes back to C1 and A		•			1.0
checks for the plate			•	•	0.5–2.5

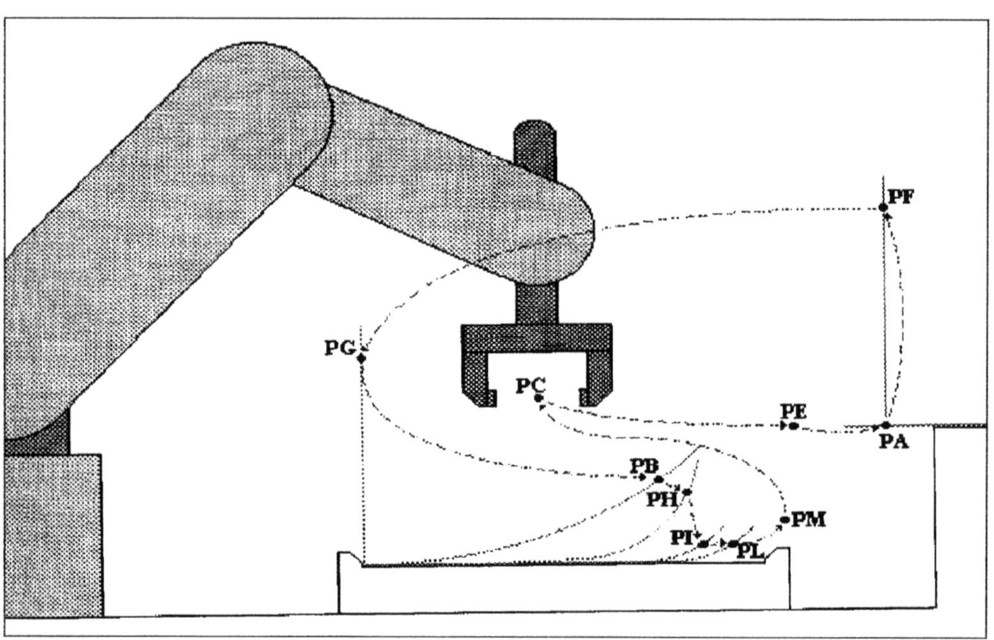

Fig. 2.6: A scheme with precision points for the manipulation of a sheet by robot arm.

The first manipulation has been characterized by the precision points shown in Fig. 2.6 together with a path of the reference point on robot extremity by combining an analysis of suitable motion for the sheet with experiments of suitable operations and programming of the robot. The motion of the sheet has been characterized by the indicated precision points that have been used for a flexible programming from a loading station at point PA to the writing plane at point PL. The size and shape (mainly the borders) of the writing plane have been designed to facilitate the deposition of the sheet also as a result of learning the robot versatility.

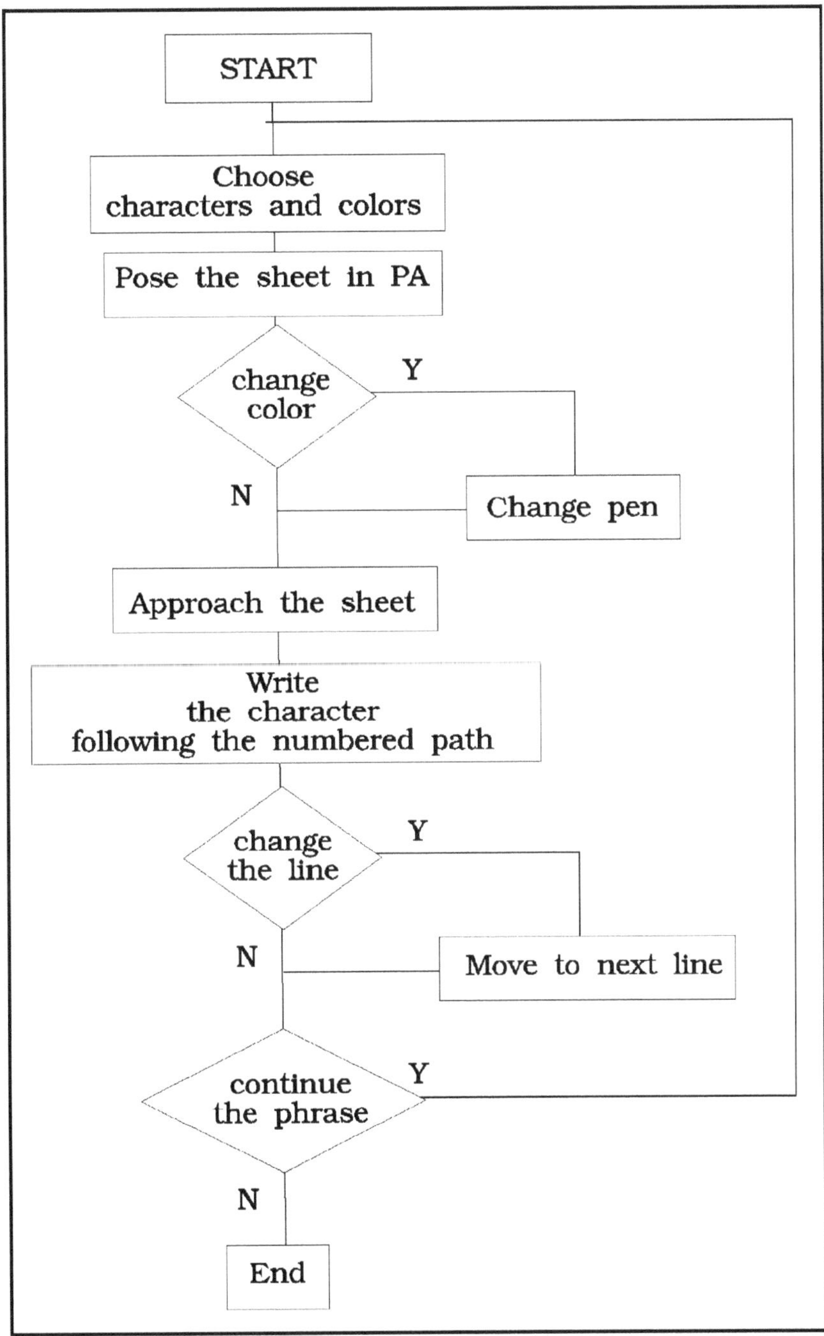

Fig. 2.7: A flowchart for all the manipulations in writing on a sheet.

The second manipulation, referring specifically to the writing, has been analyzed through the general flowchart of Fig. 2.7 by using a suitable general mapping for characters that is shown in Fig. 2.8.
The overall manipulation of the writing is described in the flowchart of Fig. 2.7 in which options for pen and colors have been considered together with a long writing that can require several lines and even several pages.
Thus, in the flowchart operations are indicated as a group of elementary actions that have been examined and represented by motions among precision points for the sheet feeding and character writing in Figs 2.6 and 2.8, respectively.
In particular, the instruction list of Fig. 2.9 refers to the writing of character R as shown in Fig. 2.8 and it has been designed as a sequence of elementary actions for linear segments and one circular arc. Thus, flexible programming is obtained by using instructions MOVELD for the straight-line segments and MOVECD for the circular segment only. In addition, although the character mapping has been designed for a specific size, the precision points have been labeled with numbers so that their coordinates can be easily changed as a function of the mapping size.

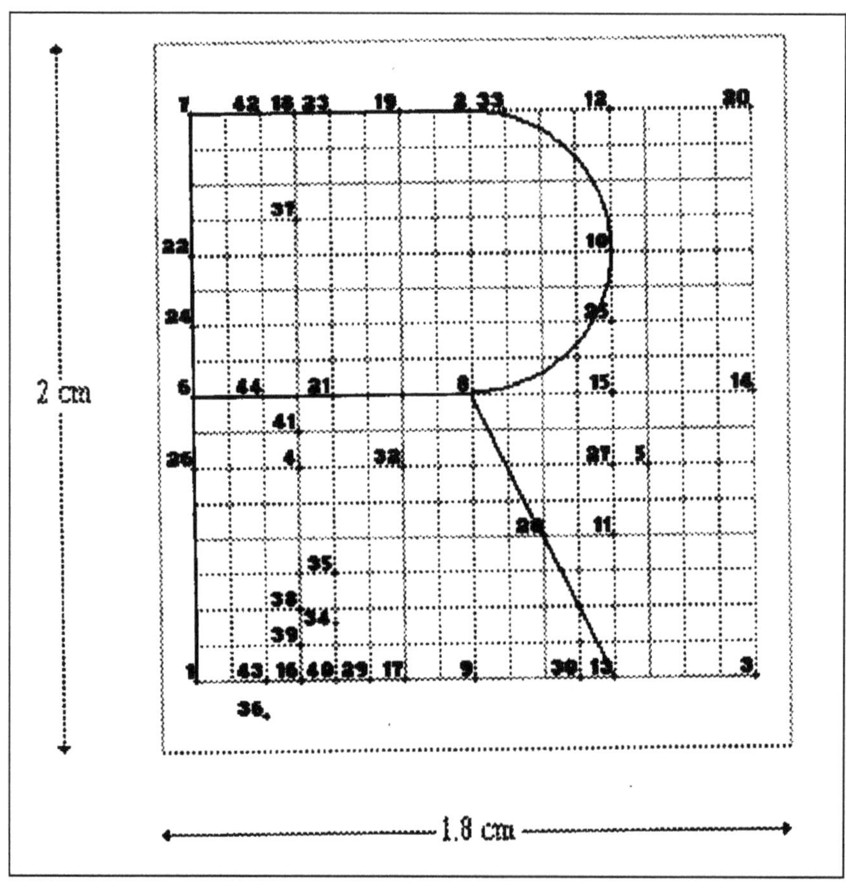

Fig. 2.8: A scheme with precision points for writing letters by using elementary actions to draw linear or circular segments.

```
SPEED 15                      MOVELD P[6]
MOVELD P[1]                   SHIFTC P[6] BY Z 200
SHIFTC P[1] BY Z 200          MOVELD P[8]
MOVELD P[1]                   SHIFTC P[8] BY Z -200
SHIFTC P[1] BY Z -200         MOVELD P[13]
MOVELD P[1]                   SHIFTC P[13] BY Z 200
MOVELD P[7]                   MOVE P[1]
MOVELD P[2]                   SHIFTC P[1] BY Z 200
MOVECD P[8] P[10]
```

Fig. 2.9: List of programming instructions for writing the character R of Fig. 2.8.

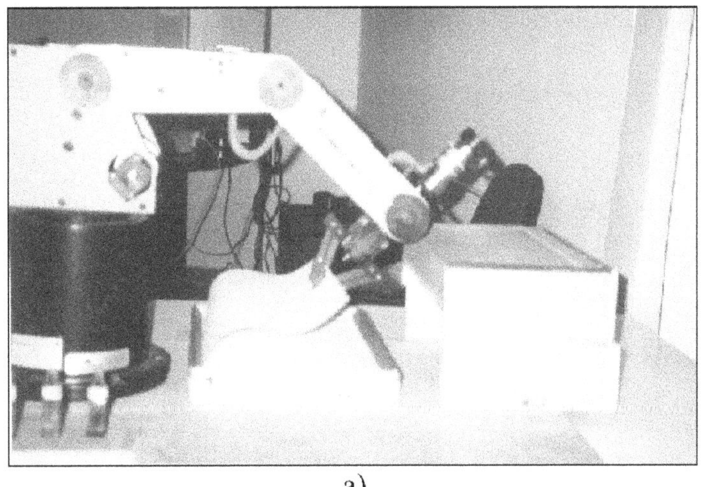

a)

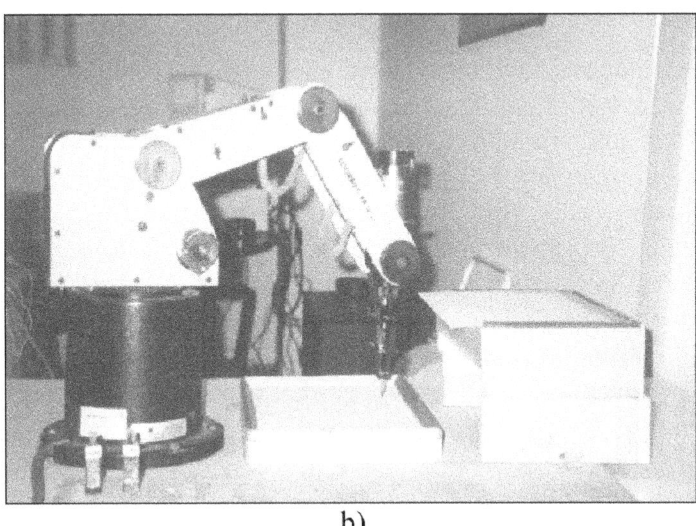

b)

Fig. 2.10: The practising layout for the manipulations of posing a sheet and writing on it by using the schemes of Figs 2.6, 2.7, and 2.8.

Fig. 2.11: Results of writing by the robot in Fig. 2.10 at different speed operation.

Table 2.2: An evaluation of the robotized manipulative actions for manipulating a sheet and writing characters of Fig. 2.11.

Action description	T	M	A	P	Time (sec.)
Robot goes to PC		•			0.5
grasps a sheet		•	•		1.0
moves to PL	•				12.5
deposits the sheet	•		•		2.0
goes to pen		•			0.5
grasps a pen		•	•		7.5
goes to sheet	•				0.5
writes character R	•		•		8.0

The manipulations have been tested and used successfully as shown in Fig. 2.10. Then in agreement with final items of the procedure of Fig. 2.1, the flexibility and versatility of the robotized writing have been tested with several conditions, and results are shown in Fig. 2.11 referring to attempts to speed up the writing. It is worthy of note that although very fast writing execution can be achieved, the quality of the writing varies considerably, as shown in Fig. 2.11 where V refers to the value of used robot speed. Thus, the student–users have chosen the speed V=3 as reference to robotized speed writing for the manipulation evaluation that is reported in Table 2.2.

The studied robotized writing process, including the sheet feeding, has been designed by using the procedure of Fig. 2.1 so that it has also been easily adapted to the PUMA robot with similar results.

The writing has been ensured with a proper mechanical design of the pen end-effector in order to avoid complex or sensored programming that should be required to consider the uncertainty of the sheet plane on which the pen writes only when in contact. A simple solution has been designed to ensure the contact of the pen with the sheet during the writing by using a suitable spring that properly pushes the pen to the sheet with a

forced contact. In addition, the pen has been properly inclined with respect to the sheet plane by suitable initial calibration.
This example mainly shows that a proper determination and use of elementary actions can give a simple robotization with suitable versatility and flexibility even for apparently complex human operations.

2.4.1.3 Intelligent packing

This example illustrates intelligent packing of batteries by using few additional sensors, but through suitable manipulations.
The intelligent packing has been designed for batteries including a test of correct production in order to pack a battery in a given box with a proper polarity direction. The process has been examined by starting from a feed on a conveyor with a loading station at which a battery can be at a vertical or horizontal posture with undetermined polarity direction. The required manipulation can be considered as intelligent when the manipulative task includes the determination of battery orientation so that the robot can pack the battery into a suitable box with a proper battery polarity.
The manipulation has been properly analyzed as shown in Fig. 2.12 where precision points have been determined together with alternative paths depending on the battery polarity and posture. In Fig. 2.12 a layout for a robotized work-cell is also indicated as a function of the feeding conveyor in A and packing station in B. The corresponding flowchart for battery manipulation is reported in Fig. 2.13.

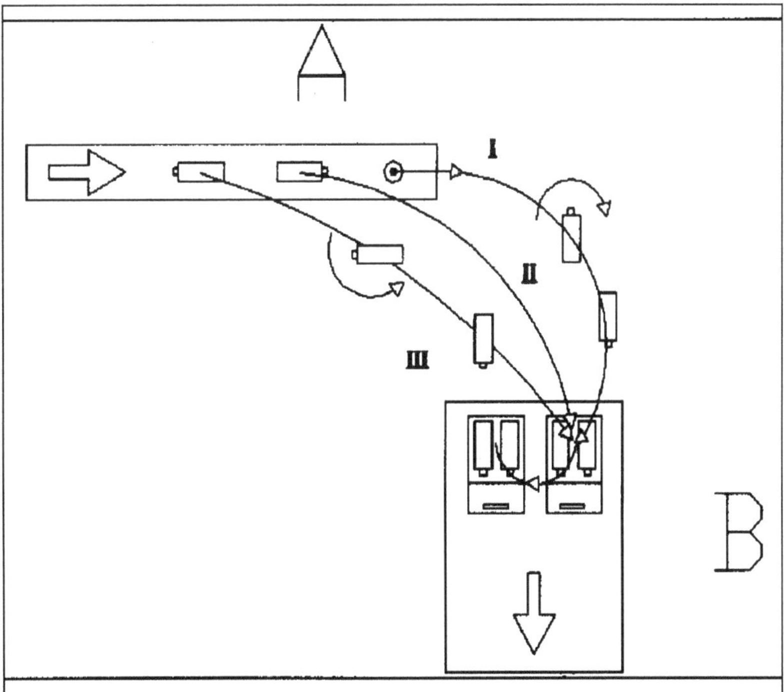

Fig. 2.12: A scheme for the analysis of manipulations for intelligent packing of batteries.

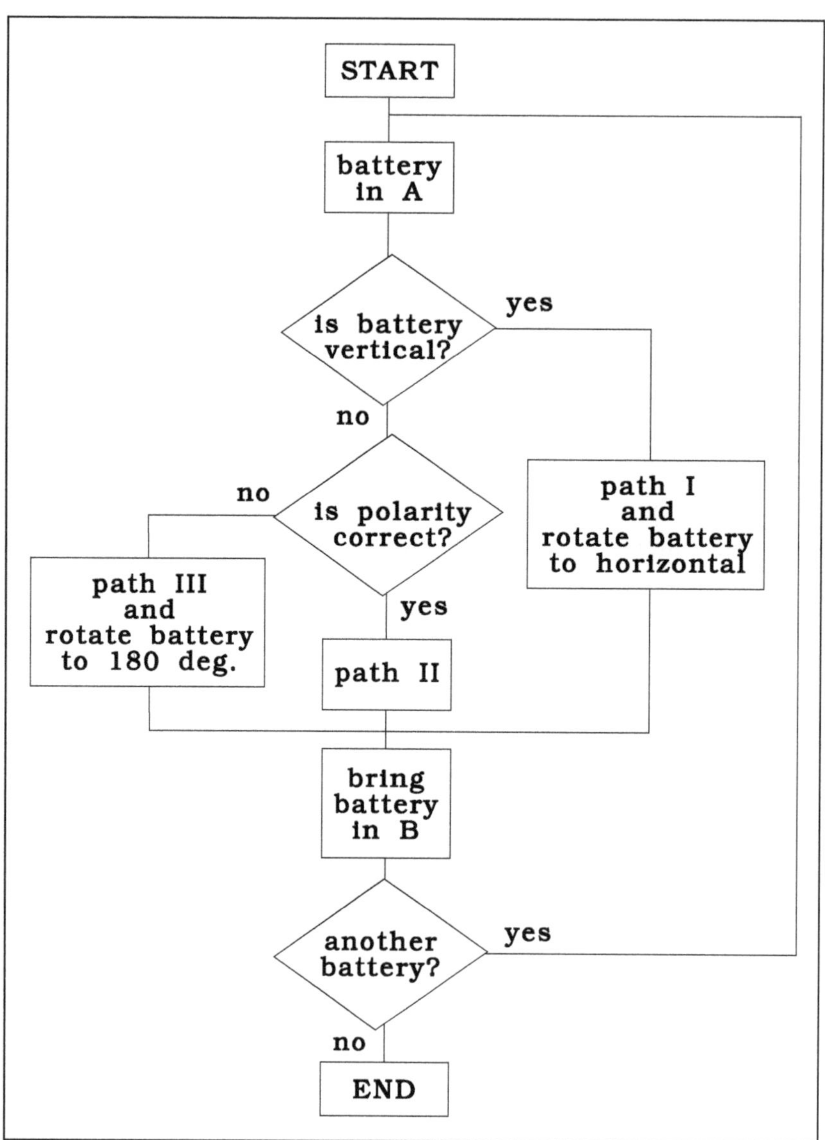

Fig. 2.13: A flowchart for the manipulation of a battery in Fig. 2.12

A scheme for a specific robotization has been designed as in Fig. 2.14 and the laboratory practising layout is shown in Fig. 2.15 by using the available robot.
The three alternative paths in Fig. 2.12 can be performed by a 5 d.o.f.s robot since three translations are required for the transfer motion from A to B, and two rotations (about vertical and horizontal axes) are necessary to rotate a battery from any posture in A to the horizontal deposition in the pack box in B with the proper polarity orientation.
The feeding conveyor has been sensored with two proximity sensors to detect the posture of a battery as in a vertical or horizontal posture. Thus, the robot can move to grasp a battery by means of a suitable mode corresponding to the current posture and quote of a battery.

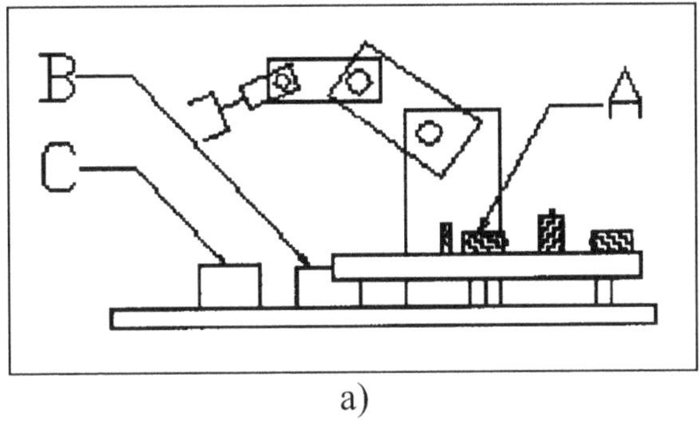

a)

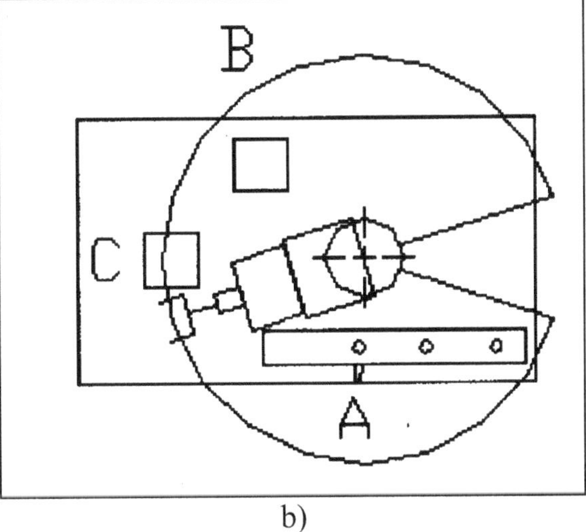

b)

Fig. 2.14 A scheme for the robotized work-cell for manipulations of Fig. 2.12: a) plane view; b) vertical view.

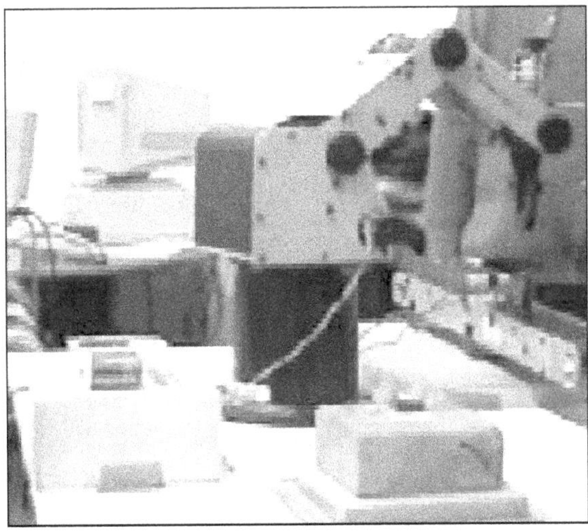

Fig. 2.15: The practising layout with robot Scorbot ER-Vplus for the manipulation of Fig. 2.12.

```
* GIVE CCORDINATES OF  PRECISION
POINTS:
* PRDO, PRV, PRH, PR1, INRIB, C, PRC, ROTA,
BATT,
* PRG1, PRG2, 1, 2, 3, 4, 11, 22 , 33, 44
FOR I =1 TO N
READ VELMAX
SET V=VELMAX*0.30
SPEED VELMAX
WAIT IN[1]=1
IF IN[2]=1
GOSUB POSV
ELSE
GOSUB POSH
ENDIF
ENDFOR
END
*PROGRAM GOSUB POSV FOR VERTICAL
BATTERY
MOVED PR0
SPEED V
OPEN
MOVELD PRV
CLOSE
MOVELD PR0
MOVEC INRIB PR1
MOVELD C
OPEN
MOVELD ROTA
MOVELD BATT
CLOSE
MOVELD ROTA
SPEED VELMAX
GOSUB  INPACK
MOVELD W
END
*PROGRAM GOSUB POSH FOR HORIZONTAL
BATTERY
MOVED PR0
SPEED V
OPEN
MOVELD PRH
CLOSE
IF IN[4]=1
GOSUB POLARITY
SPEED VELMAX
GOSUB TURNINPACK
ELSE
SPEED VELMAX
GOSUB TURNINPACK
ENDIF
MOVELD W
END
*PROGRAM GOSUB POLARITY
OPEN
MOVELD PRC
SPEED VELMAX
MOVELD PRG1
SPEED V
MOVELD PRG2
CLOSE
MOVELD PRG1
SPEED VELMAX
MOVELD PRC
END
*PROGRAM GOSUB INPACK
SPEED  V
IF I=1
MOVED 11
MOVELD 1
ELSE
IF
I=2
MOVED 22
MOVELD 2
ELSE
IF
I=3
MOVED 33
MOVELD 3
ELSE
MOVED 44
MOVELD 4
ENDIF
ENDIF
ENDIF
OPEN
CLOSE
END
*PROGRAM GOSUB TURNINPACK
SPEED  V
IF I=1
MOVECD 11 PR1
MOVELD 1
ELSE
IF
I=2
MOVECD 22 PR1
MOVELD 2
ELSE
IF
I=3
MOVECD 33 PR1
MOVELD 3
ELSE
MOVECD 44 PR1
MOVELD 4
ENDIF
ENDIF
ENDIF
OPEN
CLOSE
END
```

Fig. 2.16: Programming instructions for the robotized manipulation of Fig. 2.12.

Table 2.3: An evaluation of the robotized manipulation elementary actions of Fig. 2.12 by the robot in Fig. 2.15.

Action description	T	M	A	P	Path I time (sec.)	Path II time (sec.)	Path III time (sec.)
From home to W	•		•		0.2	0.2	0.2
Sensor signal				•	0.1-1.0	0.1-1.0	0.1-1.0
From W to A	•				3.0	3.0	3.0
Grasping battery		•	•		2.5	9.8	3.3
From A to B	•				--	5.1	5.1
From A to C	•				2.4	--	--
Deposit, check, and re-grasp	•	•	•		3.5	No check	No check
From C to B	•				2.3	--	--
Releasing into pack	•		•		1.0	1.0	1.0
From B to W		•			1.3	1.3	1.3

In order to check the battery polarity an intermediate station C has been designed accepting a horizontal battery, and it has been equipped with a proper electric circuit that gives a signal indicating to the robot controller the polarity of therein deposited battery. Consequently, the robot will transfer the battery to the pack box rotating it properly.
The intelligent manipulation has been modeled by using the signals from the feeding conveyor and intermediate polarity station as shown in the flowchart so that the robot will move depending on the orientation of the battery and can execute the three alternative paths. The corresponding program is listed in Fig. 2.16 by using the instructions of ACL language for the used Scorbot ER-Vplus robot. One can recognize the suitable sequence for the elementary actions as depending on the selected path because of the signal values. The robotized manipulation has been tested and calibrated even for future productivity improvements in agreement with the general procedure in Fig. 2.1 and results of time evaluation for elementary actions are reported in Table 2.3.
Summarizing, an intelligent manipulation can be performed by using few sensor signals in combination with a proper manipulation analysis that takes into account alternative manipulative tasks and paths depending on the different situations in the manipulative process. The success of such a manipulation strongly depends on the accuracy and depth of the manipulation analysis and particularly determination of suitable elementary actions.

2.4.2 Industrial applications

The following examples show that the same procedure used in teaching activity can also be successfully applied for analysis and design of robotized manipulations in industrial applications. In particular, the reported examples are illustrated by emphasizing the similarity with the previous examples from education practices.

2.4.2.1 Designing a robotized manipulation[3]

The need for high precision in manufacturing and great reliability in the final product characteristics for aerospace applications may require a robotized automation of the manual labor for the production of three-dimensional (3D) pieces with particular geometry. A study on a specific 3D filament winding has been carried out to design a specific robotization and a suitable robotized manipulation. Thus, the peculiarities of the manipulative task have been recognized in terms of robotized manipulation by looking at human operations and a specific study has been thought necessary for a rational design of a robotized winding manufacturing.

The human manufacturing process has been analyzed in detail through visual inspection and video recording in order to understand the basic manipulation requirements. The video means can also help to optimize the actual manual manufacturing as well as to individuate the operations to be or not to be mimicked by a robot for a suitable automation.

The material used in this manufacturing is an advanced composite material (ACM), consisting of carbon fibers impregnated into a thermosetting resin matrix.

The conventional manufacturing is performed by a manual winding of a plait PS of roving on a suitable winding support WD as shown in Fig. 2.17. The labor of a human operator is repetitive with very poor technological and decisional requirements and contributions.

Recurrent human operations can be recognized as a variable grasp from a firm to a slipping mode, and arm movements from pulling to wrapping a roving plait on the winding support. Specifically, basic operations can be modeled to be mimicked by a suitable robotic manipulator and they can be summarized as shown in Fig. 2.18 as:

- slipping of the hand along the plait direction to perform a feeding of the plait during the deposition;
- pulling, which consists of straining a plait to ensure straight-line and compact deposition;
- slipping and pulling to provide a feeding and a straining of the plait after the wrapping on curved surfaces of the winding support;
- rolling the plait about points A and B to wrap the plait on the curved surfaces of the winding support.

During the manual winding the manipulative operations of the operator have been observed as being neither regular nor continuous. Particularly, the straining of the plait is not constant during the feeding and deposition of the wound roving. The ACM gluey property seems to reduce these effects only partially. Moreover, it has been observed that a human operator cannot ensure high precision in the fiber directions during the plait deposition, as may be required by the mechanical design of the ACM product.

---

[3] This study description has been obtained from the published papers: Ceccarelli M., Iacobone F., Carrino L., Anamateros E., "On the Manipulation of a Composite Material Roving for a 3D Winding Manufacturing", Proceedings of 26th International Symposium on Industrial Robots, Singapore, 1995, pp. .597–602; Ceccarelli M., Volante G., Carrino L., Anamateros E., "A Feeding Device for Composite Material Rovings in a Robotized 3D Winding Manufacturing", Proceedings of 27th International Symposium on Industrial Robots, Milan, 1996, pp. .987–992.

A manipulation analysis has been developed on the current manual manufacturing to determine the human arm labor for a classification, as well as a numerical evaluation, as reported in Table 2.4. It has also been useful to determine the basic manipulative tasks for the robotization in agreement with the first items of the procedure shown in Fig. 2.1.

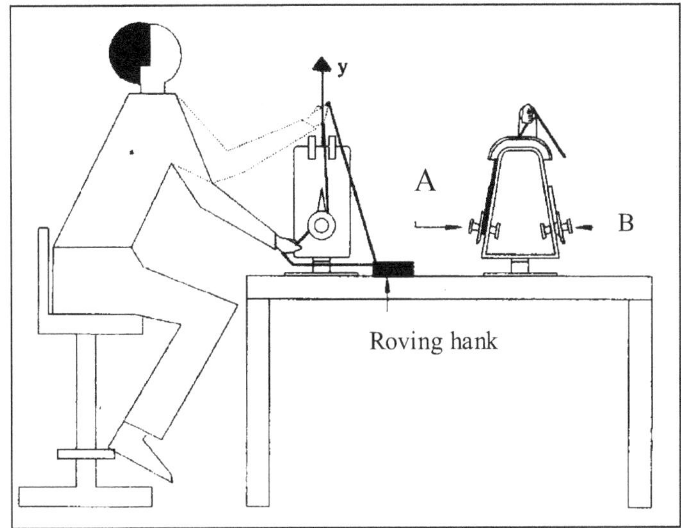

Fig. 2.17: A scheme for conventional manual manufacturing of 3D filament winding of an ACM plait of rovings and a mechanical design for the corresponding winding support.

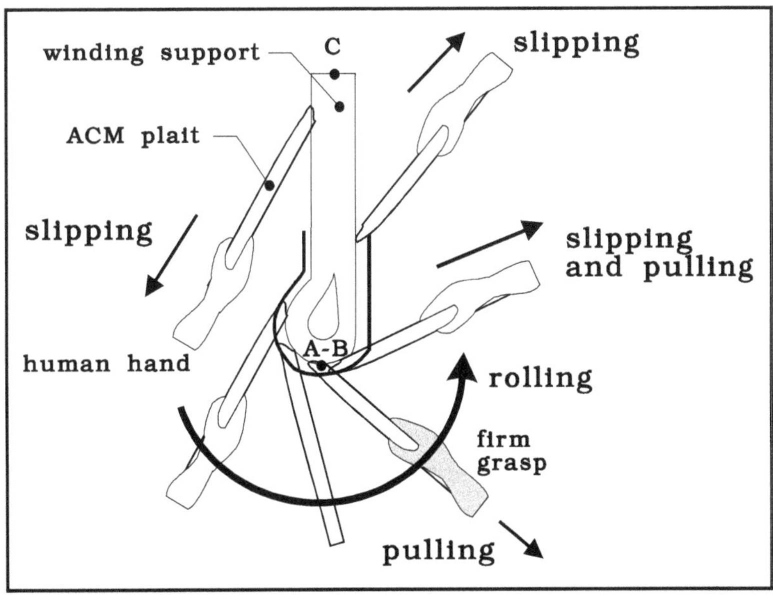

Fig. 2.18: Basic mechanical characteristics of a human manipulation of a composite material roving in a 3D winding manufacturing.

Table 2.4: Evaluation of elementary actions of human 3D filament winding manufacturing in Figs 2.17 and 2.18.
(WD is for winding support; PS is for plait of roving; FH is for feeding hank).

Left hand	**T**	**M**	**A**	**Time (sec)**	**T**	**M**	**A**	**Right hand**
1- goes to WD		•		0.3÷0.8		•		1- goes to PS
2- grasps WD	•			0.5÷0.8	•			2- grasps PS
3- holds WD			•	2÷3	•			3- winds PS up to A
4- locks wound PS	•			0.2÷0.6			•	4- holds PS
5- releases PS	•			1÷2	•			5- winds PS about A
6- grasps PS	•			0.5÷0.8			•	6- holds PS
7- locks PS	•			2÷3	•			7- unrolls PS from FH
8- hold PS and WD			•	1.5÷2	•			8- strains PS
9- releases PS	•			2÷2.5	•			9- winds PS up to C
10- wheels WD	•			1÷2			•	10- holds PS
11- releases WD	•			0.3÷0.6			•	11- strains PS
12- goes to WD		•		0.3÷0.8			•	12- strains PS
13- grasps WD	•			0.5÷0.8			•	13- strains PS
14- holds of WD			•	2÷3	•			14- winds PS up to B
15- locks wound PS	•			0.2÷0.6			•	15- holds PS
16-releases wound PS	•			1÷2	•			16- winds PS about B
17- grasps PS	•			0.5÷0.8			•	17- gets hold of PS
18- locks PS	•			2÷3	•			18- unrolls PS from FH
19- holds PS and WD			•	1.5÷2	•			19- strains PS
20- releases PS	•			2÷2.5	•			20- winds PS up to C

The study has been carried out with the aim of focusing on elementary actions, which can be reproduced by a robotic manipulator with no great complexity in robot programming to give a certain versatility and flexibility to a suitable robotization. In Table 2.4 elementary actions for both arms are individuated and listed with a performance characterization distinguishing them from a functional viewpoint as operations T, movements M, and hold on pause A, in which an end-effector or tooling action, an arm gross movement, or a pause are recognized, respectively. Results have been reported as strictly concerned with the 3D winding process of one winding path, whose execution time has been measured as variable from 20 to 35 sec.
In addition, in Table 2.4 execution times have been reported for each elementary action providing measured intervals during the operation of human operators in several occasions. Those times give measures of actions' significance and the possibility of achievable productivity improvements. In fact, Table 2.4 can be used to suggest speeding up some elementary actions more efficiently than others, as for example elementary actions n. 3, 7, 14, and 18, as well as to identify useless operations, as for example n. 4 for both hands.

In an analogous way, Table 2.4 and Fig. 2.18 can be useful for determining the elementary actions, which are needed for a better consideration both for the operator labor and manufacturing design. Thus, some elementary actions can be remarked to the operator as requiring particular attention for a manipulative path, because of fiber directions, or for straining action, because of material compactness. It is remarkable that the right hand is used much more than the left, which is mainly devoted to complementary elementary actions. Consequently, one robotic manipulator can be used to perform the basic elementary actions for the 3D winding, as for example those by right hand n. 3, 5, 9, 14, 16, and 20.
A robotization has been proposed as shown in Fig. 2.19a) by using the current winding support in order to minimize the redesigning of tools and product process. In Fig. 2.19b) a laboratory set-up is shown as it has been used for preliminary experiments.
At the moment the manufacturing is not generally for a mass production, so that the use of a robot manipulator seems to be very suitable for three-dimensional products with ACM instead of dedicated automatic machinery. This is because a robotic mechanical structure may copy the human arm to mimic human manipulative capabilities and, mostly, because its flexibility can be used for several similar manufacturing 3D winding tasks through suitable programming. In fact, suitable robot programming instructions or small groups of instructions may be individuated to perform an action with easy possibility of changes in the sequence and timing.
The models of Figs 2.18 and 2.20 have been deduced for a mechanical characterization of the manipulative process in order to design a robotized 3D filament winding as shown in Fig. 2.19. In Fig. 2.20 capital letters refer to precision points that can be identified in the manual process with the winding support, and small letters refer to precision points that can be used for robot programming of the same roving plait manipulation.
Particularly, looking at the plait deposition on a winding support, the manual 3D winding has been modeled, as in Fig. 2.20a) where since a roving plait is used there is an additional torsion due to the manipulative trajectory. This torsion seems not to be influential for the deposition trajectory but it seems to be critical for the strain properties of the final product.
However, for both robotization purpose and enhancement of product properties a 3D winding can be better provided by depositing a unique roving that can have a feeding from a force clutched spool, as shown in Fig. 2.19 with a manipulative trajectory like that in Fig. 2.20b).
Moreover, by using the scheme of Fig. 2.20b) the additional torsion in Fig. 2.20a) can be avoided. However, a proper deposition of the roving by using the winding path of Fig. 2.20a) depends on the motion of a winding eye fixed on the robot end. A winding eye is a specific tool guiding the roving in the deposition phase by giving it the proper laying plane. A specific winding eye has been designed and built as shown in Fig. 2.19b) in agreement with the desired manipulation path of Fig. 2.20.
An in-depth analysis for the manipulation characteristics and manual labor, through Figs 2.18 and 2.20 and Table 2.4, can suggest a very reduced number of elementary actions for a robotized 3D filament winding.

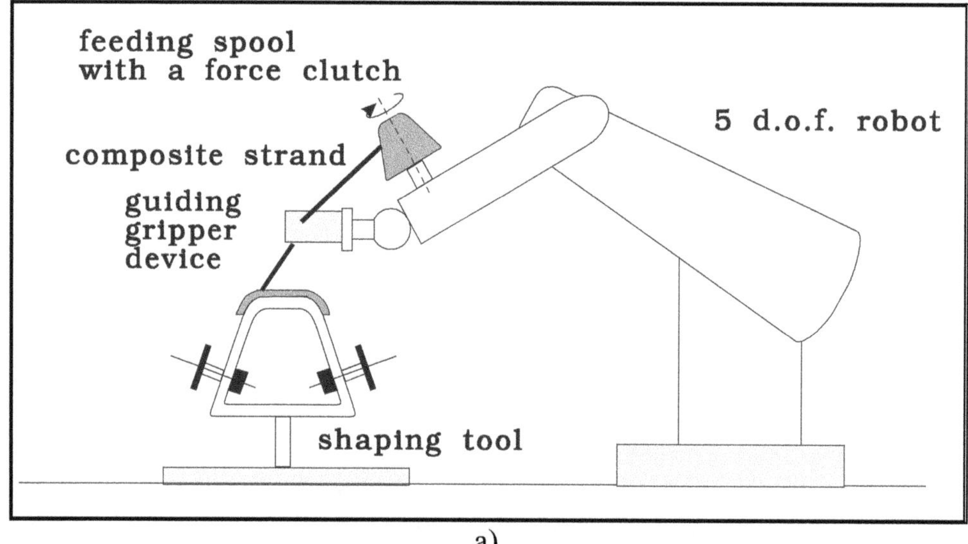

a)

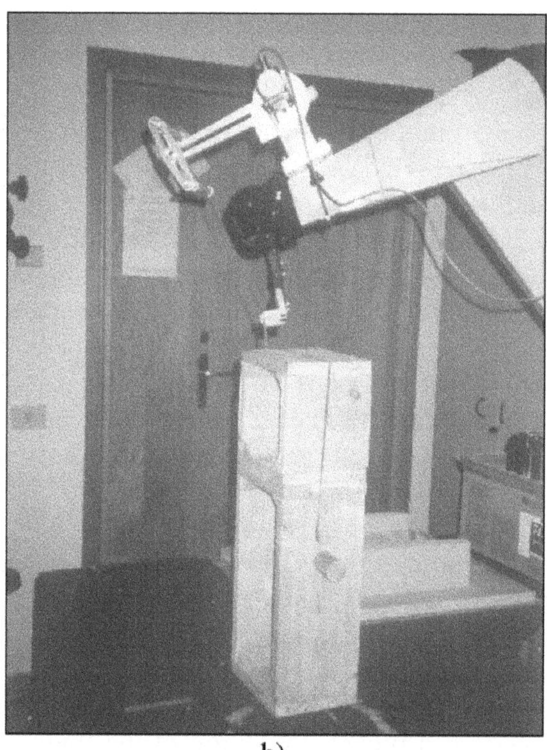

b)

Fig. 2.19: The proposed robotization layout for the 3D filament winding manufacturing of Fig. 2.18: a) a scheme; b) a laboratory set-up.

This is shown in Table 2.5, when a modified path has been synthesized to use straight-line trajectories for the robot end-effector. Thus, a robot programming can be obtained by using a proper sequence of MOVE actions only as outlined in Table 2.5.

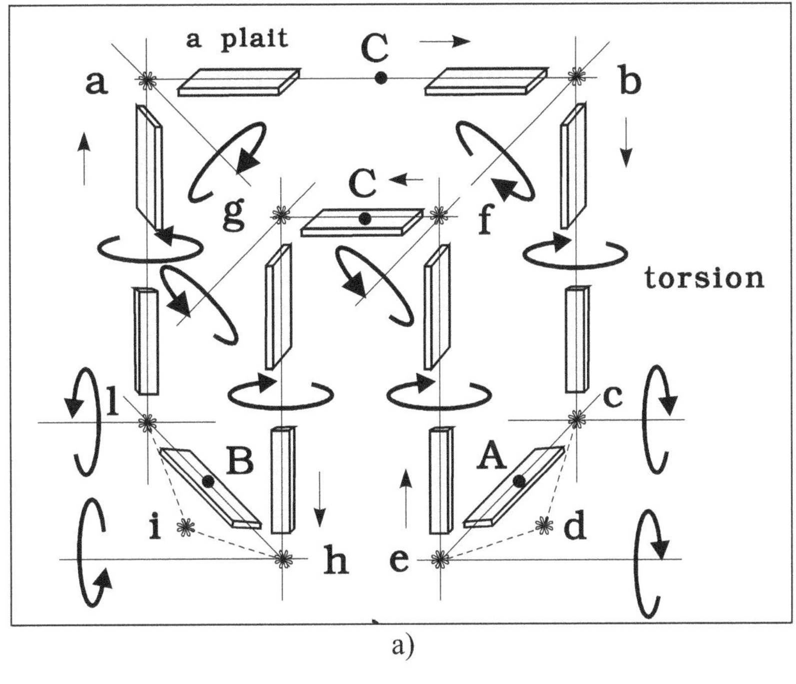

a)

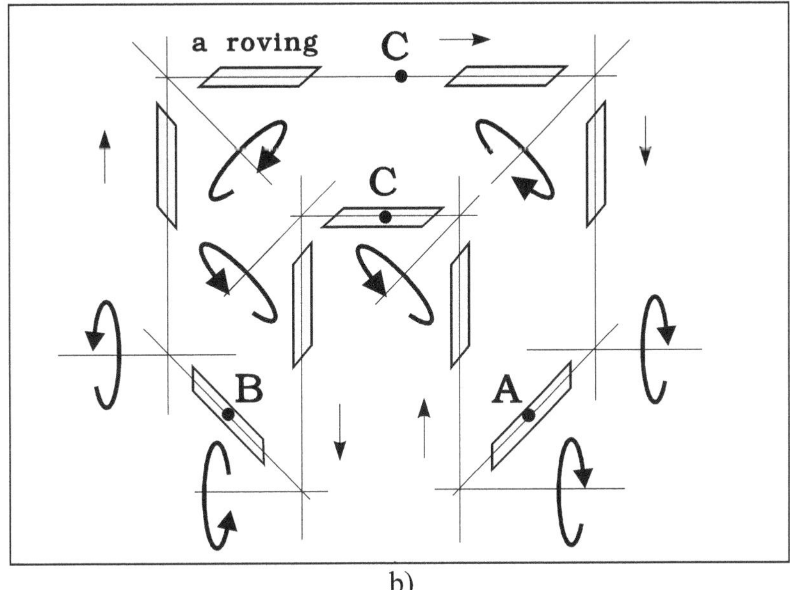

b)

Fig. 2.20: A 3D winding manipulation and basic roving motions: a) manual manufacturing by using a plait of roving; b) a roving deposition for robotization.

It is remarkable that only elementary actions of transport type can be programmed for the robotization with no dead times. Thus, the manipulative task has been identified as an end-effector path along a trajectory whose precision points are a, b, c, d, f, g, h, i, l, in Fig. 2.20a) both for a plait of rovings or a single roving. These points can be used in robot programming to specify straight-line segments for the robot movement according to MOVE elementary actions of Table 2.5. Moreover, the considerable reduction of complexity of the manipulative task has given a further great advantage for high flexibility. Such a flexibility can be achieved by using suitable changes only in the Cartesian coordinates of the above-mentioned precision points that can also be related to points A, B, and C, representing the shape and dimension of the winding support.

Table 2.5: Elementary actions of a robotized 3D filament winding manufacturing referring to Fig. 2.20a).

Robot end-effector	T	**Time** (sec)
1- goes to point b	•	0.9
2- goes to point c	•	1.3
3- goes to point d	•	0.3
4- goes to point e	•	0.2
5- goes to point f	•	1.2
6- goes to point g	•	0.9
7- goes to point h	•	1.2
8- goes to point i	•	0.3
9- goes to point l	•	0.2
10- goes to point a	•	1.2

This strategic approach has been successfully implemented by using a simple and compact programming procedure based on the VAL-II instruction 'MOVE TRANS (x, y, z, A, O, T)'. This instruction is useful to use the coordinates x, y, z of the precision points of the winding eye path, according to Fig. 2.20a) and Table 2.5, by also taking into account the orientation of the end-effector through the Euler angles A, O, T. In such a way an angle of deposition can be suitably obtained not only for the wound roving but also for the roving straining and wrist forcing.

A first considerable improvement has been obtained for one winding cycle time by reducing the time from 23 sec of the manual process to 7.7 sec of a robotized winding as obtained in laboratory experiments with the layout of Fig. 2.19b). Some laboratory experiments have been carried out and results are shown in Table 2.5 when the robot has been run at SPEED = 100. This time can be easily further decreased, speeding up the manipulator movements.

Particular attention has been devoted to a suitable roving feeding device since a feeding of continuously strained roving has been recognized as of primary importance in a successful manipulation. Thus, a prototype has been designed and built as shown in Fig. 2.21, with a clutch device for a regulation of the straining roving. It has been designed for manual adjustments through a control screw. The feeding device has been installed on the forearm of the manipulator as shown in Figs 2.19 and 2.21, near the robot wrist, with the purpose of a better manipulation of the winding eye for both payload limits and

tooling encumbrance. In fact, the installation of the feeding device and winding eye as a unique end-effector can be a great encumbrance with severe problems for manipulation programming and execution, as well as of unfeasible weight requiring stronger and larger robots.
This example illustrates how a manipulation analysis is useful also for end-effector design as is the case for the winding eye that is constrained by the deposition plane for roving trajectory of Fig. 2.20b).
This case of study is an example of how from mimicking human manipulation, a robotized solution can be obtained with better characteristics and production performances both in terms of versatility and flexibility, when the manipulation of the object and robot end-effector is properly analyzed and fully understood without any reference to human operation.

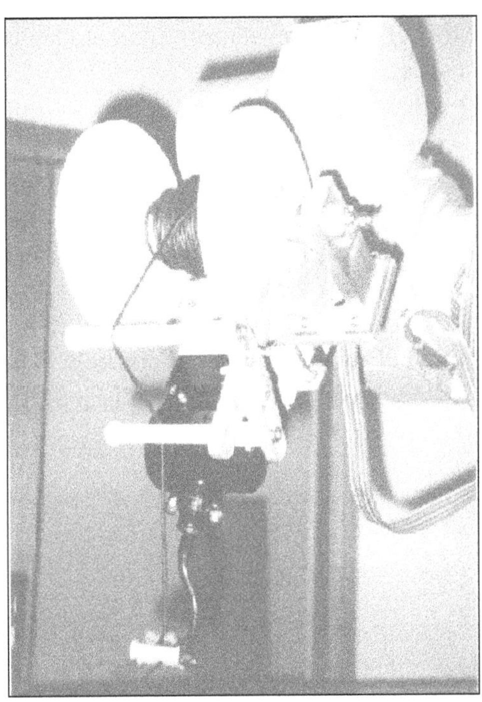

Fig. 2.21: A laboratory prototype for the feeding device for plait of ACM rovings for a robotized 3D filament winding manufacturing.

2.4.2.2 Optimizing a robotized manipulation[4]

In industrial manufacturing robots are widely used mainly in manipulative tasks to transport and transfer products and objects from one location to another. Those tasks are generally called 'pick up and place' applications. These manipulations are repetitive operations with no cultural or intellectual content so that they have been and are often

[4] This description has been obtained from the published paper: Ceccarelli M., "Analyzing a Robotized Workcell to Enhance Robot's Operation", *Journal of Robotics and Mechatronics*, 1999, Vol. 11, n.1, pp. 67–71.

automated with no doubts, often by simply substituting human operators with robots. Therefore, a robot often simply mimics human labor, to which it is devoted, since the working environment is not changed.

But problems may arise: is the robot optimally used? Can the manipulation efficiency be improved with minor changes in the work-cell?

In this example we have addressed the above-mentioned questions by referring to a case of study for an industrial pick and place application. In particular we report the application of a procedure for manipulation analysis regarding the case of a manipulation of a kinescope for a TV set production in the Videocolor SpA plant in Anagni. A robotized work-cell, which is used to rotate and transfer a kinescope from scales conveyor to a belt conveyor, has been analyzed with the aim of verifying the manipulation for productivity improvements and robot failures reduction.

The robotized work-cell is illustrated in Fig. 2.22 and a schematic model is reported in Fig. 2.23, in which the size of the system is indicated by the distances r_{LU}, r_{RU}, r_{RL}, h_B, and h_F .

A 5 d.o.f.s IRB 60/2 robot, a loading station, and an unloading station, comprises the robotized work-cell. The robot picks up a kinescope and manipulates it through a rotation and transportation to place it on a suitable plane of the unloading station.

A led sensor has been installed on the scales conveyor line to give a start signal for the robotized work-cell and particularly to the robot. Although the robot is equipped with a proper suction cup end-effector for a stable grasp of kinescopes, a safe bed has been located in the working area in order to mainly avoid the broken pieces of rarely lost kinescopes to be split everywhere in the work-cell. The suction cup end-effector is provided with a depressor sensor so that the approaching motion to a kinescope can be easily controlled by the signal of this sensor.

Fig. 2.22: A robotized work-cell for transferring and rotating a kinescope at the Videocolor plant in Anagni, Italy. (Courtesy of Videocolor SpA).

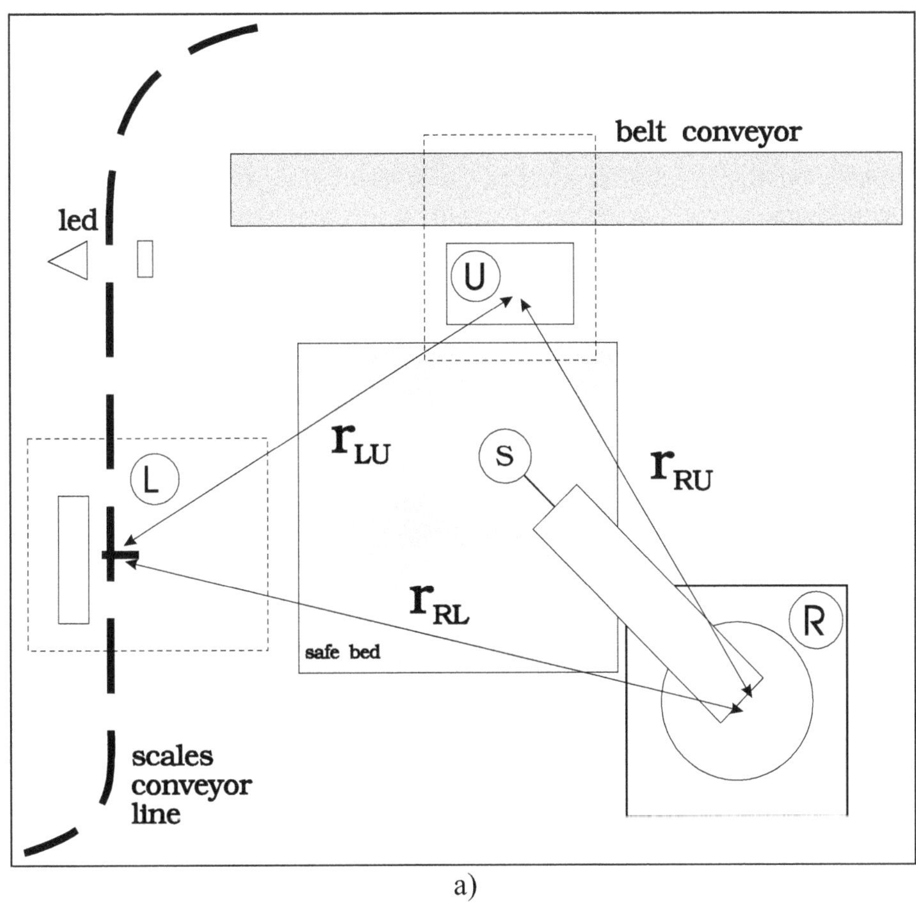

a)

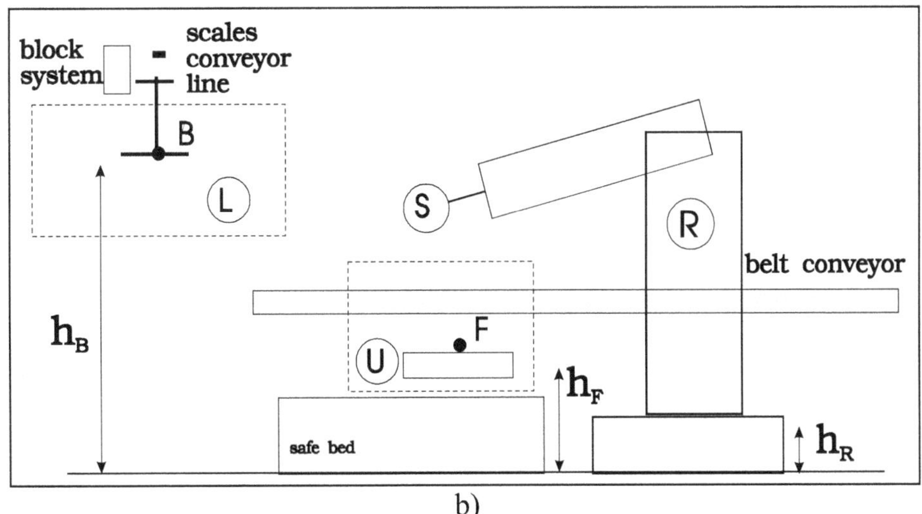

b)

Fig. 2.23: A model of the robotized work-cell of Fig. 2.22: a) horizontal view; b) vertical view. (Basic components are: R a robot; L the loading station, U the unloading station, S the suction cup robot end-effector)

The loading station is composed of mechanisms, which work to free a loaded scales conveyor from the transmission chain, to stop it for a loading operation by the robot, and to let the conveyor go back to the transmission chain. Figure 2.24 gives a simplified model of those mechanisms. The blocking system is an electropneumatically-actuated crank-sliding mechanism whose crank is fork shaped in order to fix the scales conveyor. Although this simple efficient system, the scales conveyor oscillates since oscillations can be produced previously during the path on the conveyor line.
After the robot manipulation a kinescope is placed on a suitable plane at the unloading station, Fig. 2.25. This plane is determined by four fingers of a lifting system that, after a centering operation, gives the kinescope to another eight-finger gripper. Finally, this gripper deposits the kinescope on the plate of another lifting system, which brings the kinescope to a properly shaped bed of the unloading belt conveyor.
The average weight of 200 N of a kinescope does not seem to be so critical since the maximum payload for the used robot is declared to be over 600 N . Together with the fact that an excessive speed-up of the process cannot be achieved, the robot has been thought to be under-used so that we have thought it convenient to re-think its use by looking at the performed manipulation.
Moreover, the manipulation analysis has been carried out to verify for possible small changes in the work-cell which can improve the manipulation without settling a new workcell or requiring a lengthy stop in production in order to run the suggested modifications.
The manipulation under examination can be considered as a "pick up and place" application since it consists of picking up a kinescope from a given location in the loading station and placing it at a given location in the unloading station, as shown in Fig. 2.26.

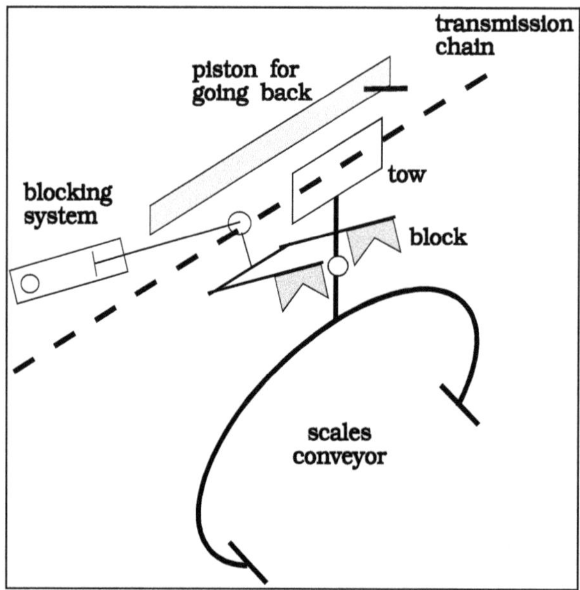

Fig. 2.24: A model for the mechanical design of the loading station.

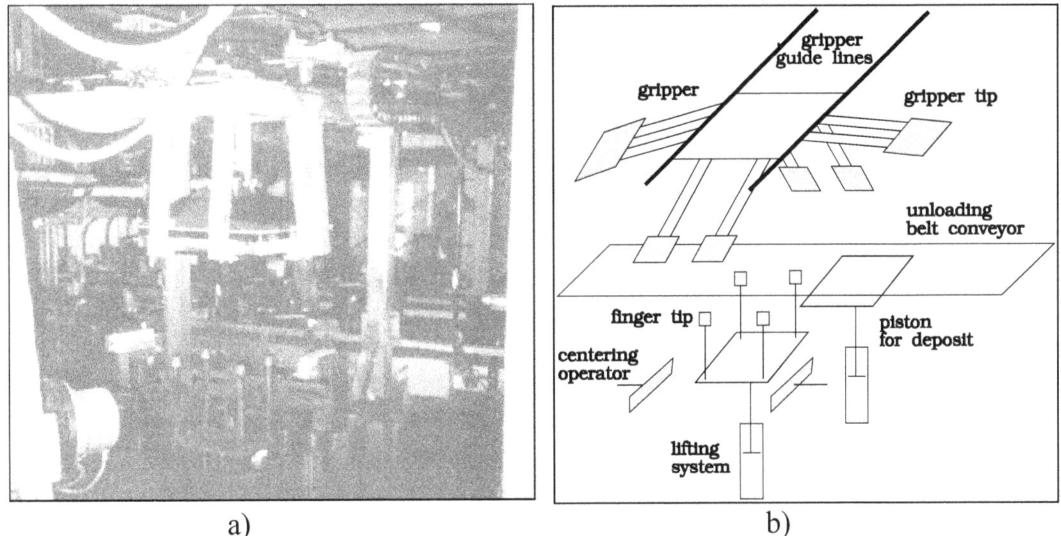

Fig. 2.25: The unloading station: a) the working layout (Courtesy of Videocolor SpA); b) a model for the mechanical design.

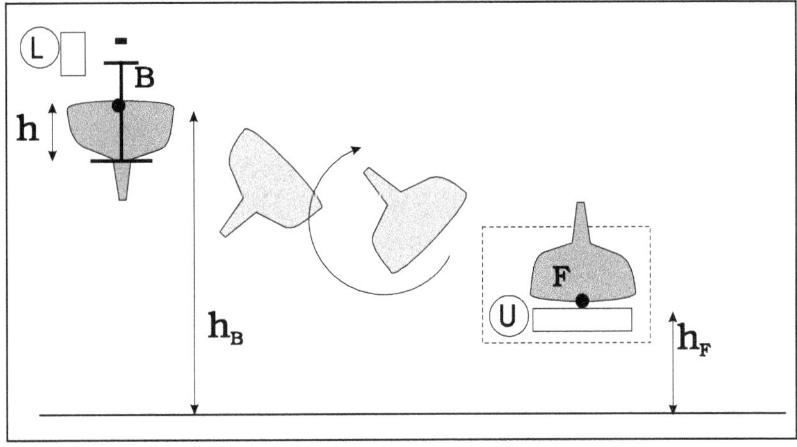

Fig. 2.26: The manipulation of a kinescope in the work-cell of Figs 2.22 and 2.23.

The term 'location' refers both to position and orientation of an object. Figure 2.26 shows the performed manipulation of a kinescope emphasizing the rotation, whose direction has been given to obtain a gyroscope torque to help in the bending motion of the robot.

From Figs 2.23 and 2.26 the degrees of freedom that are required for the manipulative task can be easily evaluated to be four, since the motion of a kinescope consists of the three spatial translations and one rotation only. Particularly, a translation along a vertical plane can even be considered pleonastic when by rearranging quotes h_B and h_F in Fig. 2.23, a faster 3 d.o.f. robot can be conveniently used.

However, an analysis of the manipulative trajectory has been carried out by observing the working and programming of the work-cell. A simplified scheme is reported in Fig.

2.27. The indicated points are reference precision points, which are used in the programming to determine the range of motion activities. Capital letters indicates precision points that are related to the manipulation given by the robot and small letters refer to the kinescope manipulation in the machinery of loading and unloading stations.
In particular, points B and F refer to the loading and unloading locations of a kinescope and the distance between them give the size of the required manipulation. Point D has been established as a point at which the robot eventually stops when the unloading station is not free or the work-cell is stopped for any other reason. The peculiarity of this position of the manipulation is the configuration of the robot with a partial rotation of the end-effector S, as shown in Fig. 2.28.
In Table 2.6 the working of the work-cell has been analyzed in terms of elementary actions referring to the manipulative trajectory of Fig. 2.27. A time measure has been reported for regular working so that the efficiency of each operation can be easily evaluated for productivity. In fact, by observing the time needed to perform an action and by considering the manipulative content, an optimization of the operations can be considered straightforward and conveniently.
Thus, if one considers the cycle time of 22 sec. to complete the kinescope manipulation from point a to l, Fig. 2.27, it is evident that a speed-up of operations with small time activity cannot give great improvements. For example the grasping at the loading station or the release by the gripper cannot improve the work-cell productivity significantly and even these can produce an additional source of accidents or improper working.
On the contrary, by addressing the attention to actions with large duration or great manipulation content, improvements can be obtained both in terms of simplification of the manipulation and speeding up the working. This is the case, for example, of the unloading which lasts 7.9 sec because the unloading plane is not located on the unloading belt conveyor.

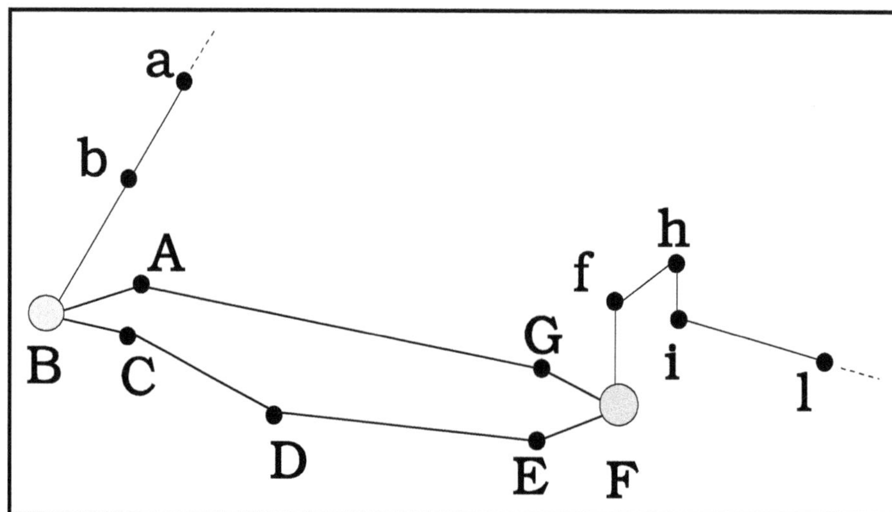

Fig. 2.27: A scheme for the manipulative trajectory of a kinescope in the robotized work-cell of Figs 2.22 and 2.23.

Fig. 2.28: The IRB 60/2 robot at the intermediate point D of the manipulative trajectory of Fig. 2.23. (Courtesy of Videocolor SpA).

Table 2.6: An evaluation of the working of the work-cell of Figs 2.22 and 2.23. (Points refer to Fig. 2.27; K is for kinescope; S is for the suction cup end-effector).

Elementary action	ref. points	**P**	**A**	**M**	**T**	time(sec.)
K approaching B	a-b				•	--
led checks for K	b-b		•			0.0
Work-cell waits	b-b	•				4.0
K in loading station	b-B				•	3.5
Conveyor is blocked	B-B		•			0.6
S approaches to grasp	B-B		•	•		0.1–0.5
S grasps	B-B		•			0.1
robot works	B-F	•	•	•	•	4.8–5.3
K in unloading plane	F-F		•			0.2
K is centered	F-F				•	1.0
K is lifted	F-f				•	1.6
gripper works	f-f		•			0.5
gripper translates	f-h				•	1.5
gripper releases K	h-h		•			0.1
depositing K	h-i				•	0.7
K is deposited	i-i		•			0.1
belt conveyor works	i-l				•	2.2

A simple solution could be suggested by adjusting the centering operation on the depositing piston, Fig. 2.25, so that time and manipulation actions are earned. Furthermore, this requires better location of the robot with respect to the loading and unloading stations. Nevertheless, this last modification can also be suggested by results of the analysis for the robot performance of Table 2.7 and by looking at the robot configurations during the working. The robot works at the loading and unloading points when it is near to its extreme reach configurations, which are related to maximum stresses both for the actuators and joints. Moreover, the approaching motions both with the unloaded and loaded robot are those with the longest duration in the robot working. Of course major care should be addressed to transfer elementary actions T because of the risk of accidental kinescope releases.

Table 2.7: An evaluation of the working of the IRB60/2 robot in the work-cell of Figs 2.22 and 2.23. (Points refer to Fig. 2.27; R is for robot; S is for suction cup end-effector; L is for loading station; U is for unloading station).

Elementary action	ref. points	**P**	**A**	**M**	**T**	time(sec.)
R waits to start	A-A	•				0–0.5
R approaches L	A-B			•		1.0
S starts	B-B		•	•		0.1–0.5
R departs from A	B-C				•	1.0
R departs from L	C-D				•	1.0
R goes across D	D-D	•			•	0–0.5
R approaches U	D-E				•	1.2
R approaches U plane	E-F				•	1.3
R deposits	F-F		•			0.2
R departs from U	F-G			•		0.7
R goes to A	G-A			•		2.0

By examining Table 2.7 and Fig. 2.28 the critical situation for the robot configuration at D can be easily recognized in the fact that the robot, once it has been stopped, starts again from an unfavorable configuration in terms of load stresses.
This case of study is an example of how a proper manipulation analysis through elementary actions can give useful results and suggestions to enhance and optimize a robotized manipulation in terms of both manipulation robot programming and robotic work-cell design.

Chapter 3:
Fundamentals of the mechanics of robots

3.1 Kinematic model

The manipulator architecture of a robot is composed of an arm mostly for movements of translation, a wrist for movement of orientation, and an end-effector for interaction with the environment and/or external objects, as shown in Fig. 3.1

Generally, the term manipulator refers specifically to the arm design, but it can also include the wrist when attention is addressed to the overall manipulation characteristics of a robot.

A kinematic study of robots deals with the determination of configuration and motion of manipulators by looking at the geometry during the motion, but without considering the actions that generate or limit the manipulator motion. Therefore, a kinematic study makes it possible to determine and design the motion characteristics of a manipulator but independently from the mechanical design details and actuator's capability.

This aim requires the determination of a model that can be deduced by abstraction from the mechanical design of a manipulator and by stressing the fundamental kinematic parameters.

The mobility of a manipulator is due to the degrees of freedom (d.o.f.s) of the joints in the kinematic chain of the manipulator, when the links are assumed to be rigid bodies.

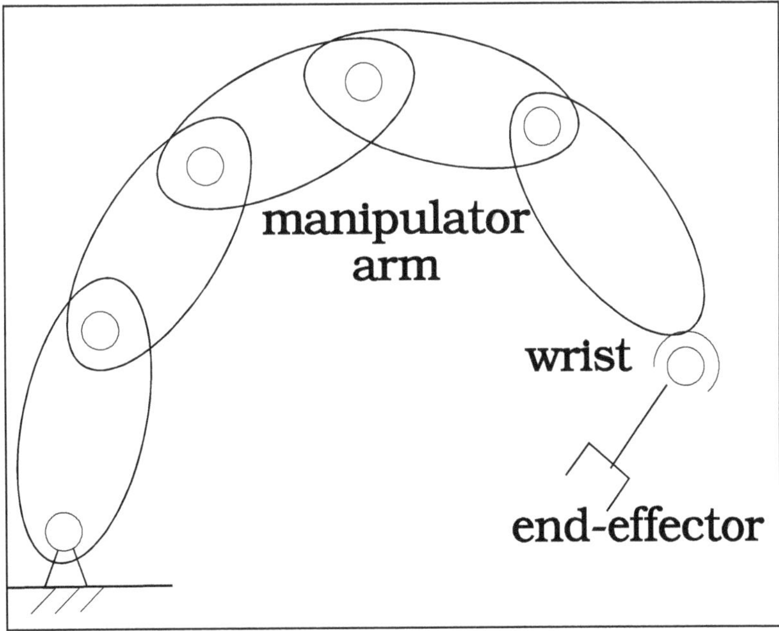

Fig. 3.1: A scheme for the manipulator architecture of a robot with the arm, wrist, and end-effector.

A kinematic chain can be of open architecture, when referring to serial connected manipulators, or closed architecture, when referring to parallel manipulators, as in the examples shown in Fig. 3.2.

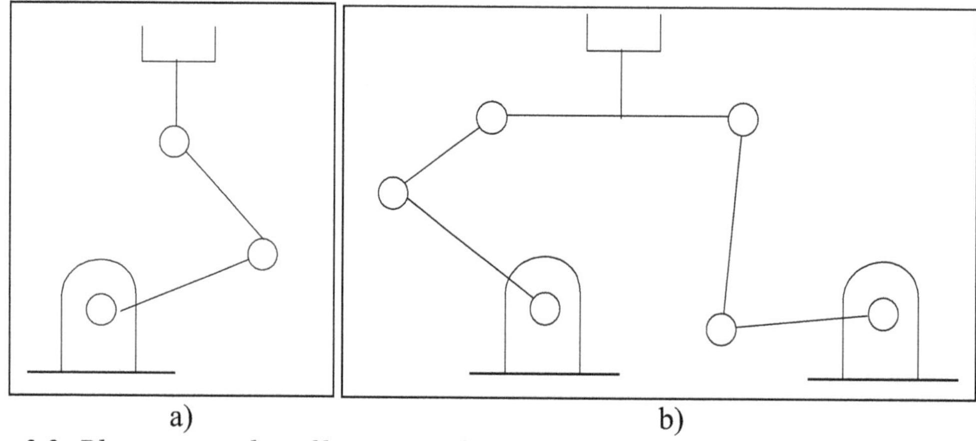

Fig. 3.2: Planar examples of kinematic chains of manipulators: a) serial chain as open type; b) parallel chain as closed type.

Of course, it is also possible to design mixed chains for so-called hybrid manipulators.
Regarding the joints, although there are several designs both from theoretical and practical viewpoints, usually the joint types in robots are related to prismatic and revolute pairs with one degree of freedom. They can be modeled as shown in Fig. 3.3.

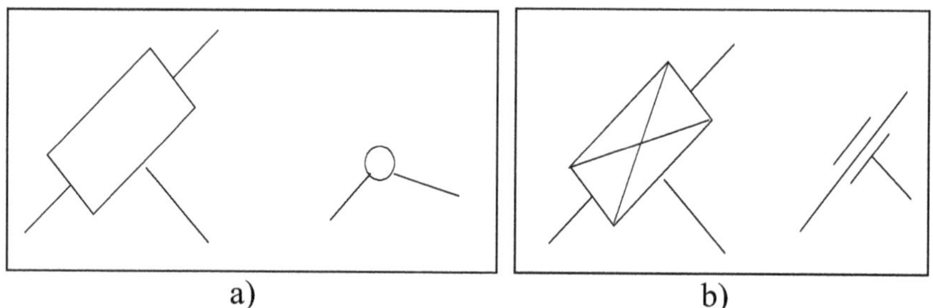

Fig. 3.3: Schemes for joints in robots: a) revolute joint; b) prismatic joint.

However, most of the manipulators are designed by using revolute joints, which have the advantage of simple design, long durability, and easy operation and maintenance. But the revolute joints also allow a kinematic chain and then a mechanical design with small size, since a manipulator does not need a large frame link and additionally its structure can be of small size in a work-cell.
In addition, it is possible to also obtain operation of other kinematic pairs with revolute joints only, when they are assembled in a proper way and sequence. For example, three revolute joints can obtain a spherical joint and depending on the assembling sequence they may give different practical spherical joints.

The most common design architectures for industrial manipulators are shown in Fig. 3.4 by using a suitable combination of revolute and prismatic joints.

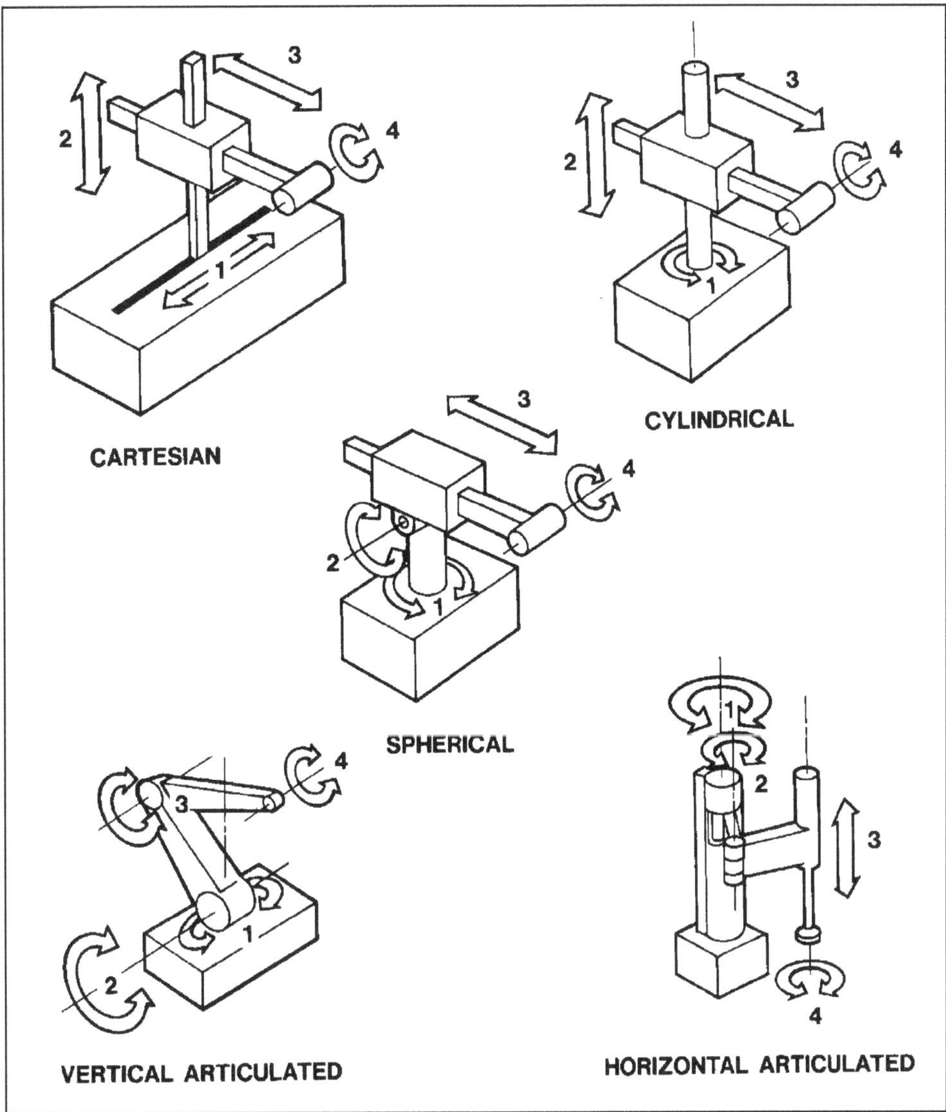

Fig. 3.4: Manipulator architectures for industrial robots.

In particular, referring to Fig. 3.4 they can be characterized as follows:

- Cartesian manipulators, which refer to a kinematic construction that makes use of the axes of a Cartesian reference frame through prismatic joints that give motions along those axes; they are characterized by a position workspace that is a bulk parallelepiped volume;
- Cylindrical manipulators, which refer to a kinematic construction that makes use of the axes of a cylindrical reference frame through one revolute joint and two prismatic

joints that give motion of an extremity point on a cylindrical surface. They are characterized by a position workspace that is a bulk cylinder volume;
- Spherical manipulators, which refer to a kinematic construction that makes use of the axes of a spherical reference frame through two revolute joints and one prismatic joint that give motion of an extremity point on a spherical surface. They are characterized by a position workspace that is a bulk spherical volume;
- Vertical articulated manipulators or so-called anthropomorphic manipulators, which refer to a kinematic human-like construction that makes use of revolute joints only; they are characterized by a position workspace that is a bulk solid of revolution volume;
- Horizontal articulated manipulators or so-called SCARA (Selective Compliance Arm for Robot Assembly) manipulators, which refer to a kinematic construction that makes use of the two parallel revolute joints and one prismatic joint; they are characterized by a position workspace that is a bulk cylindrical volume.

In Fig. 3.4 joint 4 indicates wrist capability as a common joint for any manipulation. The architectures in Fig. 3.4 are the most common in industrial robots, but many others are used, even by combining revolute and prismatic joints in alternative sequences.

Another way for manipulator identification consists of using a denomination for the sequence of joint types in the manipulator chain, when revolute joints are indicated as R and prismatic joints as P. Thus, an anthropomorphic manipulator can also be identified as a spatial RRR or 3R manipulator, and a spherical telescopic manipulator as RRP.

It is worthy of note that suitable combinations and proper design architectures with R and P joints can be equivalent to other types of kinematic pairs. For example a spherical joint can be obtained by using three revolute pairs whose joints' axes intersect in a point, and a helicoidal joint can be equivalent to a revolute pair and a prismatic pair in the form of RP or PR.

The kinematic model of a manipulator can be obtained in the form of a kinematic chain or mechanism by using the above-mentioned schemes for joints and rigid links through essential dimensional sizes for connections between two joints.

A kinematic model of a manipulator can be named as 'functional' when its scheme refers to kinematic parameters only, but also permits understanding the motion capability of the manipulator architecture.

A kinematic functional model can be determined from the mechanical design of a robot through the following step-by-step procedure:
- identification of the type of joints;
- identification of the position of each joint axes;
- identification of the geometry of the links;
- drawing of a scheme for the kinematic chain.

Figure 3.5 is an illustrative example for a kinematic functional scheme of a given mechanical design of a robot. It is worthy of note that in such a scheme the link is represented in a way that makes clear the relative motion because of the joints.

The kinematic functional model can be completed as a kinematic model by identifying the geometrical sizes and kinematic parameters. These parameters determine the relative position and orientation of a link with respect to a neighborhood one as a function of the variable coordinates of the joints that are connected through the link. Usually the variable coordinate of a joint is named as 'joint variable'.

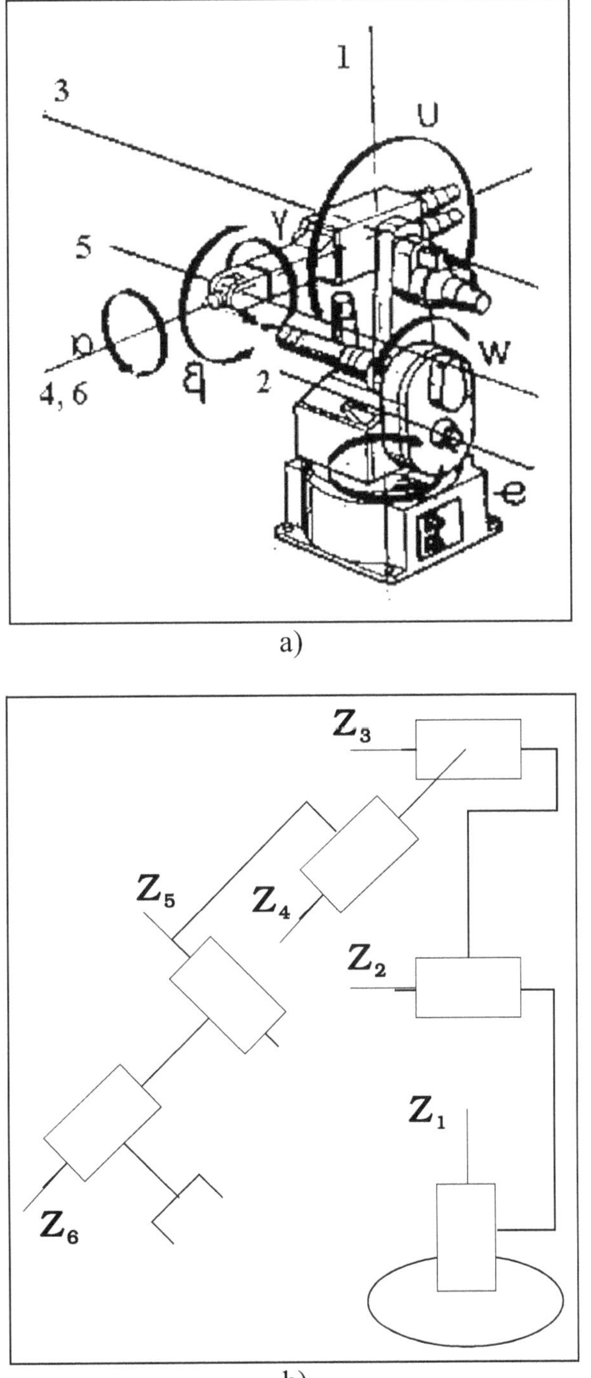

Fig. 3.5: An example of modeling an industrial robot: a) the mechanical design; b) the corresponding kinematic functional scheme.

In order to determine those geometrical sizes and kinematic parameters, one can usually refer to a scheme like that in Fig. 3.6 by using a H–D notation, in agreement with a procedure proposed by Hartenberg and Denavit in 1955.
This scheme gives the minimum number of parameters that are needed to describe the geometry of a link between two joints, but also indicates the joint variables. The joints in Fig. 3.6 are indicated as big black points in order to stress attention to the link geometry and H–D parameters.

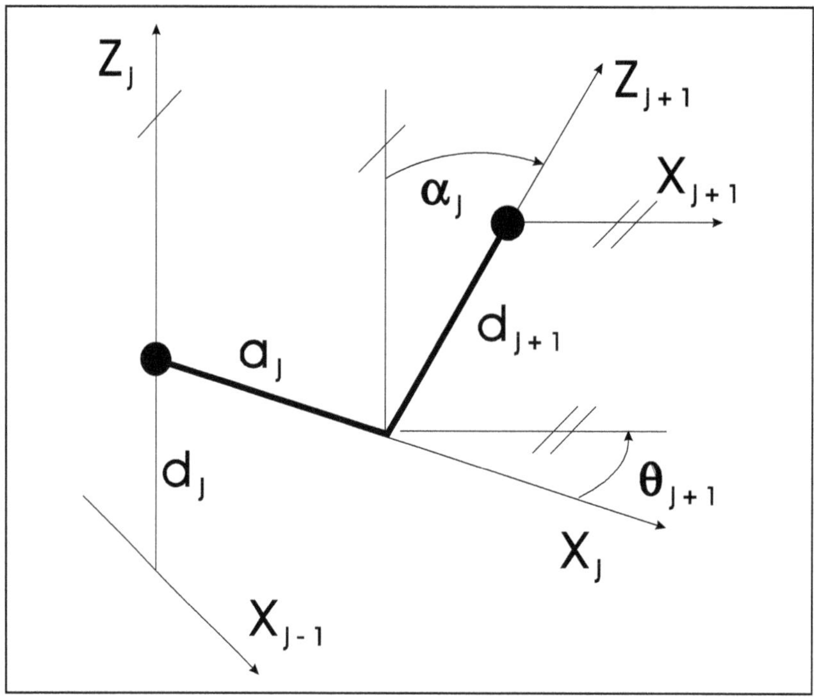

Fig. 3.6: A kinematic scheme for manipulator link parameters according to the H–D notation.

In particular, referring to Fig. 3.6 for the j-link, the j-frame $X_jY_jZ_j$ is assumed as fixed to the j-link, with the Z_j axis coinciding with the joint axis, with the X_j axis lying on the common normal between Z_j and Z_{j+1} pointing to Z_{j+1}. The origin O_j and the Y_j axis are determined to obtain a Cartesian frame as shown in Fig. 3.6.
The kinematic parameters of a manipulator can be defined according to the H–D notation in Fig. 3.6 as:

- a_j, which is named as the link length that is measured as the distance between the Z_j and Z_{j+1} axes along X_j;
- α_j, which is named as the twist angle that is measured as the angle between the Z_j and Z_{j+1} axes about X_j;
- d_{j+1}, which is named as the link offset that is measured as the distance between the X_j and X_{j+1} axes along Z_{j+1};
- θ_{j+1}, which is named as the joint angle that is measured as the angle between the X_j

and X_{j+1} axes about Zj_{j+1}

When a joint can be modeled as a rotation pair, the angle θ_{j+1} is the corresponding kinematic variable.

When a joint can be modeled as a prismatic pair, the distance d_{j+1} is the corresponding kinematic variable.

The other H–D parameters can be considered as dimensional parameters of the links.

The H–D notation is very useful for the formulation of the position problems of manipulators through the so-called transformation matrix by using matrix algebra.

The position problem of manipulators consists of determining the position and orientation of the end-effector as a function of the manipulator configuration that is given by the link position that is defined by the joint variables.

In general, the position problem can be considered from different viewpoints depending on the unknowns that one can solve in the following formulations:

- Kinematic direct problem in which the dimensions of a manipulator are given through the dimensional H–D parameters of the links but the position and orientation of the end-effector are determined as a function of the values of the joint variables;
- Kinematic inverse problem in which the position and orientation of the end-effector of a given manipulator are given, and the configuration of the manipulator chain is determined by computing the values of the joint values.

In addition, a third kinematic problem can be formulated as:

- Kinematic indirect problem (properly 'design problem') in which a certain number of positions and orientations of the end-effector are given but the type of manipulator chain and its dimensions are the unknowns of the problem.

The position problem for manipulators can be seen as an extension of the determination of a relative position and orientation between two rigid bodies and it can be studied referring to the model of Fig. 3.7 by using the reference frames that are attached at each body.

The fixed frame XYZ is attached to the body that is assumed as reference for the relative motion. The frame xyz is a moving frame that is attached to the mobile body. The relative position and orientation of the two frames XYZ and xyz describe the relative position and orientation of the two bodies. In particular, the relative position can be expressed by the position vector **h** between the origin of the two frames. The position of any point of the moving body can be computed by using its position vector **x** in the moving frame to compute the absolute position vector **X** by adding **h**, when all the vectors are computed with respect to the same frame.

The orientation of the moving frame can be evaluated though the unit vectors of xyz with respect to XYZ or by using Euler angles.

However, the three-dimensional model of Fig. 3.7 can be described in a more suitable way by using a formulation that is based on the so-called 'transformation matrix' with homogenous coordinates.

3.1.1 Transformation matrix

The so-called 'transformation matrix' with 4 rows and 4 columns is a useful analytical

means to express the relative position and motion of rigid bodies in a compact form that additionally makes possible the use of matrix algebra.
This matrix can be obtained with information on relative position and orientation between two reference frames and can express the basic relation of the relative motions shown in Fig. 3.7 in an efficient computational way in the form

$$\mathbf{X} = \mathbf{h} + {}_0\mathbf{x} \tag{3.1.1}$$

The vectors $\mathbf{X}$, $\mathbf{h}$, ${}_0\mathbf{x}$, which are expressed with respect to XYZ, represent the position vector of a reference point H and the origin of xyz from the origin of XYZ, and the position vector of H from the origin of xyz, respectively.
When computing $\mathbf{x}$ as ${}_0\mathbf{x}$ with respect to XYZ one must consider the relative orientation of xyz with respect to XYZ by using a 3 × 3 matrix whose entries are the cosine directors that are the components of the unit vectors of xyz axes along the axes of XYZ.

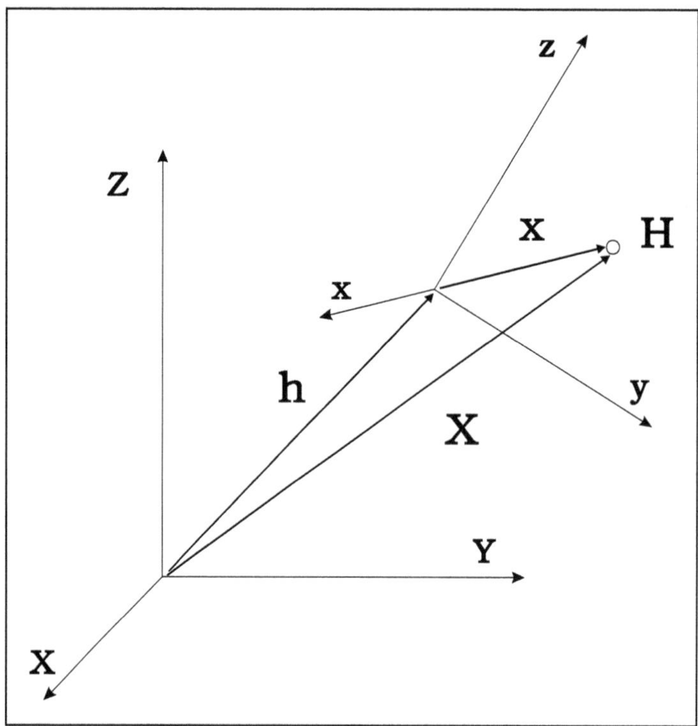

Fig. 3.7: Vectors for relative position and orientation between two reference frames.

This can be expressed as

$${}_0\mathbf{x} = \mathrm{R}\,\mathbf{x} \tag{3.1.2}$$

where R is the so-called 'rotation matrix' containing the above-mentioned cosine

directors and $\mathbf{x}$ is the position vector of H reference point with respect to xyz.
The $\mathbf{x}$ vector is described and computed with respect to xyz, while $_0\mathbf{x}$ is described with respect to xyz but is computed with respect to XYZ frame.
Therefore, the basic expression for relative motion can be written as

$$\mathbf{X} = \mathbf{h} + \mathrm{R}\mathbf{x} \tag{3.1.3}$$

When the transformation matrix is defined as

$$\mathrm{T} = \begin{vmatrix} \mathrm{R} & \mathbf{h} \\ 0 & 1 \end{vmatrix} \tag{3.1.4}$$

by using the vector $\mathbf{h}$ through its components and the cosine directors of one frame with respect to the other in the rotation matrix R, Eq. (3.1.3) can be expressed in a compact form as

$$\mathbf{X} = \mathrm{T}\,\mathbf{x} \tag{3.1.5}$$

The expression (3.1.5) requires the use of homogenous coordinates for description of geometrical and kinematic characteristics of relative motion. Thus, vectors $\mathbf{X}$ and $\mathbf{x}$ must be written in the form $\mathbf{X}$=[x, y, z, 1]^t and $\mathbf{x}$=[x, y, z, 1]^t, respectively, by using the above-mentioned Cartesian coordinates in the corresponding frames and introducing the homogenous coordinate with value one as fourth component of both vectors.
An entry R_{ij} of matrix R represents the cosine director of the i-axis with respect to the j-axis and therefore the rotation matrix R can be also expressed as

$$\mathrm{R} = \begin{vmatrix} i \cdot I & i \cdot J & i \cdot K \\ j \cdot I & j \cdot J & j \cdot K \\ k \cdot I & k \cdot J & k \cdot K \end{vmatrix} \tag{3.1.6}$$

in which the terms i, j, and k indicate the unit vectors of the axes of xyz frame and the terms I, J, and K are the unit vectors of the axes of XYZ frame. The dot operator indicates the inner product between vectors.
In addition, Eq. (3.1.5) can be understood depending on the operative meaning of matrix T that can be seen as:

1- descriptive function, when T describes a frame xyz with respect to XYZ frame, since R determines the relative orientation between the frames, and $\mathbf{h}$ gives the relative position of the frame origins;
2- mapping function, when T is used in Eq. (3.1.5) to transfer the description with vector $\mathbf{x}$ from xyz frame to an $\mathbf{X}$ representation with respect to XYZ frame;
3- operator function, when matrix T in Eq. (3.1.5) operates as an operator since it makes vector $\mathbf{X}$ from vector $\mathbf{x}$.

The convenience of expression (3.1.5) can be recognized also by taking into account the problem of determining the relative position and orientation among the N+1 link bodies, but mainly the location of the extremity link N with respect to the base frame link 0. In fact, once reference frames are fixed on the corresponding links, one can write the expression (3.1.5) for consecutive link frames in the form

$$\mathbf{X}_{N-1} = {}_{N-1}T_N \, \mathbf{X}_N \tag{3.1.7}$$

$$\mathbf{X}_{k-1} = {}_{k-1}T_k \, \mathbf{X}_k \qquad (k=1, ..., N-1) \tag{3.1.8}$$

But by considering

$$\begin{aligned} \mathbf{X}_k &= {}_kT_{k+1} \, \mathbf{X}_{k+1} \\ \mathbf{X}_{k-1} &= {}_{k-1}T_k \, \mathbf{X}_k \end{aligned} \tag{3.1.9}$$

it yields

$$\mathbf{X}_{k-1} = {}_{k-1}T_k \; {}_kT_{k+1} \, \mathbf{X}_{k+1} \tag{3.1.10}$$

so that if one will keep the notation of Eq. (3.1.5) for the expression between the references k-1 and k+1 in the form

$$\mathbf{X}_{k-1} = {}_{k-1}T_{k+1} \, \mathbf{X}_{k+1} \tag{3.1.11}$$

then the transformation matrix can be expressed by using Eqs. (3.1.10) and (3.1.11) as

$${}_{k-1}T_{k+1} = {}_{k-1}T_k \; {}_kT_{k+1} \tag{3.1.12}$$

This expression describes the condition that a resultant transformation matrix can be obtained from two matrices by pre-multiplying the initial matrix that describes the mobile frame k with respect to the fixed reference k-1.

This condition can usually be represented by a scheme like that one in Fig. 3.8 where an arrow is drawn on each line to indicate with respect to which frame a transformation matrix is formulated.

In the case that the transformation matrix describes a mobile or a new frame, then the composition of the matrices is obtained by post-multiplication of matrices.

It is worthy of note that from Eq. (3.1.12) the transformation matrices can be computed recursively by considering two frames at each step as relative motion.

Finally, the location of the extremity link N+1 can be computed with respect to the base frame as

$$\mathbf{X}_0 = {}_0T_N \, \mathbf{X}_N \tag{3.1.13}$$

and therefore the resultant transformation matrix can be obtained as

$${}_0T_N = {}_0T_1 \, {}_1T_2 \cdots {}_{N-1}T_N = \prod_{k=0}^{N-1} {}_kT_{k+1} \tag{3.1.14}$$

The last expression can be considered the fundamental typical formulation for the direct kinematic of manipulators.

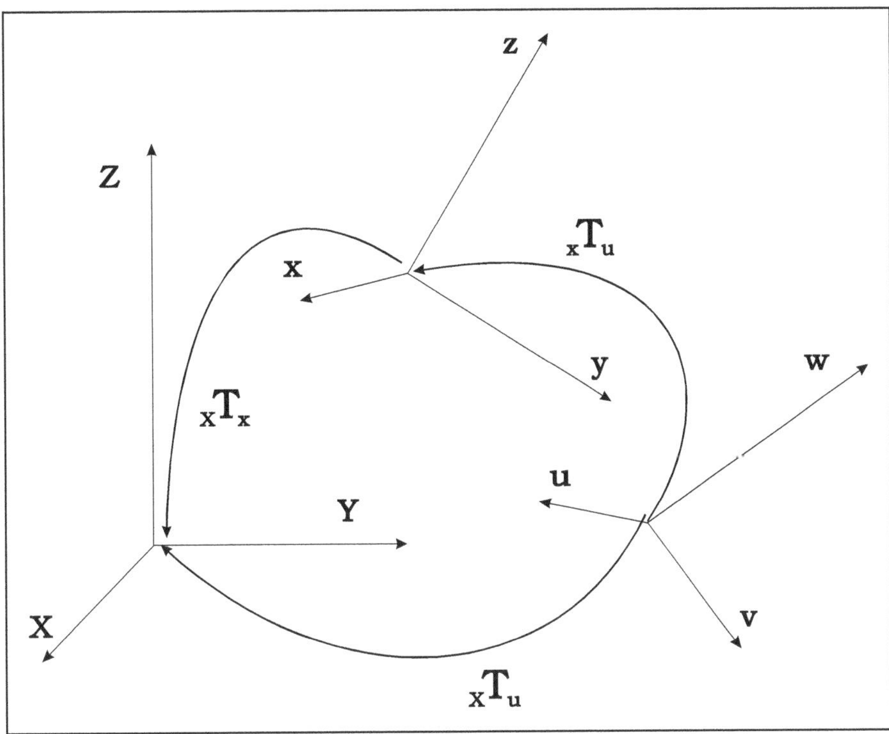

Fig. 3.8: A scheme for indicating with respect to which frames a transformation matrix is formulated.

In particular, by using the above-mentioned properties and formulation for the transformation matrix, a general expression for ${}_kT_{k+1}$ can be given by referring to the general manipulator scheme of Fig. 3.6, in the form

$${}_kT_{k+1} = \begin{vmatrix} \cos\theta_{k+1} & -\sin\theta_{k+1} & 0 & a_k \\ \sin\theta_{k+1}\cos\alpha_k & \cos\theta_{k+1}\cos\alpha_k & -\sin\alpha_k & -\sin\alpha_k \, d_{k+1} \\ \sin\theta_{k+1}\sin\alpha_k & \cos\theta_{k+1}\sin\alpha_k & \cos\alpha_k & \cos\alpha_k \, d_{k+1} \\ 0 & 0 & 0 & 1 \end{vmatrix} \tag{3.1.15}$$

as a composition of elementary matrices that describe elementary differences between reference frames that are described by the H–D parameters.
An elementary matrix corresponds to a rotation about a reference axis or a translation along such an axis.
In addition, the combination of a rotation with a translation with respect to the same axis can be described as an elementary screw motion that can be characterized by the axis itself (which is usually referred to as 'Mozzi's' axis or 'Screw axis') and its parameters.
In particular, an elementary rotation about a Cartesian axis of a reference frame gives a simple transformation matrix that in the case of a rotation Rot (Z, α) about Z axis with angle α can be expressed as

$$T_{rot} = Rot(Z, \alpha) = \begin{vmatrix} \cos\alpha & -\sin\alpha & 0 & 0 \\ \sin\alpha & \cos\alpha & 0 & 0 \\ 0 & 0 & 1 & 0 \\ 0 & 0 & 0 & 1 \end{vmatrix} \tag{3.1.16}$$

In the case of a translation Transl (Z, d) along Z axis with a distance d, the elementary transformation matrix can be written as

$$T_{trasl} = Transl(Z, d) = \begin{vmatrix} 1 & 0 & 0 & 0 \\ 0 & 1 & 0 & 0 \\ 0 & 0 & 1 & d \\ 0 & 0 & 0 & 1 \end{vmatrix} \tag{3.1.17}$$

Finally, the composition of the above-mentioned motions gives the helicoidal motion (named also as Screw motion or Mozzi's motion) that can be represented by Screw (Z, d, α) with the elementary transformation matrix T_{screw} that can be obtained as a combination of matrices of elementary rotation and elementary translation. The elementary transformation matrix T_{screw} can be formulated by using T_{rot} and T_{transl} to describe the sequence of rotation and translation of the helicoidal motion in the form

$$T_{helic} = Screw(Z, d, \alpha) = \begin{vmatrix} \cos\alpha & -\sin\alpha & 0 & 0 \\ \sin\alpha & \cos\alpha & 0 & 0 \\ 0 & 0 & 1 & d \\ 0 & 0 & 0 & 1 \end{vmatrix} \tag{3.1.18}$$

Consequently, the matrix ${}_kT_{k+1}$ can be computed by considering a motion that transports the frame k+1 onto the frame k by means of elementary rotations and translations. Thus the matrix ${}_kT_{k+1}$ can be given by the expression

$$ {}_{k}T_{k+1} = \mathrm{Rot}(X_k, \alpha_k)\,\mathrm{Trasl}(X_k, a_k)\,\mathrm{Rot}(Z_{k+1}, \theta_{k+1})\,\mathrm{Trasl}(Z_{k+1}, d_{k+1}) \qquad (3.1.19) $$

or alternatively by using elementary helicoidal motions, it is given by

$$ {}_{k}T_{k+1} = \mathrm{Screw}(X_k, a_k, \alpha_k)\,\mathrm{Screw}(Z_{k+1}, d_{k+1}, \theta_{k+1}) \qquad (3.1.20) $$

Then, the corresponding transformation matrix can be computed in the form of Eq. (3.1.15) by using the above-mentioned elementary matrices.
Similarly, one can obtain the expressions for the inverse kinematics by using the properties of the transformation matrix in agreement with the scheme of Fig. 3.8, in the form

$$ {}_{k+1}T_{k} = {}_{k}T_{k+1}^{-1} = \begin{vmatrix} {}_{k}R_{k+1}^{t} & -{}_{k}R_{k+1}^{t}\,\mathbf{h} \\ 0 & 1 \end{vmatrix} \qquad (3.1.21) $$

when the orthogonality of the Cartesian frame is considered through the expression

$$ {}_{k+1}R_{k} = {}_{k}R_{k+1}^{-1} \qquad (3.1.22) $$

3.1.2 Joint variables and Actuator Space

The kinematic variables of a manipulator refer to the joint mobility and provide the possibility to describe the state of a manipulator in terms of position, orientation, velocity, and acceleration as functions of mobility of the joints.
These kinematic variables can be considered all together as coordinates of so-called multi-dimensional reference ‘Joint Space’ that is used to represent the manipulator mobility in terms of the mobility of the joints.
In addition, the manipulator mobility can be described in so-called ‘Cartesian Space’ in the domain of Euclidean pace for the Cartesian coordinates of a reference point H on the manipulator extremity. This gives the position mobility of a manipulator, which is related to workspace characteristics.
Referring to orientation capability, a similar description can be formulated by using a Euclidean Space whose coordinates are the orientation angles of the manipulator extremity (Euler angles). This space can be named as ‘Cartesian Space’, although a more correct name is ‘Orientation Space’.
Another interesting space for mobility description is considered to be the domain of the actuator mobility, which is represented in the form of Actuator Space as a function of the actuator coordinates. These last coordinates are strongly related to the joint coordinates, and they are influenced by each other.
The practical interest of these spaces is evident for the above-mentioned variables, since the knowledge of the mobility domain permits a proper use of manipulators.
In particular, the Cartesian Space can be determined from the Joint Space by solving the

Direct Kinematics.
The Actuator Space can be determined when the transmissions connecting actuators and manipulator joints are given with their mobility range. In the following the relation between the variable of an actuator and the variable of the corresponding joint is deduced by referring to a typical linkage transmission. In general, the relation between the Actuator Space and Joint Space can be expressed by using the transmission ratio tr as

$$\psi = \text{tr}\,\theta \qquad (3.1.23)$$

where ψ represents the actuator coordinate and θ is the corresponding joint variable.
The transmission ratio tr expresses the mobility of the actuators as a function of the mobility of the joints and vice versa, by using the kinematics of the mechanical transmissions that connect actuators to the joints. The transmission ratio can assume a value that is bigger or less than 1, being constant or variable as function of ψ and θ, depending on the mechanical design of the transmission. In general, the transmissions that are used in robots can be grouped and classified referring to tr values as:
- gear systems with tr < 1 or tr > 1;
- harmonic drives with tr < 1;
- belt transmissions with tr < 1 or tr > 1;
- linkages and mechanisms with tr variable as function of ψ and θ.

The computation of tr for each system can be performed by using common schemes for mechanism analysis.
In the following, a general algorithm is outlined for the case of Fig. 3.9 in which a four-bar linkage is the transmission system between an actuator and manipulator joint.
The kinematics of a four-bar linkage can be formulated as closed-form expressions by solving the position closure equation in the form

$$\begin{aligned} &l_2 \cos\psi + l_3 \cos\gamma + l_4 \cos\theta - l_1 = 0 \\ &l_2 \sin\psi + l_3 \sin\gamma + l_4 \sin\theta = 0 \end{aligned} \qquad (3.1.24)$$

Squaring and summing them, after algebraic manipulations it yields

$$K_1 \sin\psi + K_2 \cos\psi = K_3 \qquad (3.1.25)$$

with

$$K_1 = \sin\theta\,,$$

$$K_2 = \frac{l_1}{l_2} + \cos\theta\,,$$

$$K_3 = \frac{l_1}{l_4}\cos\theta + \frac{l_1^2 + l_2^2 - l_3^2 + l_4^2}{2 l_2 l_4}$$

The solution of Eq. (3.1.25) can be expressed after suitable manipulations as

$$\mathrm{tg}\frac{\psi}{2}=\frac{K_1 \pm \sqrt{K_1^2+K_2^2-K_3^2}}{K_2+K_3} \tag{3.1.26}$$

which expresses the transmission ratio tr as the relation between ψ and θ, as function of the kinematics of the transmission mechanism. The sign ambiguity for the square root depends on the assembly configuration of the mechanism and looking at a starting mechanism configuration can solve it.

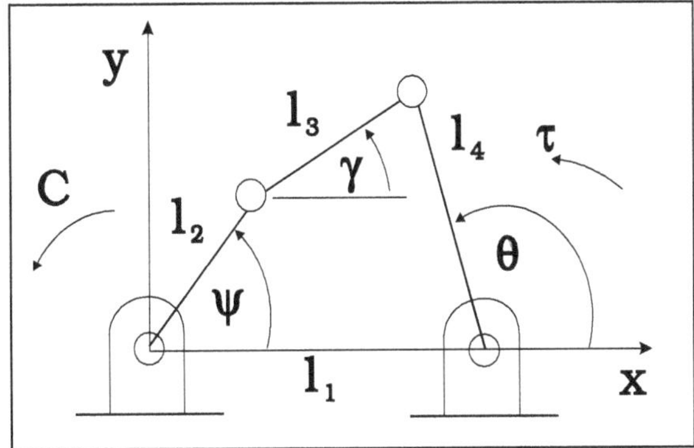

Fig. 3.9: A kinematic scheme for a four-bar linkage transmission

The above-mentioned approach for determining tr is of algebraic nature and allows us to obtain closed-form formulation when possible, or to deduce numerical procedures for more complex transmission systems. Indeed, when a transmission system is seen as a mechanism, any approach for mechanism analysis can be used to compute the transmission ratio.

3.1.3 Workspace analysis

An important subject of direct kinematics is the determination and evaluation of the workspace of manipulators.

The workspace W(H) is defined as the region of points that can be reached by a reference point H on the manipulator extremity. This is the Position Workspace and similarly the Orientation Workspace can be defined as the set of orientations that can be reached by the manipulator extremity.

Since the main task of a manipulator is recognized in moving the end-effector or a grasped object in the space, it is suitable to achieve an optimum design and location of a robot by basically taking into account workspace characteristics. Thus, a workspace formulation is of fundamental importance not only for workspace analysis and evaluation but moreover to deduce rational design criteria and algorithms for manipulators.

In this section a specific study on the workspace is discussed with the aim of giving a

formulation, which is useful to simulate and analyze basic workspace characteristics. The results can be used for a numerical classification and selection of robots, and finally to formulate procedures for design and optimization of manipulators.
Fundamental characteristics of the manipulator position workspace are recognized as:

- the shape and volume of the workspace, which is a solid of revolution for manipulators with only revolute joints, and can be a parallelepiped for manipulators with prismatic joints;
- the hole, whose existence is determined by a region of unreachable points that can individuate straight-lines surrounded by the workspace;
- the voids, which are regions of unreachable points that are buried within the workspace.

Similar characteristics can be identified for the orientation workspace that nevertheless has a different topology.
An N-revolute manipulator can be sketched as an open kinematic chain with N revolute joints as shown in Fig. 3.10. A frame X_i Y_i Z_i can be attached to each link i of the chain in a way that axis Z_i coincides with the joint axis, X_i coincides with the common normal between two consecutive joint axes, and Y_i is consequently determined to give a Cartesian frame. A base frame can be considered to be coincident with X_1 Y_1 Z_1 at an initial manipulator configuration. The geometrical parameters of a general N-R manipulator are: the link lengths $a_i \neq 0$ (i=1,...,N), the link offsets d_i (i = 2,...,N), and the twist angles $\alpha_i \neq k\ \pi/2$ (i=1,...,N-1), (k=0, ...,4). The offset d_1 may not be considered since it shifts the workspace up and down only and therefore it does not affect a workspace evaluation. The joint angles θ_i (i=1,...,N) are defined as the angles between two consecutive X_i axes. The design parameters of a serial open-chain manipulator are shown in Fig. 3.10 where the so-called 'total manipulator length' is also indicated as L.

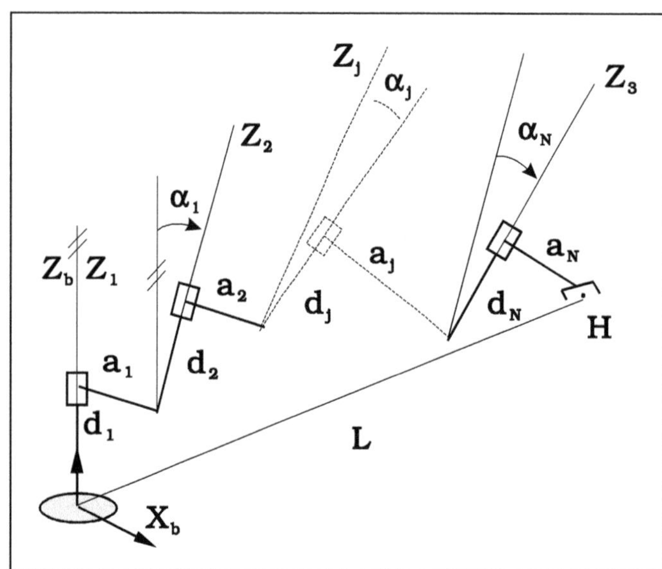

Fig. 3.10: Design parameters for a serial open-chain manipulator.

The workspace is defined as generated by a reference point H on the extremity of the manipulator chain that is moved to reach all possible positions because of the mobility ranges of the joints.
Workspace geometry of manipulators with revolute joints has been recognized as a ring topology, and this can be conveniently used to analytically describe the workspace.
Thus, the ring geometry can be used with the following modeling and reasoning by referring to Fig. 3.11 to characterize the generation of the manipulator workspace.
Revolving a torus about an axis generates a ring. Therefore, a ring volume $W_{3R}(H)$ can be thought as the union of the points swept by the revolving torus $T_{R2R3}(H)$, due to the mobility in R_2 and R_3 joints, during the θ_1 revolution about Z_1 axis. This can be expressed as

$$W_{3R}(H) = \bigcup_{\vartheta_1=0}^{2\pi} T_{R_2R_3}(H) \tag{3.1.27}$$

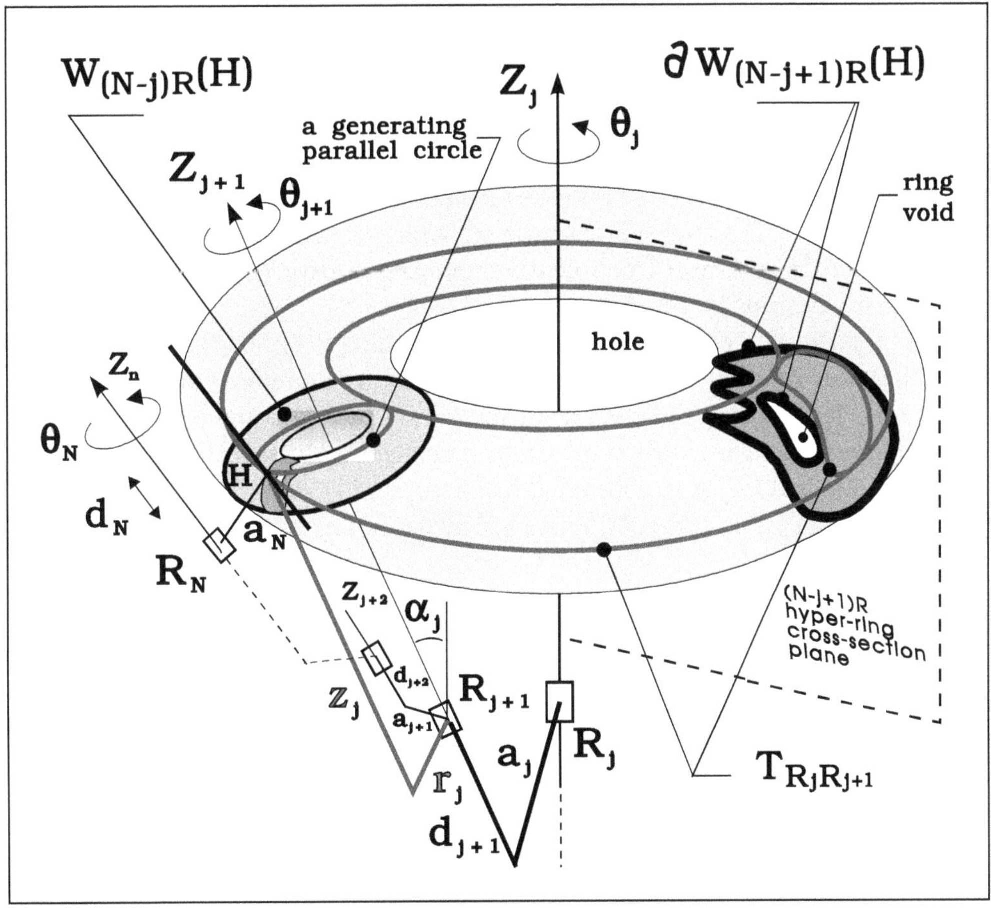

Fig. 3.11: A descriptive view of workspace generation for serial manipulators with revolute joints.

Alternatively, a ring $W_{3R}(H)$ can be considered as the union of the tori $T_{R1R2}(H)$, which are due to the mobility in R_1 and R_2 joints and are traced by all parallel circles which can be cut on generating torus $T_{R2R3}(H)$ so that a ring workspace can be formulated as

$$W_{3R}(H) = \bigcup_{\vartheta_3=0}^{2\pi} T_{R_1R_2}(H) \qquad . \qquad (3.1.28)$$

Thus, the boundary $\partial W_{3R}(H)$ of a ring can be thought as the envelope of torus surfaces generated by revolution of the generating torus or, alternatively, it can be obtained by an envelope of torus surfaces that are traced from the parallel circles of the generating torus. The latter procedure can be expressed according to Eq. (3.1.28) in the form

$$\partial W_{3R}(H) = \underset{\vartheta_3=0}{\overset{2\pi}{\text{env}}} T_{R_1R_2}(H) \qquad (3.1.29)$$

where 'env' is an envelope operator performing an envelope process.
A ring geometry has been recognized as topologically common also to hyper-rings as a 'solid hollow ring' shape on the basis of a consideration of an iterative revolving process for hyper-ring generation in N-R manipulators. In fact, referring to Fig. 3.11 a (N-j+1)R hyper-ring $W_{(N-j+1)R}$ (H) is generated by revolving a (N-j)R hyper-ring $W_{(N-j)R}$ (H) about Z_j axis. The (N-j+1)R hyper-ring, which is traced by a point H on the last link of an open chain with N-j+1 consecutive revolute pairs, is the volume swept by the (N-j)R hyper-ring during its revolution.
A basic element is a generating parallel circle, which is cut on the revolving $W_{(N-j)R}$ (H) because of a revolution about Z_{j+1} axis and whose revolution about Z_j generates a torus T_{RjRj+1} (H). The envelope of the tori of the family will give the boundary $\partial W_{(N-j+1)R}$ (H) of the hyper-ring. The chain dimensions directly involved in the process are the link sizes a_j and α_j and the axial and radial reaches r_j and z_j, whose value can be determined for each parallel circle that is individuated by a point of the boundary $\partial W_{(N-j)R}$ (H) of the revolving $W_{(N-j)R}$ (H).
The generation process of a hyper-ring is a consecutive revolving process of a circle, a torus, a ring, a 4R hyper-ring, and so on. This can be expressed through a revolution operator Rev in the form

$$W_{(N-j+1)R}(H) = \underset{\vartheta_j=0}{\overset{2\pi}{\text{Rev}}} W_{(N-j)R}(H) \qquad j = 1,...,N-1 \qquad (3.1.30)$$

where $W_{(N-j+1)R}$ (H) represents reachable points, with respect to frame X_j Y_j Z_j fixed on link j, due to (N-j+1) being the last revolute joints in the chain; and $W_{(N-j)R}$ (H) is the workspace region with respect to X_{j+1} Y_{j+1} Z_{j+1} on link j+1 due to N-j revolute pairs.
Alternatively, $W_{(N-j+1)R}$ (H) can be considered as the union of a suitable torus family

which is traced by the boundary points in the revolving torus, ring, 4R hyper-ring, and so on when they are rotated completely about the first two revolute axes in the corresponding generating sub-chain. It can be expressed in the form

$$W_{(N+1-j)R}(H) = \bigcup_{\vartheta_j, \vartheta_{j+1}=0}^{2\pi} T_{R_j R_{j+1}} \left[\partial W_{(N-j)R}(H) \right] \qquad (3.1.31)$$

where T_{RjRj+1} (H), represents a torus generated by revolutions θ_j and θ_{j+1} about the joints axes of R_j and R_{j+1}. The revolution in θ_{j+1} generates a parallel circle in $W_{(N-j)R}$ (H) and together with θ_j generates the torus T_{RjRj+1} (H).
Hence, the boundary $\partial W_{(N-j+1)R}$ (H) of a (N-j+1)R hyper-ring can be described as an envelope of the torus family traced by all the points on $\partial W_{(N-j)R}$ (H), and it can be expressed as

$$\partial W_{(N-j+1)R}(H) = \underset{\vartheta_j, \vartheta_{j+1}=0}{\overset{2\pi}{\text{env}}} \; T_{R_j R_{j+1}} \left[\partial W_{(N-j)R}(H) \right] \quad . \qquad (3.1.32)$$

Thus, a workspace boundary $\partial W_{NR}(H)$ of a general N-R manipulator can be generated by using recursively Eq. (3.1.32), to determine the tori envelopes from the ring up to the N-R hyper-ring in the chain, by computing from the extremity to the base of the manipulator chain .
The term hyper-ring has been used to stress revolution operations about more than three axes, which is the case usually referred to as the ring geometry.
A similar approach can be used for manipulators with prismatic joints as an extension of the above-mentioned formulation.
The procedure of Eqs (3.1.31) and (3.1.32) can also be conveniently adjusted for manipulators with prismatic joints, as Fig. 3.12 synthetically illustrates.
In fact, from a descriptive geometry viewpoint, the workspace of a telescopic arm also with prismatic joints can be characterized by looking at the figures, which are generated by H because of successive full movements of the joints starting from the last up to first one. The first joint is fixed to the manipulator base. A telescopic arm is a RRP manipulator whose prismatic joint is the extremity joint. Thus, movements of the sliding joint will generate a straight-line segment. Then, the second revolute pair of the chain perform a full rotation of the straight-line segment, which generates a circular cone. Indeed, depending on the orientation of the straight-line segment with respect to the revolute joint axis we may have a cylinder, a cone, or generally a hyperboloid. Finally, a full revolution of the first revolute joint will generate, by revolving the hyperboloid, a solid of revolution which can be generally depicted as a 'cylindroid ring', which contains a ring topology, as shown in Fig. 3.12. A 'cylindroid ring' shows a hollow bulk shape and its cross-section is characterized by straight-lines with possible cusps and two circular contours on the top and the bottom.
Referring to Fig. 3.12, a cylindroid ring is generated by revolving a circular cone about

axis Z_1. Therefore, workspace volume W(H) of the cylindroid ring can be thought as the union of the points swept by revolving cone $T_{R2P}(H)$, due to the mobility in R_2 and P joints, during the θ_1 revolution about Z_1 axis. This can be expressed in the form

$$W(H) = \bigcup_{\vartheta_1=0}^{2\pi} T_{R_2P}(H) \quad . \tag{3.1.33}$$

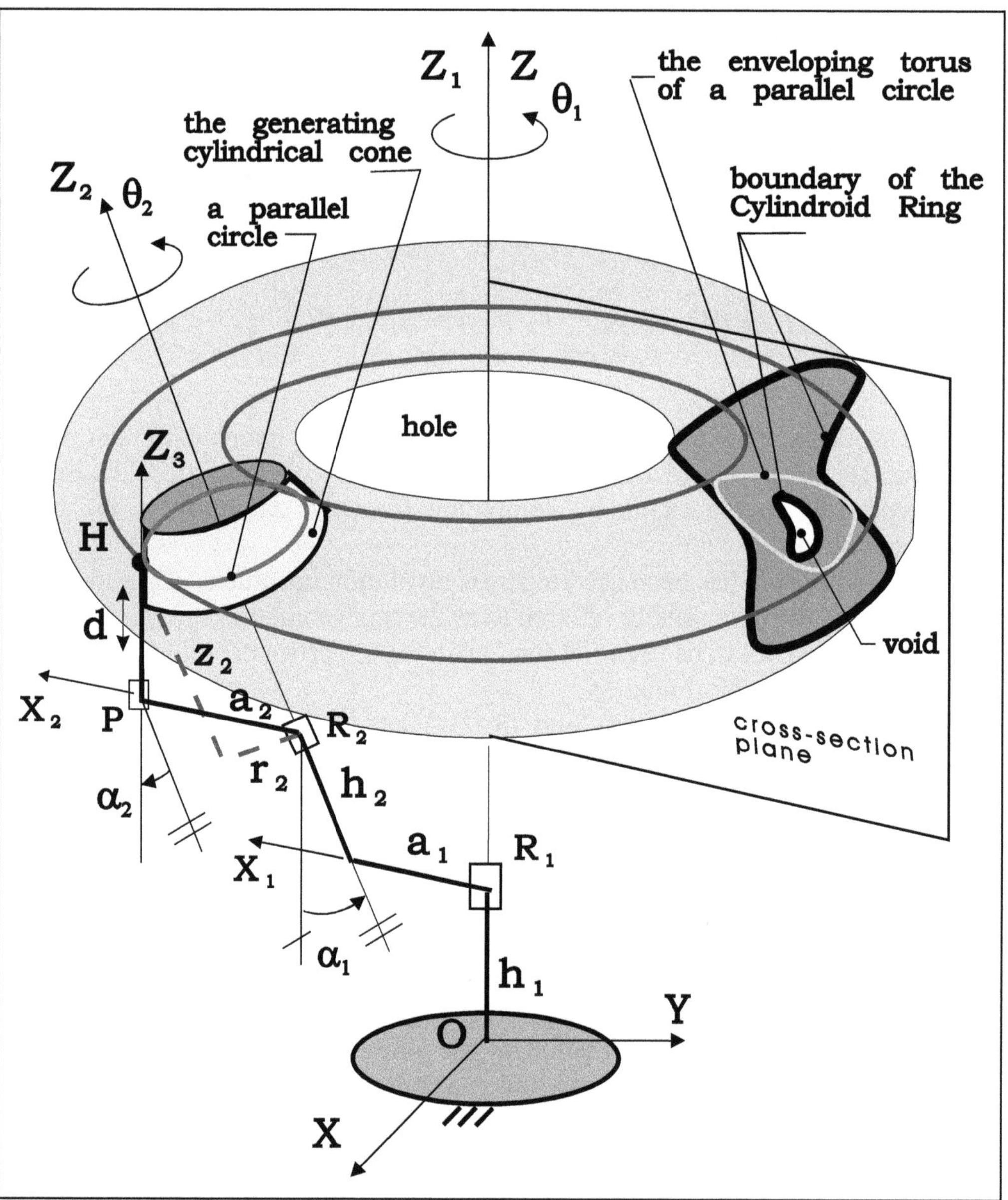

Fig. 3.12: Design parameters and topology in the generation of so-called 'cylindroid ring' workspace for telescopic manipulator arms.

Alternatively, a cylindroid ring $W_{RRP}(H)$ can be considered as the union of the tori $T_{R1R2}(H)$, which are due to the mobility in R_1 and R_2 joints and are traced by all parallel circles which can be cut on the generating cylindrical cone $T_{R2P}(H)$ so that $W_{RRP}(H)$ can be expressed in the form

$$W_{RRP}(H) = \bigcup_{d=d_{min}}^{d_{max}} T_{R_1R_2}(H) \quad . \tag{3.1.34}$$

Consequently, the workspace boundary $\partial W_{RRP}(H)$ of a cylindroid ring can be obtained as the envelope of toroidal surfaces $T_{R1R2}(H)$ which are generated by revolution of the parallel circles of the generating cone. Therefore, from Eq. (3.1.34) the boundary envelope can be expressed as

$$\partial W_{RRP}(H) = \underset{d=d_{min}}{\overset{d_{max}}{\text{env}}}\, T_{R_1R_2}(H) \tag{3.1.35}$$

where 'env' is the envelope operator performing analytically the geometric generation of an envelope.

A straightforward result of determining a workspace boundary through the above-mentioned procedure is the individuation of the workspace shape, and the existence and extension of hole and voids, which are recognized as basic characteristics for manipulators.

A hole is a region outside the ring but yet surrounded by the ring, within which it is possible to individuate at least a straight line of points not belonging to the ring. Therefore, a hole is generated when the revolving torus does not intersect or touch the revolution axis.

A void can be generally identified as an internal region within the workspace itself, which is not reachable by point H. In particular, a so-called 'ring void' is a void with ring topology, which is generated by a hole in a generating torus that does not intersect the revolution axis. A so-called 'apple void' is obtained when the hole of the generating torus intersects the axis of revolution for ring generation and a bulk apple-shaped volume characterizes it. An example of an apple void is shown in Fig. 3.13.

Particularly, referring to the case of position workspace of manipulators with revolute joints, two branches of envelope boundary contours in a cross-section of a workspace boundary are observable: an external and an internal, as shown in the example of Fig. 3.13 for 4R manipulators.

The external branch of the torus family envelope determines the external boundary of a bulk hyper-ring and it can be used for the generation of subsequent hyper-rings.

The internal branches of the torus family envelope show several separated and intricate contours in the cross-section. Nevertheless, they may indicate one ring void only. In fact, a ring void is generated by the hole in the revolving workspace. The smallest

parallel circle that is cut in the generating ring or hyper-ring generates the hole. Therefore, when this is rotated about an axis, this parallel circle also rotates and it once more generates a torus, which is the boundary surface of a ring void.
An apple void is a region of points not belonging to the workspace and containing a segment of the axis of revolution. It can be thought as generated, when no hole exists, by the volume inside a hole on the revolving workspace when the axis of revolution intersects the revolving workspace itself. In a ring it has been detected that more than one and at the most three apple voids may occur.
The common geometrical characteristics in ring and hyper-rings can be recognized in the recursive use of the ring generation. Nevertheless, each additional use introduces a multiplying effect in a sense that, for example, the number of apple voids may be one more than twice the number of voids in the revolving workspace. It seems, indeed, that each revolving operation may double the number of the envelope contours and consequently the number of apple voids may be one more of the double in revolving workspace. Thus, a 4R manipulator may have at the most seven apple voids, a 5R chain may show fifteen apple voids, and a 6R hyper-ring may present thirty-one apple voids, and so on.

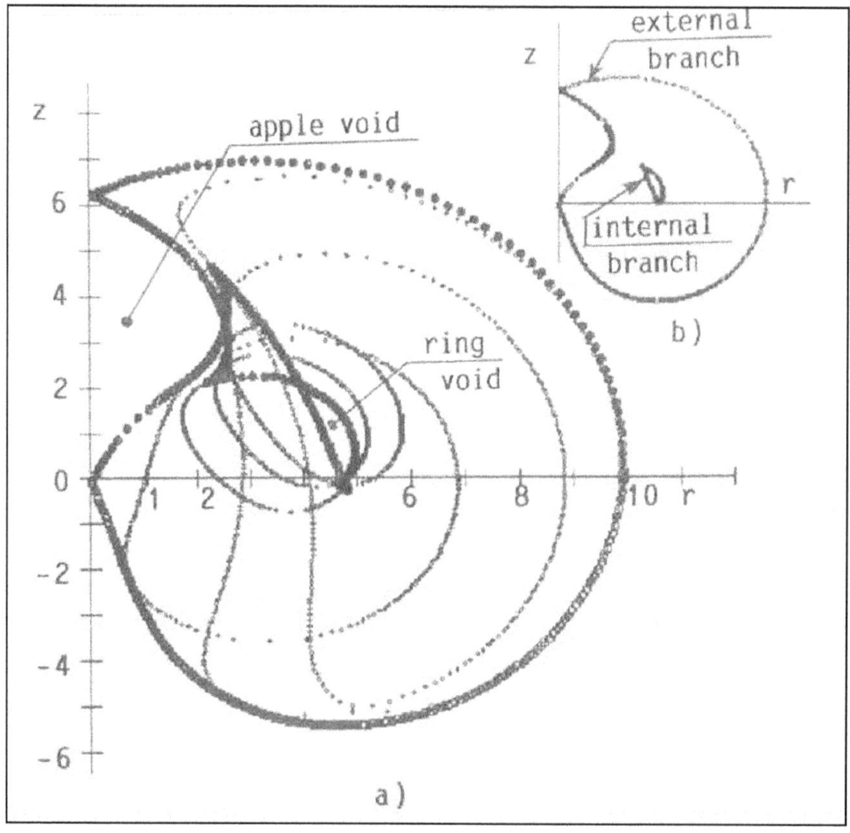

Fig. 3.13: Workspace cross-section of a 4R manipulator with a1= 3 u, a2= 4 u, a3= 2 u, a4= 5 u, α1= α2= α3= 60 deg, d2= d3= d4= 1 u. (u is the unit length)

A general shape of hyper-ring with hole may show in its cross section many inlets looking at the axis of symmetry as it may have many possible apple voids. Indeed, boundary inlets may generate apple voids and vice versa when the hyper-ring is changed to have or not have a hole. However, the peculiar shape with boundary inlets occurs when the hyper-ring boundary is near to the first revolution axis and it rapidly goes to a more regular circular shape with the increase of the minimum radial reach, i.e. a larger hole.

On the contrary, the ring void boundary shows invariable geometrical characteristics of a ring. This can be explained by taking into account that the generation process of a ring void is produced by the hole in the revolving hyper-ring and it is always due to a revolution of the circle giving the size of the generating hole.

Figure 3.13 shows an illustrative example with apple and ring voids. In Fig. 3.13 a) some tori of the generating envelope family are also shown in order to clarify the generation of the envelope boundary. In Fig. 3.13 b) the cross-section contour is indicated as internal and external branches of algebraic formulation. The internal branch always refers to the ring void.

3.1.3.1 A binary matrix formulation

It is recognized that the importance of a characterization for the workspace points and thus primary and secondary workspaces have been defined since the first studies on workspace depending on reachability and multiple reachability of points, respectively. Primary workspace can be defined as the region of reachable points without any limiting constraints, whereas secondary workspace can be defined as the region of reachable points when the robot extremity is oriented within prescribed ranges.

According to these views of the workspace, it is convenient to deduce a general procedure that is based on a numerical matrix formulation.

The workspace can be described through a binary mapping of the cross-section in the plane of radial direction r and axial line z, as shown in Fig. 3.14, according to a numerical algorithm whose basic steps are:

1. Dividing the cross-section plane r, z into I $\times$ J small rectangles of width Δi and of height Δj, where J and I are the number of divisions along the r axis and z axis, respectively. Each rectangle is individuated by P_{ij} to provide a binary image of the workspace cross-section. The values of the width Δi and of the height Δj can be properly selected as a function of the $\Delta\theta_k$ scanning intervals.
2. Initialization by setting $P_{ij}=0$ for all i and j.
3. A scanning process for each joint angle θ_k from θ_{kmin} up to θ_{kmax} with step $\Delta\theta_k$ to compute workspace point coordinates by using a matrix approach in the form of radial reach r and axial reach z as

$$r_k = \sqrt{x_k^2 + y_k^2} \qquad \qquad (3.1.36)$$
$$z_k = z_k$$

where

$$\begin{bmatrix} x \\ y \\ z \\ 1 \end{bmatrix}_k = {}_k T_{k+1} \begin{bmatrix} x \\ y \\ z \\ 1 \end{bmatrix}_{k+1} \tag{3.1.37}$$

in which ${}_k T_{k+1}$ is the transformation matrix between frame k and frame k+1, and the coordinates in k+1 frame are related to a position of extremity point H when the mobility of the joints following k joint is considered.

4. Construction of the binary map P_{ij}= 1 of the workspace cross-section by determining i and j as

$$i = \text{fix}\left[\frac{r_k}{\Delta i}\right]$$
$$j = \text{fix}\left[\frac{z_k}{\Delta j}\right] \tag{3.1.38}$$

by using the operator fix to compute the integer value of the above-mentioned ratios.

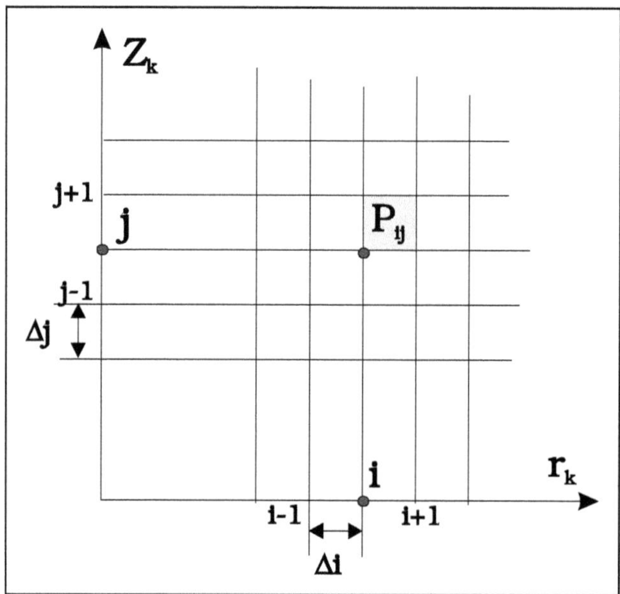

Fig. 3.14: A grid of the scanning process for binary mapping of the workspace cross-section.

Therefore, the binary mapping for the workspace cross-section is given by

$$P_{ij} = \begin{cases} 0 \text{ if } P_{ij} \notin W(H) \\ 1 \text{ if } P_{ij} \in W(H) \end{cases} \tag{3.1.39}$$

Because a binary image $P_{ij} = 1$ of the workspace has been given by Eqs (3.1.36) – (3.1.39) an efficient algorithm for determining workspace boundary can be developed by using P_{ij}. This is based on the geometry of the grid since a workspace point within the bulk of the reachable area is related to a sum equal to nine of the P_{ij} values for the points surrounding it. But when the workspace point is on the boundary contour this sum is less than nine, since there are surrounding points not belonging to the workspace with $P_{ij}= 0$. This can be analytically expressed through a variable, named as sum, given by

$$\text{sum} = \sum_{i-1}^{i+1} \sum_{j-1}^{j+1} P_{ij} \tag{3.1.40}$$

whose detection can be used to generate a binary mapping $G_{ij} = 1$ for the boundary points.

But in this numerical procedure two particular cases of possible errors may arise:

a) a rectangle with P_{ij} equal to zero is surrounded by eight rectangles with P_{ij} equal to one. This rectangle is not necessarily a void;

b) a rectangle with P_{ij} equal to one is surrounded by seven rectangles with P_{ij} equal to one and the eighth equal to zero, determining a gross boundary contour, because of a not fine resolution.

Both ambiguities can be solved by selecting the reference value for the sum variable in Eq. (3.1.40) equal to eight or less, instead of nine. It is worth noting that the numerical efficiency of the algorithm increases while its correctness does not decrease if the rectangles have small enough width Δi and height Δj.

A map of primary workspace can be obtained through the cross-section binary image P_{ij}.. Secondary workspaces are related to dexterous performances of a robot. Secondary workspaces can be defined also as regions of points that can be reached several times and it is convenient to determine regions with the same number of reaches. Thus, a measure of the reachability for each rectangle P_{ij} of a workspace grid can be introduced as a function of the number nq of times that a rectangle has been reached during the scanning process of the joints' mobility. Because of this numerical definition, a frequency matrix with entries fq_{ij} can be generated during the generation of P_{ij} itself by giving to fq_{ij} the values of the number of times that P_{ij} has been reached.

A numerical example has been reported in Fig. 3.1.5 for a COMAU robot SMART 6.100A by using the aforementioned binary matrix procedure for workspace determination and evaluation.

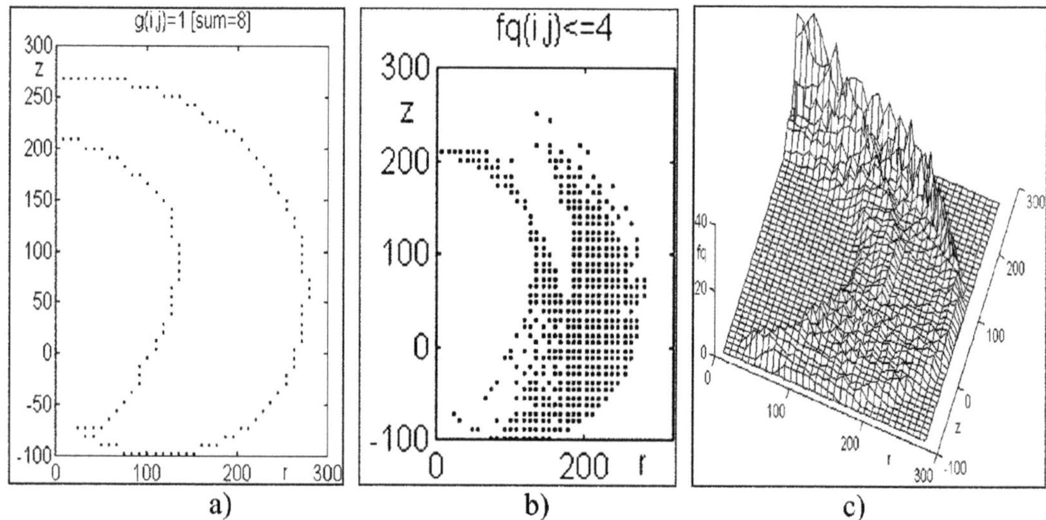

Fig. 3.15: Results of the binary matrix formulation for workspace evaluation of a robot manipulator COMAU SMART 6.100A with a_1 = 300 mm, α_1 = 90 deg., a_2 = 1,100 mm, $\alpha_2 = \alpha_3$ = 0 deg., a_3 = 1,625 mm, and $b_1 = b_2 = b_3 = 0$: a) cross-section workspace contour generated with sum = 8; b) cross-section area of secondary workspace with $fq(i,j) \leq 4$; c) a map for the frequency matrix fq(i,j).

3.1.3.2 An algebraic formulation

The boundary of an (N-j+1)R hyper-ring can be expressed algebraically when it is thought to be generated by enveloping the torus family traced by the parallel circles in the boundary of the revolving (N-j)R hyper-ring, according to Eq.(3.1.32).

An equation for a torus family can be expressed with respect to the j-th link frame, assuming $C_j \neq 0$ and $\cos\alpha_j \neq 0$, as a function of the radial r_j and axial z_j reaches, in the form

$$(r_j^2 + z_j^2 - A_j)^2 + (C_j z_j + D_j)^2 + B_j = 0 \qquad (3.1.41)$$

where the so-called torus parameters are a_j, α_j, r_{j+1}, z_{j+1}, and the coefficients are given as

$$\begin{aligned} A_j &= a_j^2 + r_{j+1}^2 + (z_{j+1} + d_{j+1})^2 \\ B_j &= -4a_j^2 r_{j+1}^2 \\ C_j &= 2a_j / s\alpha_j \\ D_j &= -2a_j (z_{j+1} + d_{j+1}) c\alpha_j / s\alpha_j \end{aligned} \qquad (3.1.42)$$

in which $s\alpha_j$ and $c\alpha_j$ designate $\sin\alpha_j$ and $\cos\alpha_j$, respectively.

Particular cases with $C_j = 0$ or $\cos\alpha_j = 0$ are not represented by Eq. (3.1.41) and specific

formulation can be developed when the torus boundary is not generated as an envelope of the revolving circle. When $a_j = 0$, the torus degenerates in an ellipsoid surface with major axis r_{j+1}. When $s\alpha_j = 0$, the torus degenerates in a circular rim of width r_{j+1}; and when $c\alpha_j = 0$, a 'common' or 'symmetrical-offset torus' is obtained from the revolving circle of radius r_{j+1}.
Indeed, a torus family can be thought to be generated by considering r_{j+1} and z_{j+1} as variables, which can be calculated from the boundary points of the revolving (N-j)R hyper-ring. Particularly, r_{j+1} and z_{j+1} can be expressed recursively by using Eq. (3.1.41) up to the last revolving torus workspace, or they can be formulated as a function of the revolute joint angles θ_{j+1}, θ_{j+2}, or finally, they can be a function of the last revolute joint angle by using an iterative formulation of the revolving boundary workspace and its torus family up to the extreme ring workspace for the manipulator chain.
Reaches r_{j+1} and z_{j+1} can be usefully expressed as a function of the last kinematic joint angle θ_N in the N-R chain to have a single variable formulation. Moreover, θ_N can be indeed considered as a torus family parameter since for each value a torus of the enveloping family can be determined and points of the workspace boundary can be generated.
The envelope equations of a torus family can be obtained from Eq. (3.1.41) and its derivative with respect to the torus family parameter θ_N. After some algebra for which $C_j \neq 0$ and $E_j \neq 0$ are needed, the so-called ring boundary equations can be obtained in the form

$$r_j = \left[A_j - z_j^2 + \frac{(C_j z_j + D_j) G_j + F_j}{E_j} \right]^{1/2}$$

$$z_j = \frac{-F_j G_j \pm Q_j^{1/2}}{(E_j^2 + G_j^2) C_j} - \frac{D_j}{C_j} \tag{3.1.43}$$

where the so-called ring coefficients are given as

$$\begin{aligned} E_j &= R_{j+1} + S_{j+1} \\ F_j &= -2 a_j^2 R_{j+1} \\ G_j &= -2 a_j z'_{j+1} c\alpha_j / s\alpha_j \\ Q_j &= -E_j^2 \left[F_j^2 + B_j \left(E_j^2 + G_j^2 \right) \right] \end{aligned} \tag{3.1.44}$$

with

$$R_{j+1} = \left[(C_{j+1} z_{j+1} + D_{j+1})(E_{j+1} G'_{j+1} - G_{j+1} E'_{j+1}) + F'_{j+1} E_{j+1} - F_{j+1} E'_{j+1}\right] / E^2_{j+1}$$
$$+ G_{j+1} E_{j+1} (C_{j+1} z'_{j+1} + G_{j+1}) / E^2_{j+1} + E_{j+1} - 2 z_{j+1} z'_{j+1} ,$$

$$S_{j+1} = 2\left(z_{j+1} + d_{j+1}\right) z'_{j+1} \tag{3.1.45}$$

$$z'_{j+1} = \left[\left(\pm 0.5\, Q'_{j+1} Q^{-1/2}_{j+1} - F_{j+1} G'_{j+1} - G_{j+1} F'_{j+1}\right)\left(E^2_{j+1} + G^2_{j+1}\right)\right.$$
$$\left. + 2\left(F_{j+1} G_{j+1} \mp Q^{1/2}_{j+1}\right)\left(E_{j+1} E'_{j+1} + G_{j+1} G'_{j+1}\right)\right] / \left(E^2_{j+1} + G^2_{j+1}\right)^2 C_{j+1} - G_{j+1} / C_{j+1}$$

The symbol ' represents the derivative operator with respect to the torus family parameter θ_N.
Equations (3.1.45) can be computed through the equation coefficients E_{j+1}, F_{j+1}, G_{j+1}, Q_{j+1} and their derivatives E'_{j+1}, F'_{j+1}, G'_{j+1}, Q'_{j+1}, whose expressions and values can be deduced from r_{j+1} and z_{j+1} formulation. Particularly, the derivatives of the ring coefficients can be calculated from Eqs (3.1.44), as

$$\begin{aligned} E'_j &= R'_{j+1} + S'_{j+1} \\ F'_j &= -2a^2_j R'_{j+1} \\ G'_j &= -2a_j\, z''_{j+1}\, c\alpha_j / s\alpha_j \\ Q'_j &= -2E^2_j \left[F_j\left(F'_j + E^2_j + G^2_j\right) + B_j (E_j E'_j + G_j G'_j)\right] + 2Q_j E'_j / E_j \end{aligned} \tag{3.1.46}$$

The derivatives R'_{j+1}, S'_{j+1}, and z''_{j+1} can be computed by means of a further derivative operation from Eqs (3.1.45) and by using, additionally, the second derivatives of the ring equation coefficients (3.1.44). The radial r_{j+1} and axial z_{j+1} reaches, the ring coefficients and derivatives of the (N-j)R workspace boundary may be expressed in iterative form by means of the envelope development from the (N-j-1)R hyper-ring boundary up to the extreme ring workspace due to the last three revolute joints in the chain. Finally, last ring workspace can be expressed in an algebraic form as a function of the torus family parameter θ_N.
The algebraic formulation of Eqs (3.1.42)–(3.1.46) can be used to generate numerically the workspace boundary of an N-revolute chain according to Eq. (3.1.32), from the extremity to the base of the manipulator when the geometrical sizes of the links are given. It is worthy of note that the generating workspaces can also be evaluated for the computations.
The coefficients R_{j+1}, S_{j+1}, and z'_{j+1} can be calculated through E'_{j+1}, F'_{j+1}, G'_{j+1}, Q'_{j+1}, as well as these can be computed through R'_{j+2}, S'_{j+2}, and z''_{j+2}, and so on, since the implicit expressions (3.1.44) of E_{j+1}, F_{j+1}, G_{j+1}, Q_{j+1}, as far as explicit expressions of E_{n-2}, F_{n-2},

G_{n-2}, Q_{n-2} can be computed. This iterative computation can be expressed, according to Eqs (3.1.46), in a general iterative form

$$\begin{aligned} E^k_{j+1} &= R^k_{j+2} + S^k_{j+2} \\ F^k_{j+1} &= -2a^2_{j+1} R^k_{j+2} \qquad\qquad k = 0, 1,\dots, j;\ j = 0,1,\dots, N\text{-}4 \\ G^k_{j+1} &= -2a_{j+1}\, z^{k+1}_{j+2}\, c\alpha_{j+1}/s\alpha_{j+1} \end{aligned} \tag{3.1.47}$$

and, according to Eqs (3.1.45), as

$$\begin{aligned} R^k_{j+2} &= f^k\left(E^{k+1}_{j+2}, F^{k+1}_{j+2}, G^{k+1}_{j+2}\right) \\ S^k_{j+2} &= g^k\left(E^{k+1}_{j+2}, F^{k+1}_{j+2}, G^{k+1}_{j+2}\right) \qquad k = 0, 1,\dots, j;\ j = 0,1,\dots, N\text{-}4 \\ z^{k+1}_{j+2} &= h^k\left(E^{k+1}_{j+2}, F^{k+1}_{j+2}, G^{k+1}_{j+2}\right) \end{aligned} \tag{3.1.48}$$

where f^k, g^k, and h^k, represent the k-derivatives of the functions f, g, and h which are those expressing R_{j+1}, S_{j+1} and z'_{j+1} in the form of Eqs (3.1.45), respectively.
However, all the above-mentioned computations can be numerically evaluated by starting from the computation of derivatives of E_{N-2}, F_{N-2}, G_{N-2}, Q_{N-2} up to (N-3) order. In fact, the ring coefficients can be algebraically expressed from the ring equations as

$$\begin{aligned} E_{N-2} &= -2a_N(d_{N-1}\, s\alpha_{N-1} c\theta_N + a_{N-1}\, s\theta_N) \\ F_{N-2} &= 4a^2_{N-2} a_N (a_N\, s^2\alpha_{N-1} s\theta_N c\theta_N + a_{N-1} s\theta_N - d_N s\alpha_{N-1} c\alpha_{N-1} c\theta_N) \\ G_{N-2} &= 2a_{N-2} a_N\, c\alpha_{N-2} s\alpha_{N-1} c\theta_N / s\alpha_{N-2} \end{aligned} \tag{3.1.49}$$

where from the geometry of the manipulator chain it holds

$$\begin{aligned} r_{N-1} &= \left[(a_N c\theta_N + a_{N-1})^2 + (a_N s\theta_N c\alpha_{N-1} + d_N s\alpha_{N-1})^2\right]^{1/2} \\ z_{N-1} &= d_N\, c\alpha_{N-1} - a_N\, s\theta_N\, s\alpha_{N-1} \end{aligned} \tag{3.1.50}$$

An easy numerical algorithm can be developed to compute the radial r_j and axial z_j reaches of all envelopes for j from N-2 to 1, which is from the extreme workspace ring up to NR hyper-ring. This can be obtained by scanning the joint angle θ_N from 0 to 2π and calculating at each j the coefficients A_j, B_j, C_j, D_j, E_j, F_j, G_j, R_j, S_j, and z'_j, and finally r_j, z_j when the j derivatives of E_j, F_j, G_j, are evaluated by using previous calculations for R_{j+1}, S_{j+1} and z'_{j+1}.
A numerical example of the workspace determination by using the aforementioned algebraic formulation has been reported in Fig. 3.16 in which the envelope contours of the generating workspace are shown together with the final results that are obtained by

neglecting the internal branches not representing void boundary.

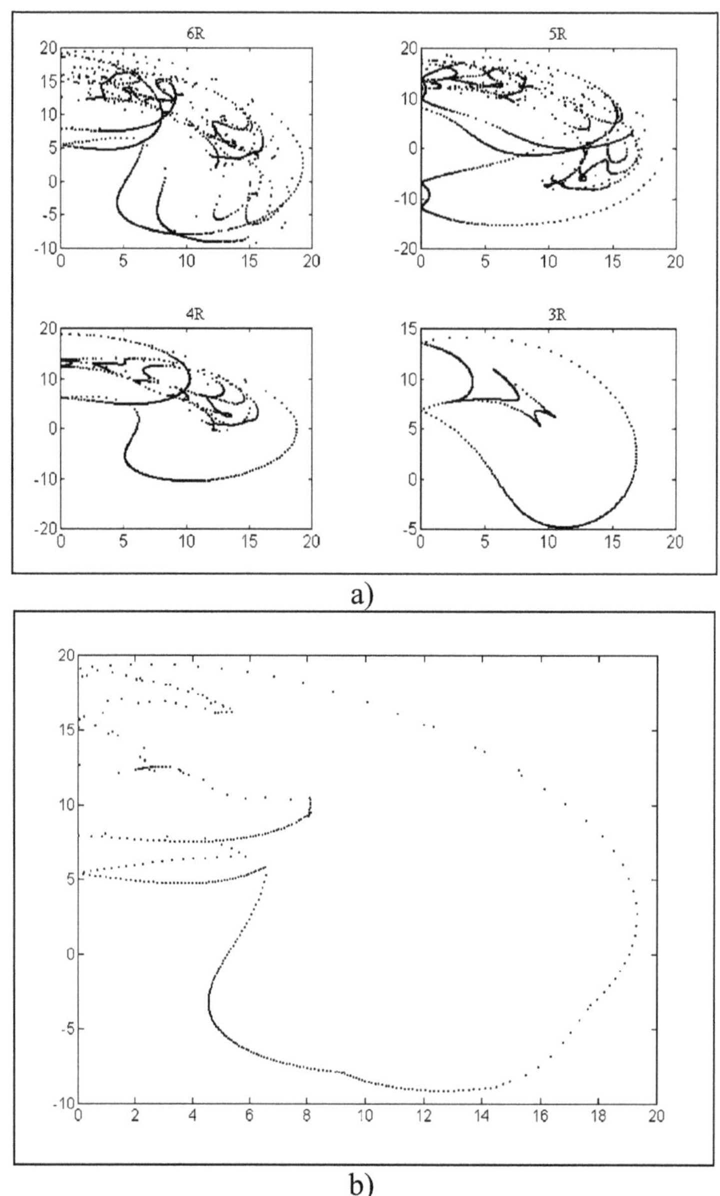

b)

Fig. 3.16: Workspace cross-sections of a 6R manipulator with $a_i = i$ u ($i = 1,...,6$), $d_1 = 0$ u, $d_i = (i-1)$ u ($i = 2,...,5$), and $\alpha_i = \pi/4$ ($i = 1,...,5$). *(u is the unit length): a) the envelope boundary in the generating workspaces; b) the cross-section of primary workspace.*

A similar algebraic formulation can be deduced for the workspace of manipulators with prismatic joints.
In particular, by referring to the case of a telescopic arm in Fig. 3.12, the procedure of Eq. (3.1.35) can be formulated as in the following.
Figure 3.12 shows that the workspace boundary of the cylindroid ring is composed of

two different geometrical topologies: envelope segments and toroidal surfaces. The envelope segments are located in the lateral sides of the cross-section representation, and two toroidal surfaces are the top and bottom covers, respectively. This cross-section shows a shape that justifies once more the name cylindroid ring which we have given to workspace volume of telescopic arms, since it recalls the geometry both of a cylinder and a ring.

Consequently, the workspace boundary $\partial W(H)$ of a cylindroid ring can be obtained as the envelope of toroidal surfaces T_{R1R2} (H), which are generated by revolution of the parallel circles of the generating cylindrical cone, and straight-line segments of the envelope contour.

In particular, each point of the straight-line segment is individuated through a value of stroke d. The corresponding torus may be written assuming $h_1 = 0$, with the hypothesis $\sin \alpha_1 \neq 0$ and with respect to OXYZ, by using the radial reach r and the axial reach z in the form

$$\left(r^2 + z^2 - A\right)^2 + (C\,z + D)^2 + B = 0 \tag{3.1.51}$$

The so-called structural coefficients are expressed as

$$\begin{aligned}
A &= a_1^2 + r_2^2 + (z_2 + h_2)^2 \\
B &= -4\,a_1^2\,r_2^2 \\
C &= 2\frac{a_1}{\sin \alpha_1} \\
D &= -2\,a_1\,(z_2 + h_2)\frac{\cos \alpha_1}{\sin \alpha_1}
\end{aligned} \tag{3.1.52}$$

where r_2 and z_2 are the radial and axial reaches with respect to $O_2X_2Y_2Z_2$. The distances r_2 and z_2 can be given from the geometry of the chain and they can be expressed as

$$\begin{aligned}
r_2 &= \sqrt{a_2^2 + d^2 \sin^2 \alpha_2} \\
z_2 &= d \cos\alpha_2
\end{aligned} \tag{3.1.53}$$

where the independent variable is the stroke parameter d.

Equations (3.1.51) to (3.1.53) can be used to determine the workspace volume for a given telescopic arm by scanning the stroke interval from a minimum value d_{min} to a maximum value d_{max} to perform Eq. (3.1.34).

Indeed, a torus is traced by using Eq. (3.1.51) to determine r by assuming z. This is obtained by scanning z within its range from minimum to maximum values, which can be calculated by using the geometry of the telescopic arm in Fig. 3.12 from

$$z = r_2 \sin\theta_2 \sin\alpha_1 + (h_2 + d\cos\alpha_2)\cos\alpha_1 + d_1 \tag{3.1.54}$$

From this computation the toroidal segments can be determined as the top and bottom of the workspace boundary.
The envelope segments can be determined by a formulation by using Eq. (3.1.35) to obtain a more efficient computational algorithm for workspace determination. In fact, enveloping all the tori traced by H during the scanning process generates the envelope segments. An analytical expression of the envelope can be obtained from Eq. (3.1.51) and its derivative with respect to the envelope parameter, which is the stroke d.
Thus, differentiating Eq. (3.1.51) with respect to d one obtains

$$-\left(r^2 + z^2 - A\right)A' + (Cz + D)D' + B' = 0 \tag{3.1.55}$$

where the symbol ' indicates d-derivative operator.
By using Eqs (3.1.52) and (3.1.53) analytical expression for derivatives of the structural coefficients can be obtained in the form

$$\begin{aligned} A' &= 2\left(d + h_2 \cos\alpha_2\right) \\ B' &= -4a_1^2\, d \sin^2\alpha_2 \\ D' &= -2a_1 \cos\alpha_2 \frac{\cos\alpha_1}{\sin\alpha_1} \end{aligned} \tag{3.1.56}$$

Finally, by using Eqs (3.1.51) and (3.1.56) with suitable algebraic manipulations, the radial and axial reaches of points of envelope boundary segments can be expressed with the hypothesis $A' \neq 0$ as

$$\begin{aligned} z &= \frac{-B'D' \pm A'\sqrt{-\left[B'^2 + B\left(A'^2 + D'^2\right)\right]}}{\left(A'^2 + D'^2\right)C} - \frac{D}{C} \\ r &= \sqrt{\frac{B' + (Cz + D)D'}{A'} + A - z^2} \end{aligned} \tag{3.1.57}$$

The sign ambiguity exists to give two envelope segments.
A numerical example of the workspace determination by using the aforementioned algebraic formulation has been reported in Fig. 3.17 in which the boundary contour of the workspace of a telescopic manipulator is shown together with contours of the

enveloping toroidal surfaces.
The family of the enveloping tori also determines the top and bottom covers of the workspace as part of the extreme tori.

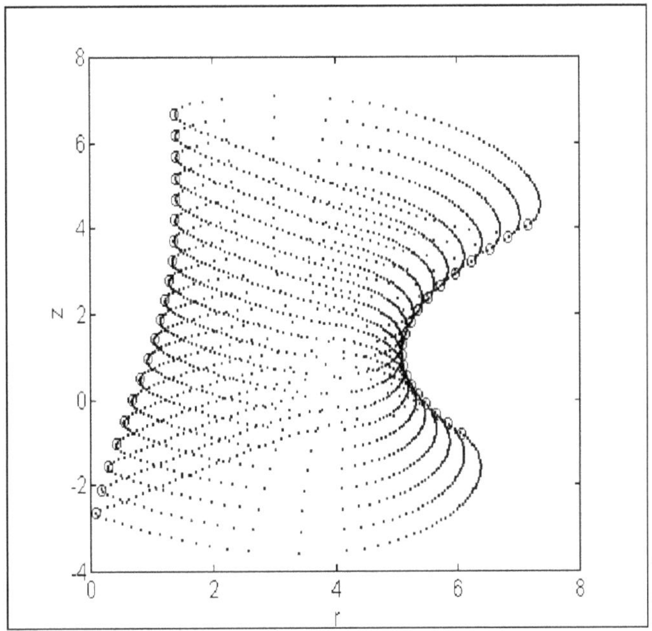

Figure 3.17: Cross-section of a workspace boundary computed as tori envelope segments (circle points) and toroidal covers (dotted points) for a telescopic manipulator arm with $a_1 = 3$ u, $a_2 = 2$ u, $h_1 = 0$ u, $h_2 = 2$ u, $\alpha_1 = 30$ deg., $\alpha_2 = 30$ deg., $d_{min} = -5$ u, $d_{max} = 5$ u (u is a unit length).

3.1.3.3 A workspace evaluation

An evaluation of the workspace can be obtained by means of numerical simulations and/or experimental tests.
The fundamental characteristics of manipulator workspace for a numerical evaluation can be identified for both position and orientation capabilities as:

- shape and value of cross-section areas;
- shape and value of workspace volume;
- shape and extension of hole and voids;
- reach distances and reach ranges.

As explained in 3.1.3.1, it is also convenient to evaluate numerically the repeatability measure through frequency matrix plots.
The above-mentioned characteristics can be determined numerically by using the binary matrix procedure or algebraic formulation in 3.1.3.2 through general or specific algorithms for simulation.
Once the workspace points (both in position and orientation) are determined, one can use them to perform an evaluation of the above-mentioned workspace characteristics.

In particular, a cross-section area can be determined by selecting from the computed workspace points those that lay on the cross-section plane under examination, as shown in the examples of Figs 3.15, 3.16, and 3.17. Thus, the shape can be illustrated by the result in the plot form. The computation of the value of a cross-section area can be obtained by using a grid evaluation or an algebraic formula.
By referring to the scheme of Fig. 3.14 for a grid evaluation, one can calculate the area measure A as a sum of the scanning resolution rectangles over the scanned area as

$$A = \sum_{i=1}^{I} \sum_{j=1}^{J} \left({}_{A}P_{ij}\ \Delta i\ \Delta j \right) \tag{3.1.58}$$

by using the ${}_{A}P_{ij}$ entries of the binary matrix P_{ij} that are related to the cross-section plane for A.
Alternatively, one can use the workspace points of the boundary contour of a cross-section area that can be determined from the algebraic formulation in 3.1.3.2 or by using the entries of the binary matrix G_{ij} from the scanning procedure through Eq. (3.1.40). Thus, referring to the scheme of Fig. 3.18 and by assuming as computed the coordinates of the cross-section contour points as an ordinate set (r_j, z_j) of the contour points H_j with j=1, …, N, the area measure A can be computed as

$$A = \sum_{j=1}^{N} \left(z_{1,j+1} + z_{1,j} \right) \left(r_{1,j} - r_{1,j+1} \right) \tag{3.1.59}$$

By extending the above-mentioned procedures, the workspace volume V can be computed by using the grid scanning procedure in a general form

$$V = \sum_{i=1}^{I} \sum_{j=1}^{J} \sum_{k=1}^{K} \left[P_{ijk}\ \Delta i\ \Delta j\ \Delta k \right] \tag{3.1.60}$$

in which P_{ijk} is the entry of a binary representation in a 3D grid.
By using the boundary contour points through the Pappus-Guldinus Theorem the workspace volume V can be computed within the binary mapping procedure in the form

$$V = 2\pi \sum_{i=1}^{I} \sum_{j=1}^{J} \left[P_{ij}\ \Delta i\ \Delta j \left(i\,\Delta i + \frac{\Delta i}{2} \right) \right] \tag{3.1.61}$$

or within the algebraic formulation in the form

$$V = \frac{\pi}{2} \sum_{j=1}^{N} \left(z_{1,j+1} + z_{1,j} \right) \left(r_{1,j}^{2} - r_{1,j+1}^{2} \right) \tag{3.1.62}$$

when the workspace volume is a solid of revolution.
Otherwise the workspace volume can be computed by using the algebraic expressions for the corresponding topology.
Therefore, it is evident that the formula of Eq. (3.1.60) has a general application, while Eq. (3.1.61) is restricted to serial open-chain manipulators with revolute joints.

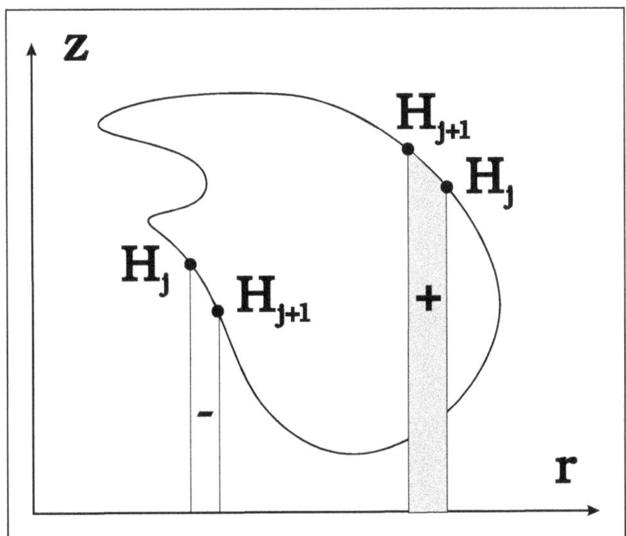

Fig. 3.18: A scheme for the computation of workspace volume of serial open-chain manipulators with revolute joints.

Similarly, hole and void regions can be numerically evaluated by using the formulas of Eqs (3.1.58) to (3.1.62) to obtain the value of their cross-sections and volumes, once they have been preliminarily determined.
Orientation workspace can be similarly evaluated by considering the angles in a Cartesian frame representation.
Reach distances can be defined as those distances that are related to workspace points that are at the minimum and maximum distances from the base axis of the manipulator.
In particular, radial reach r can be defined as the maximum distance of the extremity point H as measured from the base axis of a manipulator along a radial direction. Axial reach z is defined as the maximum distance of the extremity point H as measured from the base of a manipulator along the direction of the base axis. A maximum global reach can also be defined as the maximum distance of the extremity point H as measured from the origin of the base frame of a manipulator.
In a similar way, minimum values can be evaluated for minimum radial reach, axial reach, and global reach.
The corresponding reach ranges can be computed as a difference between maximum and minimum values. Those workspace reach ranges can be used to represent a parallelepiped volume that can identify the workspace limits. In addition, other suitable topology can be used to represent the workspace limits, but they require the

computation of suitable reach performance in agreement with the topology under examination.
Similar evaluation can be carried out for the orientation workspace when the axes of a Cartesian frame of workspace representation are related to the orientation angles for the manipulator orientation performances.
Experimental tests for workspace evaluation are usually performed to detect mainly the reach characteristics that can be used to check the soundness of numerical evaluation of the workspace performances.
Indeed, the so-called 'kinematic calibration' refers mainly to the experimental identification of the H--D parameters that can be used in numerical algorithms for evaluating workspace performances.
Specific tests can be carried out both in calibration and certification procedures to check the feasibility of the manipulator motion and performance within its workspace.

3.1.4 Manipulator design with prescribed workspace

Workspace design data can be prescribed as shown in Fig.3.19 by means of workspace limits in term of r radial reach and z axial reach. Thus, a general design problem can be formulated as finding the dimensions of a manipulator whose workspace cross-section is within or is delimited by the given axial and radial reaches r_{min}, r_{max}, z_{min}, z_{max}.
The algebraic formulation for workspace boundary can be conveniently used for synthesis purposes when the workspace is prescribed by means of workspace limits, as in Fig. 3.19, to give precision points for workspace boundary. In this case, in fact, the algebraic formulation can be used to build a system of algebraic design equations where the axial and radial reaches are given by the prescribed workspace data and the design unknowns are the ring or hyper-ring coefficients. Once these coefficients are solved, the chain parameters can be calculated by using the definitions of the coefficients.
In order to outline a design procedure, the case of a 3R manipulator is discussed in detail.

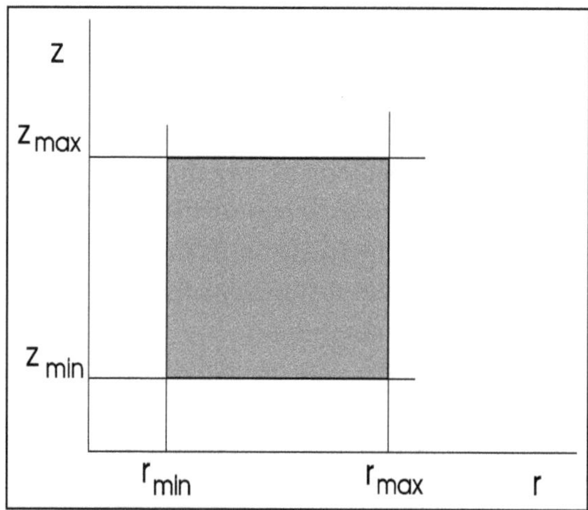

Fig. 3.19: A general scheme for prescribing workspace limits of a manipulator.

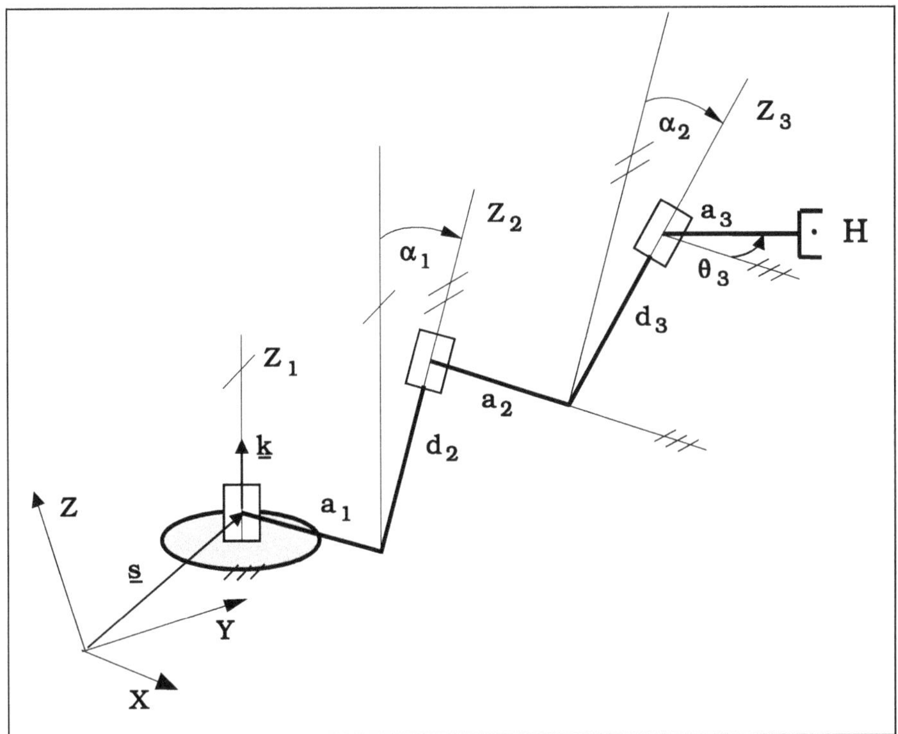

Fig. 3.20: The kinematic chain of a general 3R manipulator and its design parameters.

A general open-chain 3R manipulator with three revolute joints is sketched in Fig. 3.20, in which the design parameters are represented as the H–D parameters a_1, a_2, a_3, d_2, d_3, α_1, α_2, and θ_3 is the Z_3 joint variable.

With the hypothesis $\alpha_1 \neq k\ \pi/2$ (k = 0,1,...,4), the workspace boundary of a three-revolute open-chain manipulator can be expressed as a function of the radial reach r_1 and axial reach z_1 with respect to the manipulator base frame $X_1Y_1Z_1$ as

$$r_1 = \left[A_1 - z_1^2 + \frac{(C_1 z_1 + D_1) G_1 + F_1}{E_1} \right]^{1/2}$$

$$z_1 = \frac{-L_1 \pm Q_1^{1/2}}{K_1 C_1} - \frac{D_1}{C_1} \tag{3.1.63}$$

with so-called 'structural coefficients' as

$$A_1 = a_1^2 + r_2^2 + (z_2 + d_2)^2$$

$$B_1 = -4 a_1^2 r_2^2 \tag{3.1.64}$$

$$C_1 = \frac{2a_1}{s\alpha_1} \tag{3.1.64}$$

$$D_1 = -2a_1(z_2 + d_2)\frac{c\alpha_1}{s\alpha_1}$$

in which the distance of reference point H from the Z_2 axis is valuable from the geometry of the chain as

$$r_2 = \left[(a_3 c\theta_3 + a_2)^2 + (a_3 s\theta_3 c\alpha_2 + d_3 s\alpha_2)^2\right]^{1/2}$$
$$z_2 = d_3\, c\alpha_2 - a_3\, s\theta_3\, s\alpha_2 \tag{3.1.65}$$

The remaining structural coefficients can be computed by using the general procedure that is outlined in section 3.1.3.2 through the following expressions

$$E_1 = -2a_3(d_2\, s\alpha_2 c\theta_3 + a_2\, s\theta_3)$$
$$F_1 = 4a_1^2 a_3(a_3\, s^2\alpha_2 s\theta_3 c\theta_3 + a_2 s\theta_3 - d_3 s\alpha_2 c\alpha_2 c\theta_3)$$
$$G_1 = 2a_1 a_3\, c\alpha_1 s\alpha_2 c\theta_3 / s\alpha_1$$
$$K_1 = G_1^2 + E_1^2 \tag{3.1.66}$$
$$L_1 = F_1 G_1$$
$$Q_1 = L_1^2 - K_1\left(F_1^2 + B_1 E_1^2\right)$$

Thus, the workspace boundary of a general three-revolute open-chain manipulator can be evaluated by scanning the angle θ_3 and plotting r_1, z_1. This gives numerically the workspace boundary shape with its hole and voids, and its volume. It should be noted that the workspace formulation of Eqs (3.1.63)–(3.1.66) is an algebraic function of the θ_3 joint variable and all design parameters.
Assuming as additional design parameters the position and orientation vectors **s** and **k** of the manipulator base with respect to the fixed world frame XYZ, Fig. 3.20, the workspace design equations (3.1.63) can be modified to include the reference change. To accomplish this we take **x** as the position vector of a boundary point with respect to XYZ, and we use the expression

$$\mathbf{x}_1 = R\mathbf{x} - \mathbf{s}_1 \tag{3.1.67}$$

where r_1, z_1 are the components of $\mathbf{x}_1$ with respect to the manipulator base frame $X_1Y_1Z_1$; R is the 3 × 3 rotation matrix describing the fixed frame with respect to the manipulator base frame; the position vector of robot base is $\mathbf{s}_1$ as measured in $X_1Y_1Z_1$ and $\mathbf{s} = R^t\, \mathbf{s}_1$ (R^t is the transpose of R) as measured in XYZ.

Introducing Eq. (3.1.67) into Eqs (3.1.63), the result can be expressed, after some algebraic manipulations, in a vector form as

$$E_1(\mathbf{x}\cdot\mathbf{x}-2\mathbf{s}\cdot\mathbf{x}+\mathbf{s}\cdot\mathbf{s}-A_1)-\{C_1[\mathbf{r}_3\cdot(\mathbf{x}-\mathbf{s})]+D_1\}G_1+F_1=0 \tag{3.1.68}$$

and

$$\mathbf{r}_3\cdot(\mathbf{x}-\mathbf{s})=\frac{-L_1\pm Q_1^{1/2}}{K_1C_1}-\frac{D_1}{C_1} \tag{3.1.69}$$

where $\mathbf{r}_3$ is the third row vector of R; the Z_1 component of $\mathbf{s}_1$ can be computed as $s_{1z} = R\,\mathbf{k}_1 \cdot R^t\,\mathbf{s}_1 = \mathbf{r}_3 \cdot \mathbf{s}$; and $\mathbf{k}_1$ is the orientation vector of robot base as measured in $X_1Y_1Z_1$. Consequently, a general synthesis problem can be formulated by using Eqs (3.1.63) to (3.1.69), when a certain number of workspace boundary points are given and the design unknowns are represented by the link sizes, the structural coefficients A_1, B_1, C_1, D_1, E_1, F_1, G_1, and the manipulator base location vectors **s**, **k**.
When the manipulator base location is known, assuming seven given workspace boundary points, whose coordinates can be expressed directly with respect to $X_1Y_1Z_1$, the aforementioned unknown structural coefficients can be calculated through the Newton–Raphson technique from a set of equations which express the workspace boundary points. This set is formulated through the second of Eqs (3.1.63) and another one which is deduced from Eqs (3.1.63) with the hypothesis $E_1 \neq 0$ in the form

$$E_1K_1(r_1^2+z_1^2-A_1)-(-L_1\pm Q_1^{1/2})G_1-F_1K_1=0 \tag{3.1.70}$$

Once Eq. (3.1.70) is used with five boundary points to solve A_1, E_1, F_1, G_1, and B_1 through Q_1 as defined in Eqs (3.1.66), the remaining structural unknowns C_1 and D_1 can be evaluated by means of the second of Eqs (3.1.63) written for two more workspace boundary points.
In a similar way, in the case of the unknown vectors **s** and **k**, the Eqs (3.1.68) and (3.1.69) can be conveniently expressed to obtain two decoupled sets of design equations. The first one can be deduced from Eq. (3.1.68) by using Eq. (3.1.69) to obtain

$$E_1K_1(\mathbf{x}\cdot\mathbf{x}-2\mathbf{s}\cdot\mathbf{x}+\mathbf{s}\cdot\mathbf{s}-A_1)-(-L_1\pm Q_1^{1/2})G_1-F_1K_1=0\,. \tag{3.1.71}$$

By writing Eq. (3.1.71) for eight given workspace boundary points the unknown A_1, E_1, F_1, G_1, Q_1, and **s** can be solved when the ambiguity in the sign of the radical Q_1 is resolved, taking into account that the upper branch of the envelope boundary is related to the positive sign and the bottom part to the negative sign. Once the first set of equations is solved, Eq. (3.1.69) can be used with the remaining four boundary points to give C_1, D_1, and $\mathbf{r}_3$ taking into account that $\mathbf{r}_3$ represents Z axis unit vector with respect to $X_1Y_1Z_1$, and therefore its components must satisfy the expression

$$r_{31}^2 + r_{32}^2 + r_{33}^2 = 1 \tag{3.1.72}$$

Since we are interested in **k**, $\mathbf{r}_3$ together with the orthonormal unit vector constraints can be used to evaluate the R matrix, whose third column represents **k** unit vector with respect to XYZ.
Once the structural coefficients are numerically determined, assuming $w = \sin^2\alpha_1$, Eqs (3.1.64) can be inverted to give

$$\begin{aligned} a_1 &= 0.5\, C_1 w^{1/2} \\ r_2^2 &= -\frac{B_1}{w C_1^2} \\ z_2 + d_2 &= -\frac{D_1}{C_1 (1-w)^{1/2}} \end{aligned} \tag{3.1.73}$$

and only the parameter w needs to be solved.
Substituting Eqs (3.1.73) into the first of Eqs (3.1.64), with the position $w = y + (1 + 4 A_1 / C_1^2) / 3$, it yields

$$y^3 + 3\,p\,y + 2\,q = 0 \tag{3.1.74}$$

where

$$\begin{aligned} p &= \frac{4}{9C_1^2}\left[A_1 - \frac{C_1^2}{4} - \frac{12}{C_1^2}\left(\frac{B_1 + D_1^2}{4} + \frac{A_1^2}{3}\right)\right] \\ q &= \frac{1}{27C_1^2}\left(1 + 4\frac{A_1}{C_1^2}\right)\left(10A_1 - 16\frac{A_1^2}{C_1^2} - 18\frac{B_1 + D_1^2}{C_1^2} - C_1^2\right) + 2\frac{B_1}{C_1^4}\,. \end{aligned} \tag{3.1.75}$$

Depending on the discriminant term

$$D = q^2 + p^3 \tag{3.1.76}$$

Equation (3.1.74) can be solved algebraically using Cardano's formula as function of the discriminant D:
- when $D > 0$, one real solution is expressed as

$$y = [\, q + D^{1/2} \,]^{1/3} + [\, q - D^{1/2} \,]^{1/3} \tag{3.1.77}$$

- when $D = 0$ two real solutions are expressed as

$$y_1 = 2\, q^{1/3}, \qquad y_2 = -\, q^{1/3}; \tag{3.1.78}$$

- when $D < 0$ three real solutions are expressed as

$$\begin{aligned} y_1 &= 2\, p^{1/2} \cos(u/3), \\ y_2 &= 2\, p^{1/2} \cos(u/3 + 2\,\pi/3), \\ y_3 &= 2\, p^{1/2} \cos(u/3 + 4\pi/3), \end{aligned} \tag{3.1.79}$$

where

$$u = \cos^{-1}(q/p^{3/2}) \tag{3.1.80}$$

Equation (3.1.74) can be solved algebraically to give one, two, or three solutions depending on the discriminant D by using Eq. (3.1.75). Looking at the formulae (3.1.77) to (3.1.79) it is observable that Eq. (3.1.74) gives one or two solutions of w according to the condition $0 < w < 1$ since at least one of the Eqs (3.1.79) gives a negative value.
Once Eq. (3.1.74) is solved, we can compute the parameters a_1, r_2, and $(z_2 + d_2)$ from Eqs (3.1.73) and calculate the α_1 angle as

$$\alpha_1 = \pm \sin^{-1} w^{1/2} \tag{3.1.81}$$

so that each solution for w corresponds to two manipulators distinguished at this step by the α_1 sign and to two more manipulators taking into account the supplementary values of α_1.
Successively, with the hypothesis that $\theta_3 = 0$, inverting Eqs (3.1.65) and (3.1.66) the remaining chain parameters can be obtained as

$$\begin{aligned} d_2 &= -\frac{E_1 a_1}{G_1 \tan \alpha_1} \\ d_3 &= \frac{z_2}{\cos \alpha_2} \end{aligned} \tag{3.1.82}$$

$$a_3 = \frac{G_1 \tan\alpha_1}{2a_1 \sin\alpha_2} \tag{3.1.82}$$

$$a_2 = \left(r_2^2 - z_2^2 \tan^2 \alpha_2\right)^{1/2} - a_3$$

where α_2 needs to be previously solved. The angle α_2 can be obtained by substituting Eqs (3.1.73) into the expression of the two-link length $d_3^2 + a_2^2 + a_3^2 = r_2^2 + z_2^2$ to give, with the hypothesis $\alpha_2 \neq k\ \pi/2$, $k = 0,1,...,4$, and after some manipulations, the expression

$$\alpha_2 = \pm \sin^{-1}\left[\frac{U^2 + r_2^2 \pm\left[\left(U^2 - r_2^2\right)^2 - 4U^2 z_2^2\right]}{2\left(r_2^2 + z_2^2\right)}\right]^{1/2} \tag{3.1.83}$$

where U is a short form for the quantity $(G_1 \tan\alpha_1)/(2a_1)$.

Equation (3.1.83) provides none, two or four solutions depending on the values of U, r_2, and z_2. Two or four more solutions can be obtained by taking into account supplementary values of the calculated α_2.

It is worthy of note that each numerical solution for the structural coefficients and the manipulator base location corresponds to sixty-four different manipulator parameter sets at the most, depending on the number of solutions for α_1 and α_2. However, since the above-mentioned values for the twist angles α_1 and α_2 may give negative values for the length sizes but they do not correspond to alternative design solutions, meaningful solutions can be considered only as the sixteen sets, which can be synthesized for $-\pi/2 < \alpha_1 < \pi/2$ and $-\pi/2 < \alpha_2 < \pi/2$.

The case of the design of 3R manipulators has been approached by inverting the formulation for the workspace boundary, but a general extension to n-R manipulators by using the algebraic formulation in 3.1.3.2 seems to be unpractical and numerically too laborious. Therefore, a general design problem for n-R manipulators can be formulated as an optimization problem with prescribed workspace characteristics.

The design parameters can be considered as the link time-independent sizes of the chain, i.e. for n revolute-connected manipulators a_i, d_i, $i = 1,...,n$, and α_i, $i = 1,..., n-1$, and the vectors **b** and **k**, which represent respectively the position and orientation of the robot base with respect to a fixed frame XYZ, as shown in Fig. 3.21.

An optimum design of manipulators can be formulated as an optimization problem in the form

$$\min f \tag{3.1.84}$$

subject to

$$\mathbf{x}_j \geq \mathbf{X}_j \qquad (j = 1,...,J) \tag{3.1.85}$$

$$V \geq V_0 \tag{3.1.86}$$

where f is the objective function; $\mathbf{X}_j$ (j=1,...,J) represent given precision points, with respect to the fixed frame, to be reached as workspace points $\mathbf{x}_j$; V_0 is a minimum value for a desirable workspace volume.
The robot base location can be included into the design parameters through the **b** vector expressed by means of the radial a_0 and axial d_0 components, Fig. 3.21. Particularly, the angles α_0 and θ_0 can be considered as describing the orientation of the manipulator base as Z_b axis with respect to Z-axis and X_b with respect to X, respectively. In addition, the base frame can be conveniently assumed to be parallel with the frame $X_1Y_1Z_1$, fixed on the first link of the manipulator chain, at a starting motion configuration. Therefore, the dimensional parameters for the optimal design problem will also include the parameters α_0, θ_0, a_0, and d_0 as representing the vectors **b** and **k**.
The workspace precision points can be considered inside as well as on the boundary of the workspace volume. Nevertheless, they are assumed, at most, as limiting points for the workspace design capability and, consequently, it is usually convenient to think of them as workspace boundary points in agreement with the scheme of Fig. 2.19.
In addition, it is also possible to prescribe workspace characteristics as voids and hole, which can be useful in practical applications as it saves regions for equipment and personnel in an automated environment.
Therefore, a workspace description through determination of its boundary is needed and $\mathbf{x}_j$ is assumed to be determinable from an analytical expression of the workspace boundary.

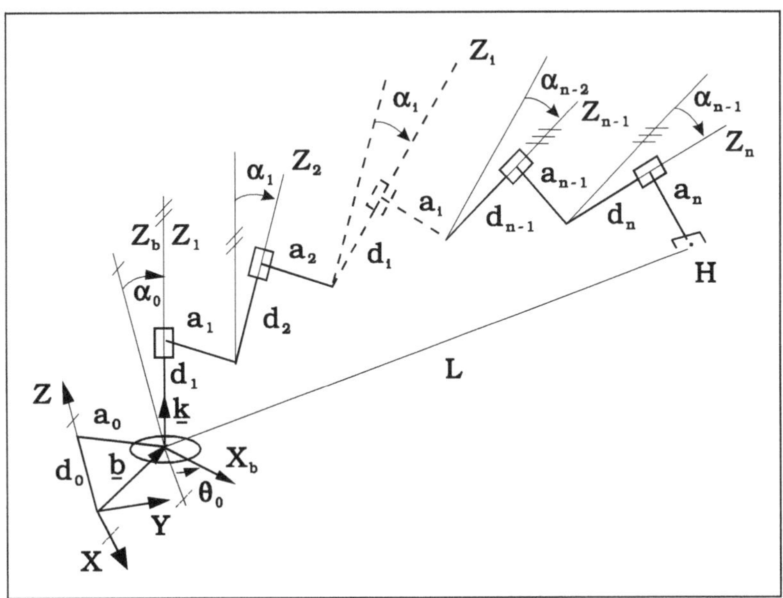

Fig. 3.21: Design parameters for a general n-R manipulator.

Once workspace boundary points are given, the manipulator solution must satisfy exactly the restriction with its workspace boundary and the sign equal is to be considered. Otherwise, a delimiting region can be assigned within which the workspace may be outlined and a weak restriction can be used to give larger design possibilities.
The constraint on the workspace volume can be of a determinant significance, since this restraint may have an effect on the robot size to counterbalance the minimization of manipulator size in order to ensure a non-degenerated solution of the manipulator chain.
The proposed optimization formulation refers to the dimensional synthesis of manipulator chains in a sense that the type of manipulator structure cannot be a direct result of this optimal design, but is better considered as a datum. However, this usually can be obtained by considering the number of degrees of freedom, which is required for a specific manipulation task. Nevertheless, a certain design optimal selection can be achieved by comparing different manipulator structures through the results of the same optimization problem.
Since the main task of a manipulator is recognized in moving the end-effector or a grasped object in the space, it is suitable to achieve an optimal design and location of a robot taking into account basically workspace characteristics.
Moreover, also the size of the manipulator is recognized as an important issue since smaller and faster machinery and robots are desirable in modern industrial applications. Nevertheless, the two design features are not independent and are even in contradiction, but a unique meaningful objective function in an optimization design problem can be formulated taking into account the following considerations.
It can be thought intuitively that the volume V of a manipulator workspace is related to the manipulator length L, in a sense that the manipulator is larger when obtainable workspace volume is larger . This can be expressed synthetically in the form

$$V = c\,L^{\beta} \tag{3.1.87}$$

where β is a constant and c is a function of the chain parameters; the manipulator length L can be defined as

$$L = \sqrt{\Sigma_1^n (a_i^2 + d_i^2)} \tag{3.1.88}$$

In addition, main characteristics as shape and volume of the workspace of revolute pairs' connected manipulators depend on the link ratios and dimensions, respectively. Therefore, it can be useful to express the manipulator length L in the form

$$L = a_1 \sqrt{\Sigma_1^{2n+1} k_i^2} \tag{3.1.89}$$

where k_i , i=1,...,2n+1, are the link ratios of a_i and d_i , i=1,...,n, with respect to a reference dimension as for example a_1. Consequently, the workspace volume can be computed by introducing Eq. (3.1.88) in the expression for a volume of revolution to give

$$V = \frac{2\pi}{\sqrt[3]{\sum_{1}^{2n+1} k_i^2}} L^3 \oint_{\partial W} r^{*2}\, dz^* \qquad (3.1.90)$$

where r* and z* are the radial and axial reaches normalized with respect to a_1, and δW is the workspace cross-section boundary along which the integral is calculated. In Eq. (3.1.90) it is possible to recognize both a shape factor and a ratio factor influencing the workspace shape. The quantity L^3 represents a scaling factor giving the size of the workspace volume when the manipulator dimension is given through a_1 in Eq. (3.1.89). From the above expressions it can be observed that the size and shape of a manipulator workspace are not completely independent in a sense that variations of link dimensions give rise to different changes of the manipulator length and workspace volume.
Nevertheless, if the manipulator size increases the workspace volume will also increase. But sometimes the workspace volume may vanish even if L does not, since the workspace ring volume can degenerate into a toroidal surface for particular values of the link parameters, as it has been stressed for the workspace ring geometry of three-revolute chains. Furthermore, the minimum of the workspace volume can be the null value, but it is not related to the minimum manipulator size since it depends on the size of the degenerated toroidal surface. In this last case the two-revolute manipulator will be the optimal solution since its toroidal workspace can satisfy the same number of conditions for the workspace boundary surface, as in the case of three-revolute manipulators, and probably it also occurs in the case of n-revolute manipulators.
Consequently, it is thought convenient to assume as an objective function the so-called 'manipulator global length' L_{tot}, which can be defined in the form

$$L_{tot} = \sum_{0}^{n} (a_i + d_i) \qquad (3.1.91)$$

where the robot base location has been included through the **b** vector expressed by means of the radial a_0 and axial d_0 components.
Several workspace characteristics can be used to formulate the objective function but however workspace volume and manipulator length are usually preferred. These characteristics can also be conveniently combined to formulate a performance index for manipulators.
The general optimization formulation of Eqs (3.1.84) to (3.1.86) can be better illustrated by referring to a specific case for a manipulator optimum design.
Thus, the optimum design of a general three-revolute manipulator deals with the synthesis of the parameters a_1, a_2, a_3, d_2, d_3, α_1, α_2, (d_1 is not meaningful since it shifts up and down the workspace only), Fig. 3.20, when the workspace characteristics are considered to fulfil the design requirements which are expressed in Fig. 3.19. A specific optimum design can be formulated as an optimization problem in the form

$$\min -\frac{V^*}{L^3} \tag{3.1.92}$$

subject to

$$\begin{aligned} &\min (z) \geq z_{min} \\ &\max (z) \leq z_{max} \\ &\min (r) \geq r_{min} \\ &\max (r) \leq r_{max} \end{aligned} \tag{3.1.93}$$

where V* indicates a measure of the workspace volume; L is the total dimension of the manipulator given as in Eq. (3.1.88); z and r are the axial and radial reaches of the boundary points of the manipulator workspace.
Equation (3.1.92) has been used to express the maximization of manipulator workspace in terms of volume V referring to the design dimension L of the manipulator. Indeed, the ratio V* / L gives a measure of the workspace maximization since the volume V* is compared to a spherical volume achievable with a single link arm of size L with a base spherical joint.
Equations (3.1.93) express the constraints of Fig. 3.19 in a general way to include several cases for which the workspace is only prescribed within a given volume. Indeed, those disequations permit an easier solution of the design problem with respect to the case of equality constraints, since they give a wider field of feasible solutions.
A way to perform numerically in an efficient procedure requires a suitable analytical formulation for the involved quantities of the manipulator workspace. Because of the optimum formulation, a formulation for the manipulator workspace can be conveniently expressed in terms of the workspace boundary.
The ring equations in 3.1.2.1 are useful to express analytically the constraint Eqs (3.1.93) as explicit functions of the design parameters. In addition, these equations are useful to analytically calculate the derivatives, which are involved in the numerical optimization process. These derivatives can be computed in a closed-form as a function of the design parameters from the above-mentioned expressions. Thus, for example, the first derivatives with respect to the design parameter a_1 can be expressed in the form

$$\begin{aligned} \frac{\partial r_{1,j}}{\partial a_1} = \frac{1}{2r_{1,j}}\left[\frac{\partial A_1}{\partial a_1} - 2z_{1,j}\frac{\partial z_{1,j}}{\partial a_1} - \frac{(C_1 z_{1,J} + D_1)G_1 + F_1}{E_1^2}\frac{\partial E_1}{\partial a_1}\right] + \\ + \frac{1}{2r_{1,j}E_1}\left[\frac{\partial F_1}{\partial a_1} + \frac{\partial G_1}{\partial a_1}(C_1 z_{1,J} + D_1) + G_1\left(\frac{\partial D_1}{\partial a_1} + z_{1,J}\frac{\partial C_1}{\partial a_1} + C_1\frac{\partial z_{1,j}}{\partial a_1}\right)\right] \end{aligned} \tag{3.1.94}$$

$$\frac{\partial z_{1,j}}{\partial a_1}=-\frac{z_{1,j}}{C_1}\frac{\partial C_1}{\partial a_1}-\frac{-L_1 \pm Q_1^{1/2}}{K_1^2 C_1}\frac{\partial K_1}{\partial a_1}+\frac{1}{K_1 C_1}\left(\pm\frac{1}{2Q_1}\frac{\partial Q_1}{\partial a_1}-\frac{\partial L_1}{\partial a_1}-K_1\frac{\partial D_1}{\partial a_1}\right) \quad (3.1.95)$$

$$\frac{\partial u_j}{\partial a_1}=\mp\frac{1}{2}\left[\frac{\partial A_1}{\partial a_1}-\left(\frac{-L_1 \pm Q_1^{1/2}}{E_1^2 K_1}+\frac{F_1}{E_1^2}\right)\frac{\partial E_1}{\partial a_1}-\frac{-L_1 \pm Q_1^{1/2}}{K_1^2 E_1}\frac{\partial K_1}{\partial a_1}+\frac{1}{E_1}\frac{\partial F_1}{\partial a_1}\right]$$
$$-\frac{1}{2K_1 E_1}\left(\frac{\partial L_1}{\partial a_1}\pm\frac{1}{2Q_1^{1/2}}\frac{\partial Q_1}{\partial a_1}\right) \quad (3.1.96)$$

where $r_{1,j}$ and $z_{1,j}$ are the reach coordinates of $\mathbf{x}_{1,j}$ with respect to the link 1 frame in the constraints equations (3.1.85) and (3.1.86); u_j, v_j, and w_j are Cartesian components of $\mathbf{x}_{1,j}$; and the derivatives of the structural coefficients can be evaluated by means of the definition equations (3.1.64) and (3.1.66). Higher derivatives can be obtained by further differentiating the above expressions and using the algebraic formulation in 3.1.3.2.
A further useful development of the algebraic approach is concerned with the volume calculation. In fact, the ring geometry of a workspace of revolute-joint manipulators allows an efficient numeric expression for volume evaluation by considering a limited number of workspace boundary points through Eq. (3.1.62) in 3.1.3.3.
Moreover, the derivative of the workspace volume with respect to the design parameters can be computed by using Eqs (3.1.94)–(3.1.96). Thus for example, the analytical derivative of V with respect to a_1 can be obtained in the form

$$\frac{\partial V}{\partial a_1}=\Sigma_1^N\frac{\pi}{2}\left[\left(\frac{\partial z_{1,n+1}}{\partial a_1}+\frac{\partial z_{1,n}}{\partial a_1}\right)\left(r_{1,n}^2-r_{1,n+1}^2\right)+2\left(z_{1,n+1}+z_{1,n}\right)\left(r_{1,n}\frac{\partial r_{1,n}}{\partial a_1}-r_{1,n+1}\frac{\partial r_{1,n+1}}{\partial a_1}\right)\right] \quad (3.1.97)$$

in which the radial and axial reach derivatives with respect to a_1 can be calculated for the boundary discretization in θ_3, by using once more Eqs (3.1.94)– (3.1.96).
The design problem for three-revolute manipulators has been formulated in the form of an optimization problem as a function of seven design variables a_1, a_2, a_3, d_2, d_3, α_1, α_2. The involved expressions are nonlinear functions of the design variables, which have been expressed as algebraic functions of the data because of the algebraic formulation for the workspace. This makes possible and indeed convenient the use of available commercial software packages for the calculations of optimization problems. The numerical procedure can have an advantage for the formulation for the workspace boundary and the Sequential Quadratic Programming technique can be useful for the non-linear optimization problems. This numerical procedure works in such a way that at each step k a solution is found along a search direction $\boldsymbol{\delta}_k$ with a variable update ψ_k . The iteration continues until the variable vectors converge. However, the procedure can be developed so that the formulation for the workspace can be conveniently included

within the solving procedure for the optimization problem. Of course the initial guess manipulator affects the optimum solution. The algebraic formulation for the workspace can be useful to obtain optimum design also when the initial workspace of the guess manipulator is far away from the prescribed limits, or it even violates some of the prescribed requirements, due to the explicit above-mentioned function formulation.
An example is illustrated in Fig. 3.22. The optimization problem involves nonlinear constraint functions of the design parameters and a Sequential Quadratic Programming technique has been used to solve the design problem. The numerical procedure starts with an initial guess for the manipulator design solution. In each iteration, a quadratic programming problem is solved to give a search direction, and a minimization of the objective function is performed to give a step size so that the design parameters are updated until they converge. The optimization is started with an initial manipulator whose workspace is far enough to satisfy the design requirements, which are indicated with limits for the working area and black circles in Fig. 3.22. In particular, the black circles indicate points to be crossed or be overpassed by the workspace boundary and the limit gross lines of a prescribed working area give the region within which the workspace boundary must be drawn. The ground has been assumed on zero axial coordinate so that the workspace is required to be outlined not below $z_b = 0$, with a workspace volume near as possible to 1000 a_1^3. In this case, the robot base has been considered as given to stress the efficiency of the algebraic formulation. The results for some iterations are shown by a thin dotted line, and a heavier line reports the final result. It can be noted that although the starting manipulator workspace does not satisfy any one of the design requirements, nevertheless the algorithm reaches a feasible solution with few iterations and it optimizes the workspace with a manipulator length of 11.89 a_1 and a volume of 1000.34 a_1^3.

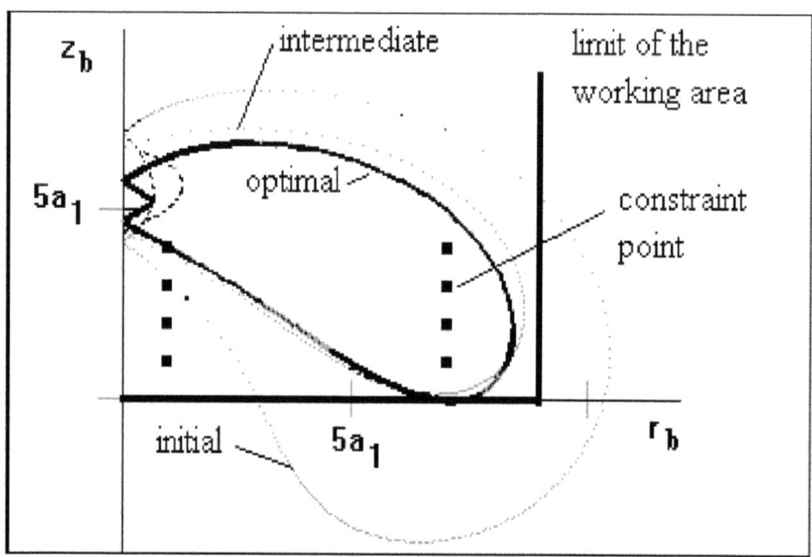

Fig. 3.22: An example of optimum design evolution for a 3R manipulator with prescribed workspace.

3.2 Inverse kinematics and path planning

The problem of inverse kinematics consists of solving the kinematic joint variables of a manipulator as function of a manipulator configuration and particularly depending on the position of a reference point in the manipulator extremity.
Indeed, this problem is strongly related to the subject of path planning and joint trajectory since a manipulator movement requires the determination of joint motion and finally actuator action, as outlined in the scheme of Fig. 3.23.
In general, mapping between Joint Space and Actuator Space is not a complicated problem because mechanical transmissions are used with well-known kinematics.
The problem of the mapping between Cartesian Space and Joint Space is a high non-linear problem for general manipulators and it requires specific attention that is usually addressed as inverse kinematics of manipulators.

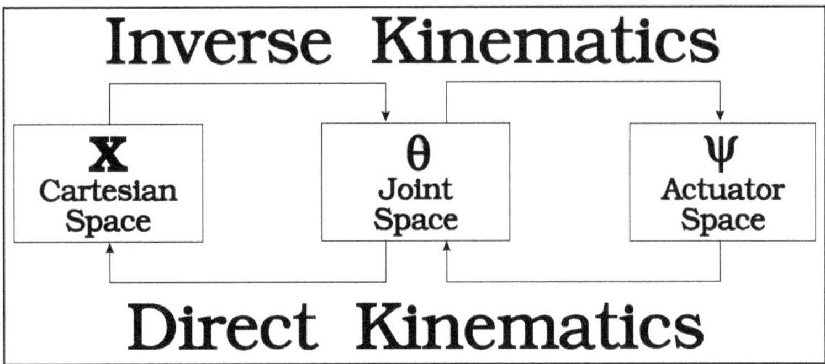

Fig. 3.23: Mapping of direct and inverse kinematics.

3.2.1 A formulation for inverse kinematics

From a numerical viewpoint the problem of inverse kinematics can be formulated to determine the joint coordinates q_i (i=1, ..., N) as functions of given manipulator configurations.
The numerical solution can be obtained by attaching the problem through:

- reverse formulation of closed-form expressions for direct kinematics;
- matrix formulation to determine suitable equations;
- geometrical–analytical formulation to determine suitable equations.

However, any formulation must take into account numerical aspects that are related to solution existence and multiple solutions.
Solution existence refers to the fact that when a given configuration is not within the manipulator mobility range, the numerical solution of the inverse kinematics cannot be obtained within the manipulator workspace. This aspect requires checking for mobility range of the joints and workspace limits before attaching the numerical solution of inverse kinematics. From a mathematical viewpoint satisfaction of workspace limits means that solutions should be obtained as real numbers.
Multiple solutions can be obtained due to the high nonlinear nature of the position kinematics of manipulators. Multiple solutions refer to the several configurations by which a manipulator can reach an extremity position. The fact that a manipulator can

reach a position with several configurations may cause problems because the robot has to be able to choose one. Therefore, it is important to have a characterization of the multiple solutions and a definition of a criterion for the configuration selection during the motion.
Figure 3.24 shows typical configurations of elbow up and elbow down that can reach the same workspace point. Those configurations are the basic geometry for understanding and distinguishing multiple solutions of the inverse kinematics problem.
A procedure for solving the inverse kinematics problem can be defined by using the definition of inverse kinematics. Since inverse kinematics refers to the inverse mapping of the direct kinematics, one can formulate the inverse kinematics by reversing the expressions for the direct kinematics. Thus, a reverse formulation can be obtained by reversing the expressions for Cartesian coordinates of the reference point H on the manipulator extremity and orientation angles of the manipulator extremity.
Even when closed-form expressions are obtained for direct kinematics, it is not always possible or numerically convenient to deduce the reverse formulation for inverse kinematics, since the high nonlinearity and transcendent functions will not always give suitable closed-form expressions for computational purposes both in control algorithms and simulation procedures.

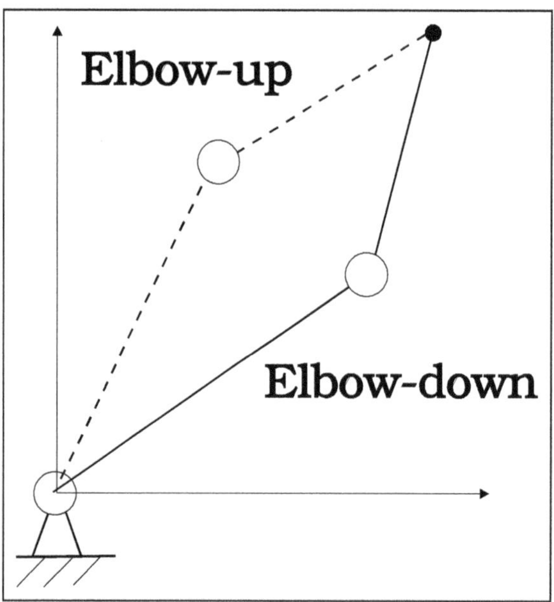

Fig. 3.24: Multiple solutions for inverse kinematics as represented by elbow up and elbow down configurations.

By taking into account the matrix formulation for direct kinematics the inverse problem can be formulated in a general way so as to solve the system of equations

$$ {}_0\mathbf{T}_N = {}_0\mathbf{T}_1(q_1)\ {}_1\mathbf{T}_2(q_2) \ldots\ {}_{N-1}\mathbf{T}_N(q_N) \tag{3.2.1}$$

for the unknowns $q_1, q_2, \ldots, q_N$ when ${}_0T_N$ is given for a suitable number of manipulator configurations in order to have as many equations as unknowns.
But Eq. (3.2.1) gives a system of equations that are highly nonlinear and coupled. Thus, a general closed-form solution is not possible and they have been obtained only for specific manipulator architectures.
A general procedure by using Eq. (3.2.1) can be outlined as in the following:
- computing matrices

$$ {}_j\mathbf{T}_{j+1}^{-1}\ {}_0\mathbf{T}_N = {}_0\mathbf{T}_j\ {}_{j+1}\mathbf{T}_N \tag{3.2.2}$$

  in which one or few joint variables are separated in the elements of the left hand;
- solving the expressions equating elements of the resulting matrices in both sides of Eq. (3.2.2) so that one or few equations are obtained with one or few unknown variables.

This procedure is continued by computing Eq. (3.2.2) with all the possible matrices ${}_jT_{j+1}$ that can help to find matrix elements suitable for separate solving of the unknown variables.
Geometrical–analytical formulation for solving inverse kinematics is based on algebraic manipulation of basic equations for manipulator kinematics. The final results are algebraic equations that can be solved with no great efforts. Two procedures can be identified as based on geometric interpretation or algebraic manipulation of manipulator configurations, respectively.
In a geometric approach the spatial geometry of a manipulator is decomposed into several plane geometry problems. The geometry problems are identified as related with triangle geometry, which is described by using basic expressions and law of cosines. Thus, once a planar triangle is identified, looking at the expressions of triangle geometry in which it is involved solves the unknown joint variable. The convenience of the procedure consists of the simplicity of the computations, although the modeling and identification can be complex and cumbersome. However, the geometric study can also give the opportunity to have a direct interpretation of manipulator behavior.
In an algebraic approach the solution for inverse kinematics is obtained by manipulating algebraically basic expressions for the kinematics of a given manipulator.
Indeed, the algebraic approach can be seen as part both of the matrix formulation and geometric approach. By considering it separately, the algebraic approach is a mathematization of the inverse kinematics problems, in a sense that the attention is focused only on the equations without any reference to their meaning for manipulator behavior. But the kinematic interpretation is again considered when a solution is found.

3.2.1.1 An example

The above-mentioned procedures for inverse kinematics can be better illustrated by referring to an example for a planar 3R manipulator, as shown in Fig. 3.25.
From the geometry of the manipulator one can deduce the kinematic equations straightforwardly from equating matrix ${}_0T_3$ to matrix ${}_0T_T$, which describes tool frame 3

at the end-effector with respect to the base frame 0. Matrix ${}_0T_3$ is expressed as

$$
{}_0T_3 = \begin{vmatrix} c_{123} & -s_{123} & 0 & a_1c_1 + a_2c_{12} \\ s_{123} & c_{123} & 0 & a_1s_1 + a_2s_{12} \\ 0 & 0 & 1 & 0 \\ 0 & 0 & 0 & 1 \end{vmatrix} \tag{3.2.3}
$$

and matrix ${}_0T_T$ can be given as

$$
{}_0T_T = \begin{vmatrix} \cos\varphi & -\sin\varphi & 0 & h_x \\ \sin\varphi & \cos\varphi & 0 & h_y \\ 0 & 0 & 1 & 0 \\ 0 & 0 & 0 & 1 \end{vmatrix} \tag{3.2.4}
$$

Thus, by equating the two matrices the following set of equations can be determined

$$
\begin{aligned}
\cos\varphi &= c_{123} \\
\sin\varphi &= s_{123} \\
h_x &= a_1c_1 + a_2c_{12} \\
h_y &= a_1s_1 + a_2s_{12}
\end{aligned} \tag{3.2.5}
$$

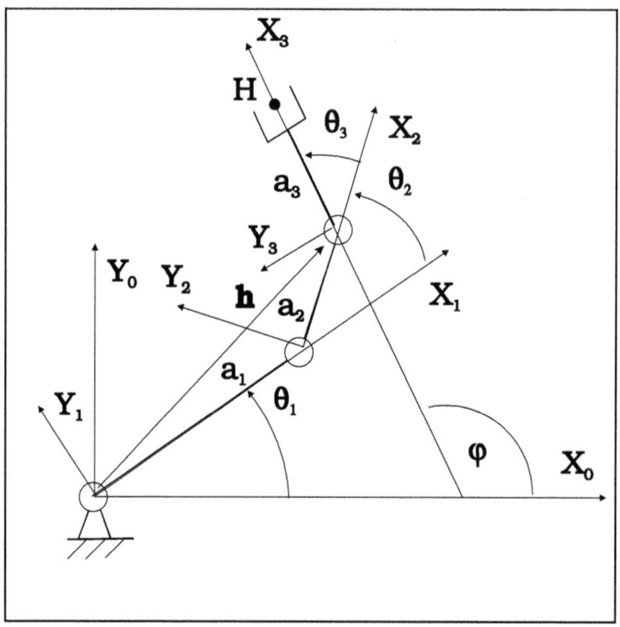

Fig. 3.25: A planar 3R manipulator and its parameters.

in which c_{12} is for $\cos(\theta_1+\theta_2)$, s_{12} is for $\sin(\theta_1+\theta_2)$, and so on. The left terms are given as functions of the manipulator configuration and the unknowns are the joint angles θ_1,θ_2,and θ_3.
Then, Eqs (3.2.5) can be manipulated algebraically to obtain closed-form solutions.
In fact, adding the square of the last two equations in (3.2.5), one can obtain the expression

$$h_x^2 + h_y^2 = a_1^2 + a_2^2 + 2a_1 a_2 c_2 \tag{3.2.6}$$

from which θ_2 joint angle can be solved as

$$\theta_2 = \cos^{-1} \frac{h_x^2 + h_y^2 - a_1^2 - a_2^2}{2a_1 a_2} \tag{3.2.7}$$

Because of the cos function, two configurations are feasible, namely the elbow up and the elbow down, as in Fig. 3.24. If also sin θ_2 is computed, the arctangent function can be used to obtain one solution only.
Once θ_2 is solved, Eqs (3.2.5) can be used again to compute the solution for θ_1 by writing

$$\begin{aligned} h_x &= Ac_1 - Bs_1 \\ h_y &= As_1 + Bc_1 \end{aligned} \tag{3.2.8}$$

with

$$\begin{aligned} A &= a_1 + a_2 c_2 \\ B &= a_2 s_2 \end{aligned}$$

From the first equation (3.2.8) one can obtain

$$s_1 = \frac{Ac_1 - h_x}{B} \tag{3.2.9}$$

that once substituted into the second equation permits solution of

$$c_1 = \frac{Ah_x + Bh_y}{A^2 + B^2} \tag{3.2.10}$$

Alternatively, from the second equation one can obtain

$$c_1 = \frac{Bs_1 + h_x}{A} \tag{3.2.11}$$

that once substituted into the second equation permits solution of

$$s_1 = \frac{Ah_y - Bh_x}{A^2 + B^2} \tag{3.2.12}$$

Then, a unique value for θ_1 can be obtained by using solutions for c_1 and s_1 into the two-argument arctangent function Atan2 as

$$\theta_1 = A\tan 2(s_1, c_1) \tag{3.2.13}$$

The outlined procedure for obtaining θ_1 is one way to manipulate algebraically the kinematic equations and several other ways can be carried out to gain algebraic solution of the involved equations.
Finally, θ_3 can be solved by using the equality

$$\theta_1 + \theta_2 + \theta_3 = \varphi \tag{3.2.14}$$

from the first two expressions in Eqs (3.2.5) but also the geometry of Fig. 3.25.
Alternatively, looking at the geometry of the manipulator configuration can give the equations and the way to solve them.
Thus, once a triangle is determined with a_1, a_2, and the segment joining O_0 with O_3, the law of cosines can be applied to give

$$h_x^2 + h_y^2 = a_1^2 + a_2^2 - 2a_1a_2\cos(\pi - \theta_2) \tag{3.2.15}$$

from which θ_2 can be solved. Again both elbow-up and elbow-down configurations can be a solution.
The angle θ_1 can be obtained considering the slope angle γ of **h** direction and the angle between the line of **h** and line of a_1 link to obtain

$$\theta_1 = \gamma - \psi \tag{3.2.16}$$

The slope angle γ can be computed from the components of **h** as

$$tg\gamma = \frac{h_x}{h_y} \tag{3.2.17}$$

and the slope angle ψ can be given by applying again the law of cosines as

$$a_2^2 = a_1^2 + h_x^2 + h_y^2 - 2a_1\left(h_x^2 + h_y^2\right)^{1/2} \cos\psi \tag{3.2.18}$$

Of course the ambiguity in cos ψ will give sign ambiguity for the summing of ψ in Eq. (3.2.16).
Finally, the orientation angle φ of the last link is the sum of the three joints angles

$$\theta_1 + \theta_2 + \theta_3 = \varphi \tag{3.2.19}$$

that gives the solution for θ_3 .

3.2.2 Trajectory generation in Joint Space

By inverse kinematics one can also compute the joint variables corresponding to locations of the manipulator extremity. In many applications those locations are of fundamental significance while the motion among them is not required with particular performances but smooth enough for the task. Those locations are usually named 'precision points', although they refer both to position and orientation. The corresponding set of joint variables can be named as precision points. They are named precision points since it is required that the motion law passes through them precisely or even the motion stops there.
The smooth motion of a joint and consequently a link can be obtained by using a continuous function with continuous first derivative. Sometimes a continuous second derivative is also desired in order to avoid vibrations and wear in joint connections.
Therefore, one can consider a trajectory in Joint Space as the problem to determine the path for each joint between an initial value θ_0 at the time $t_0 = 0$ and a final value θ_f at the time t_f. The simplest interpolating motion refers to a linear scheme that is given by

$$\theta(t) = k_0 + k_1 t \tag{3.2.20}$$

in which t is the time parameter, and k_0 and k_1 are motion coefficients that can be determined by using the known boundary constraints for the joint motion as

$$k_0 = \theta_0 \tag{3.2.21}$$

$$k_1 = \frac{\theta_f - \theta_0}{t_f} \tag{3.2.22}$$

In the case of linear interpolation for the joint motion, only position sensors on the joints are needed to measure the values θ_0 and θ_f. But the motion is not smooth since it is not possible to take into account velocity constraints for continuous position and velocity in a trajectory with several precision points.

A smooth path can be obtained by adding a parabolic blend segment at the extremity of the joint path. During the blend path portions, constant acceleration is usually imposed to spline it together with the linear joint function of Eq. (3.2.20). Assuming that the constant acceleration will give a velocity at the end of the blend portion that is equal to that of the substituted linear portion, symmetric function is ensured by using halfway time parameter t_h and joint position θ_h, as shown in Fig. 3.26.

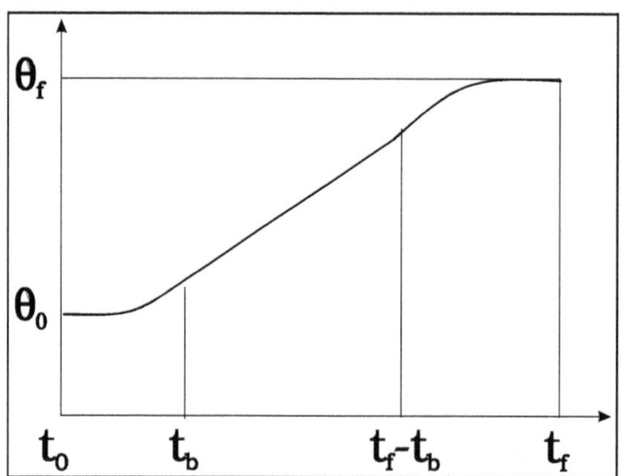

Fig. 3.26: Linear interpolation joint function with parabolic blend at extremities of the path.

Thus, the parabolic blend portions are characterized by an acceleration value that is given by

$$\ddot{\theta}_b = \frac{\theta_f - \theta_0}{(t_h - t_b)t_b} \tag{3.2.23}$$

and blend motion duration is expressed by the value θ_b of the joint angle that is computed as

$$\theta_b = \theta_0 + \frac{1}{2}\ddot{\theta}_b t_b^2 \tag{3.2.24}$$

In order to obtain smooth motion, a single function for joint motion can be formulated by using a polynomial, and particularly a cubic function can be introduced as

$$\theta(t) = k_0 + k_1 t + k_2 t^2 + k_3 t^3 \tag{3.2.25}$$

whose coefficients can be determined by using four constraint conditions that are given by the known values for initial and final positions and velocities, as

$$k_0 = \theta_0$$
$$k_1 = \dot{\theta}_0$$
$$k_2 = 3\frac{\theta_f - \theta_0}{t_f^2} - 2\frac{\dot{\theta}_0}{t_f} - \frac{\dot{\theta}_f}{t_f} \tag{3.2.26}$$
$$k_3 = -2\frac{\theta_f - \theta_0}{t_f^3} + \frac{\dot{\theta}_f + \dot{\theta}_0}{t_f^2}$$

The smooth motion is ensured by the continuous functions for velocity and acceleration that are expressed as derivatives of Eq. (3.2.25) as

$$\dot{\theta}(t) = k_1 + 2k_2 t + 3k_3 t^2 \tag{3.2.27}$$
$$\ddot{\theta}(t) = 2k_2 + 6k_3 t$$

In the case of cubic interpolating joint motion, position and velocity sensors on the joints are needed to measure the values of the given constraint condition at the extremities of the joint path as indicated by Eqs (3.2.26). Usually, this is in the case of industrial robots.

Higher polynomials can be used when suitable constraint conditions are considered to determine the polynomial coefficients. Those constraint conditions are identified from the operation of the joints, which can be described from a kinematic viewpoint by position, velocity, and acceleration. Thus in this last case six constraint conditions can be formulated to compute six polynomial coefficients that correspond to a quintic polynomial function. But in this case acceleration sensors must also be equipped on the joint axis and this sensorization can make complications both in the mechanical design and operation of the joint.

Several other functions and algorithms are available to obtain smooth joint motion and they can be identified by using suitable constraint conditions that are similar to the above-illustrated ones for the case of polynomial interpolating functions.

In addition, an interpolating joint motion function is usually used to connect precision points at which a manipulator comes to rest. But there are cases in which the manipulator motion passes through some points within the interpolated path without stopping. Those points are usually named 'via points' and they can be used to split the joint motion in several interpolated segments that are splined together.

3.2.3 A formulation for path planning in Cartesian coordinates

In many applications the movements of a manipulator are planned manually in an *ad hoc* manner so that robots do not perform the tasks with optimal path. Repetitive processes are often applied by using robots and they require optimal path planning that can be achieved with suitable offline computations. But even in many other applications optimal path can be necessary to perform manipulation tasks with suitable

performances.
An optimal path can be designed as a function of an index of merit whose value is maximized or minimized during the path.
Thus, an optimal path can be obtained as a solution of an optimization problem in which the objective function is a formulation of desired index of merit, and constraint functions are considered to take into account characteristics of manipulator operation and task performances. Many aspects of manipulation operation can be considered for formulating suitable objective functions. The most used aspects can be recognized in:
- mechanical energy in joint actuation;
- mechanical energy of performing a task;
- velocity of performing a path;
- acceleration of performing a path;
- obstacle avoidance;
- time duration of the path.
The above-mentioned aspects can be used separately but also in conjunction through suitable weighted formulation. However, a common frame can be considered for the formulation of an optimization problem whose solution gives an optimal path in terms of the used index of merit.
In the following an optimal path planning is presented as related to basic kinematic formulations that also include a solution for the inverse kinematic problem.
In Fig. 3.27 a robotic manipulator is sketched as an open kinematic chain with N+1 rigid links connecting consecutive revolute or prismatic joints. The manipulator chain can be modeled by using the so-called 'natural coordinates' as a natural way to describe mechanisms through some reference points in Cartesian coordinates. These points can be selected according to a few basic rules, i.e. a reference point in each revolute joint and two reference points for each prismatic joint to give the stroke length and the actual position of the joint. Some additional reference points may be assumed as points of interest for the chain behavior, as for example point $H = P_m$ on the end-effector. Thus, an N d.o.f.s manipulator may be described through the Cartesian coordinates of $m \geq N+1$ reference points $P_i = (x_i , y_i , z_i)$ $(i = 0, 1, ..., m)$ and the link sizes can be given as the distances l_i $(i = 1,...,N)$ between consecutive reference points P_i and P_{i-1}. Here, P_i represents the reference point i and its position vector with respect to the base frame is given as P_0P_i , by assuming P_0 fixed on the base frame.
Therefore a manipulator configuration C_j can be represented by the positions of P_i reference points for the joints. More generally, a manipulator configuration will be described also by the interest reference points so that a manipulator configuration C_j may be represented as a vector in a multi-dimensional space by taking into account all the position vectors of the reference points as $C_j = (P_0P_1{}^j , P_0P_2{}^j , ... , P_0P_m{}^j)$ when $P_0P_i{}^j$ will represent the position vector in Cartesian coordinates of reference point P_j at the manipulator configuration j. For path problems of manipulators P_0 position may be excluded as a configuration variable because it is constant. A mobile robotic manipulator can also be considered when P_0 point is also considered as a variable with respect to a reference point on a fixed frame.

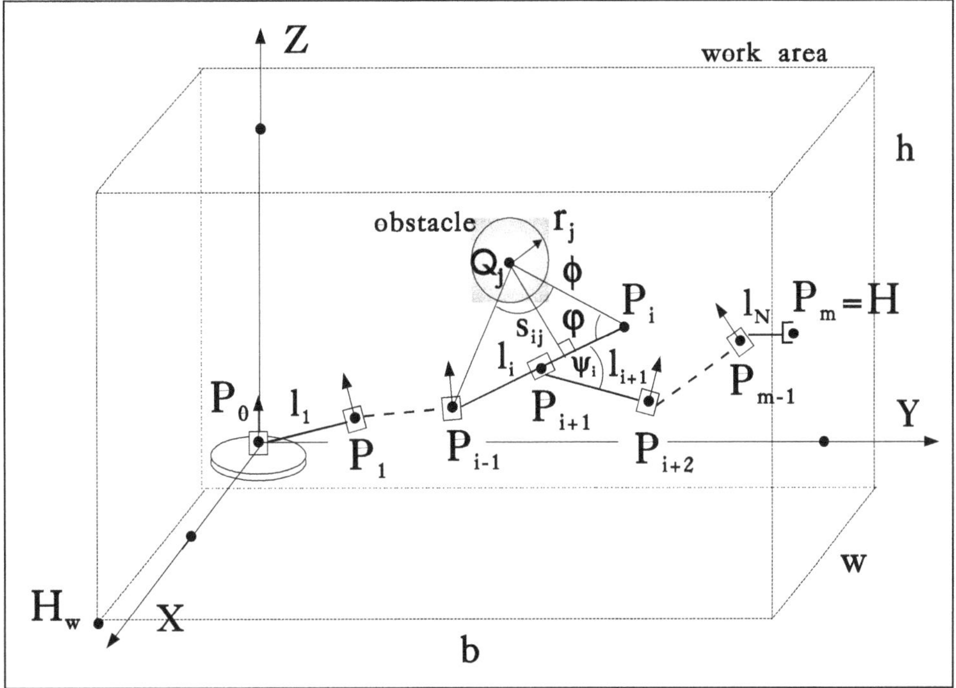

Fig. 3.27: Natural reference points P_i $i = 0,1, \ldots, m$ ($m \geq N+1$) fixed on manipulator links, and a spherical model for an obstacle.

Thus, a path planning problem is concerned with determining a suitable sequence of possible C_j configurations.
However, the Cartesian coordinates of the joint reference points must comply with congruence constraints due to the rigidity of links and kinematic characteristics of joints. Particularly, a rigid body condition for a link i can be expressed as

$$\left| P_0P_{i-1} - P_0P_i \right| = l_i \tag{3.2.28}$$

where | - | is the module operator for vectors and l_i is the link length.
A correct model for a joint on link l_i needs, Fig. 3.27, that point P_{i+1} be on link l_i and it is expressed as

$$\left| P_{i-1}P_{i+1} \times P_{i-1}P_i \right| = 0 \tag{3.2.29}$$

and the orientation of link l_{i+1} be maintained as formulated in the form

$$P_{i-1}P_i \cdot P_{i+1}P_{i+2} = l_i \, l_{i+1} \cos\psi_i \tag{3.2.30}$$

where ψ_i is a constant structural angle between links l_i and l_{i+1} (generally it is assumed equal to $\pi/2$). Moreover, a prismatic joint may slide along the link segment l_i between a minimum value l_i^m and a maximum value l_i^M so that the rigid body condition (3.2.8) for a link l_i with a prismatic joint must be replaced by

$$l_i^m \leq | P_0P_{i-1} - P_0P_i | \leq l_i^M \tag{3.2.31}$$

An obstacle within the manipulator workspace can be modeled as a sphere with radius r_j and center Q_j, as shown in Fig. 3.27, that can be determined as a container of the obstacle itself. A free-collision configuration may be evaluated by looking at the distances among the obstacles and the manipulator links. These distances may be computed from the geometry of Fig. 3.27 as s_{ij} ($i = 1,...,m$ and $j = 1,...,v$) given as

$$s_{ij} = \begin{cases} P_i Q_j \sin \varphi & \Phi \geq \Phi_{min} \\ \min(P_i Q_j , P_{i-1} Q_j) & \Phi < \Phi_{min} \end{cases} \tag{3.2.32}$$

with

$$\varphi = \cos^{-1} \frac{P_i Q_j^2 - P_{i-1} Q_j^2 + l_i^2}{2 P_i Q_j l_i}$$

$$\Phi_{min} = \sin^{-1} \frac{l_i}{\min(P_i Q_j , P_{i-1} Q_j)}$$

where Φ is the angle at Q_j under which link l_i is seen.

In order to generate a path to a given end configuration C_f it can be convenient to calculate a suitable Map of Feasible Configurations in which the problem of obstacle avoidance path will then be solved. The feasible configurations for the Map can be computed by assuming that the movements of end-effector point H be parallel to the axes of a Cartesian base frame and the configurations during the motion be adjacent to each other.

The adjacent condition between configurations C_r and C_s with H motion constraints can be analytically expressed as

$$\min \| C_r - C_s \|$$

subject to (3.2.33)

$$\left| P_0P_m^r - P_0P_m^s \right|_{x,y,z} = \Delta_{x,y,z}$$

$$s_{ij} > 0$$

where $\| - \|$ indicates a norm of $C_r - C_s$, $| - |_x$ represents the X component of H displacement between the configurations r and s, and Δ_x is a discretization size along X axis of the manipulator work area within the workspace, as well as for Δ_y and Δ_z . Usually the discretization sizes will be taken equal to a common value Δ.

The optimization problem (3.2.33) states that given a manipulator configuration we may calculate adjacent configurations related with possible motions of the end-effector reference point along the coordinate axes. Particularly, although several criteria may be adopted to formulate the norm $\| C_r - C_s \|$, from a kinematic viewpoint it can be of practical convenience to consider the criterion concerning with a total displacement of the manipulator. A total displacement of the manipulator can be given as the sum of Euclidean square displacements of all the reference points, i.e. in an analytical form

$$\| C_r - C_s \| = \sum_{i=1}^{m} \left| P_0P_i^r - P_0P_i^s \right|^2 . \tag{3.2.34}$$

This formula takes into account both the joint displacements and area swept by the manipulator links during the motion between adjacent configurations.
Thus, starting from a given configuration the problem can be solved to give 4 adjacent configurations for planar cases and 6 for spatial manipulations. An algorithm can be deduced to automatically generate a Map of Adjacent Configurations by means of a ramification procedure through formulation (3.2.33) in a discretized work area within the manipulator workspace for a given end configuration. In particular, once adjacent configurations have been computed the ramification procedure provides an update of the problem (3.2.33) by assuming a new C_r from the computed configurations with the aim of filling up the workspace area.
Then, a suitable numbering can also be arranged so that a path will be synthetically described by a sequence of numbered configurations joining a start one to a final one. If a configuration will correspond to a point out of the workspace it will not be considered. The work area within the workspace is assumed to be given through a left bottom corner point H_w and a parallelepiped region with size b, h, and w, as shown in Fig. 3.27.
A free-collision path configuration may be formulated by means of a refinement of Eqs (3.2.33) and (3.2.34) for adjacent configurations by considering the obstacle avoidance through Eqs (3.2.32) and the link constraints (3.2.28) to (3.2.31).
Summarizing the problem of path planning for manipulators can be formulated as a sequence of manipulator free-collision configurations, with the s one obtained from a previous r one by solving the problem

$$\min \sum_{i=1}^{m} \left| P_0P_i^r - P_0P_i^s \right|^2 . \tag{3.2.35}$$

subject to

$$\left| P_0P_{i-1} - P_0P_i \right| = l_i \tag{3.2.36}$$

$$l_i^m \leq \left| P_0P_{i-1} - P_0P_i \right| \leq l_i^M \tag{3.2.37}$$

$$H_w \, P_m \leq (b, h, w) \qquad (3.2.38)$$

$$s_{ij} > r_j \qquad (3.2.39)$$

with i = 1, ... , m ; j = 1, ... , v; when all the quantities are expressed in terms of Cartesian coordinates.

Equations (3.2.36) or (3.2.37) are applied alternatively to the links with revolute or prismatic joints. Equation (3.2.38) represents workspace constraints prescribing P_m to be inside the parallelepiped work area description. Finally, Eq. (3.2.39) states the obstacles avoidance for all the links.

It is worth noting that interference among the links can be avoided by considering adjacent configurations, in a sense that starting with a given suitable configuration with link positions far away from interference the adjacent configurations formulation (3.2.33) may ensure no collisions among links. A remarkable characteristic of the formulated optimization problem with Eqs (3.2.35) to (3.2.39) is that both the objective function and constraints have analytical derivatives, because of the Cartesian coordinates description, and they help to find a solution in the adopted numerical procedure.

Once a Map of Free-collision Adjacent Configurations has been generated by solving the problem of Eqs (3.2.35) to (3.2.39), a graph may be determined to obtain a path between an initial configuration C_i and a final configuration C_f , or vice versa.

It is worthy of note that, although the procedure has been formulated to connect a given final configuration to a given initial one, the generation of the Map of Configurations is useful to obtain a free-collision path between any other two configurations of the Map if it is possible to determine a graph connecting them. This means that the first item with great amount of computations can be developed once for several applications of different manipulative tasks requiring updating only the path planning graph determination.

The Cartesian coordinates representation of a manipulator can also be used to solve straightforwardly the inverse kinematic problem since a joint variable can be computed from positions of reference points for joints. In fact, a displacement of a prismatic joint can be given from the distance of P_{i+1} from P_i or P_{i-1}, as well as a rotation of a revolute joint can be evaluated as the angle between $P_{i-1} P_i$ and $P_i P_{i+1}$.

Therefore, this approach can also be used to determine suitable motion of the joints when a graph is prescribed, since the adjacent configuration condition for the manipulator also ensures continuity in joint performances.

3.2.3.1 Illustrative examples

The path planning problem formulated in the previous section can be recognized to be a constrained nonlinear optimization problem. Thus, since algorithms, which are based on Sequential Quadratic Programming, can solve this kind of optimization problem, one can successfully use those algorithms in commercial software packages.

The example in Fig. 3.28 illustrates the free-collision path obtained with 46 configurations for a planar 4R1P manipulator (m = 5) with l_1 = 10 u, l_2 = 7 u, 1 u $\leq l_3 \leq$ 7.5 u, l_4 = 0 u, l_5 = 7 u, from P_4^{START} (13, 16) to P_4^{END} (20, 20) assuming a C_f = (6, 8, -0.2, 10.8, 0, 12, 6, 15.2, 13, 16). The work area with H_w (0,0) and b = h = 20 u has been discretized by Δ = 1 u, and the obstacles, that are shown in gray, have been considered as a union of v = 35 equal circles with r_j = 1 u. All the distances have been expressed in term of u unit length.

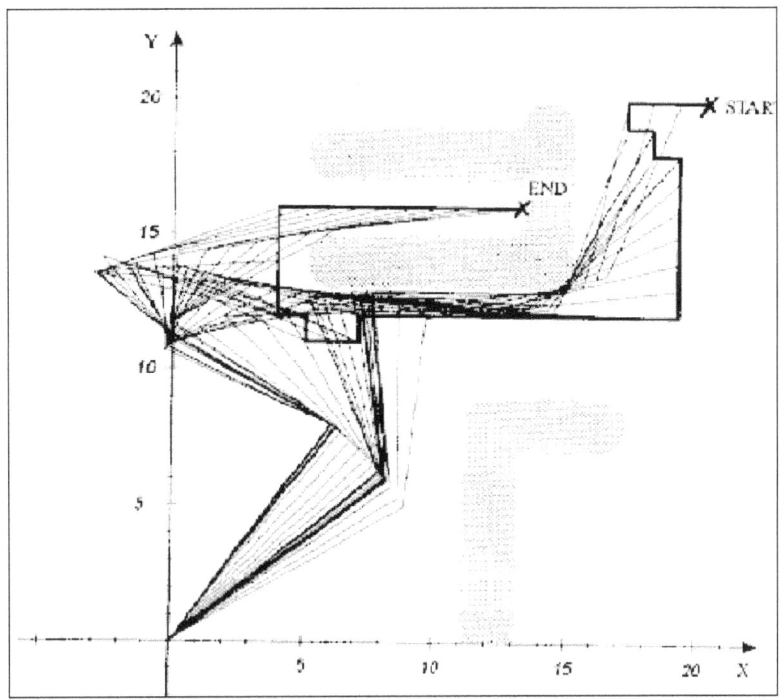

Fig. 3.28: A numerical example: free-collision path of the end-effector of a planar 4R1P manipulator among spherical modeled obstacles.

The 46 configurations have been obtained from a Map of 1,504 adjacent manipulator configurations. The obtained smooth motion is appreciable looking at the shown sequence of manipulator configurations. The good quality of the manipulator motion reflects the adopted optimal criteria on the total displacement. It is also of note that, although the path of the reference point on the end-effector does not seem to be the shortest, it is obtained by manipulator configurations wrapping the obstacle without touching them. It is also of remark that by assuming very small discretization size Δ the motion may not be constrained to be along the coordinate axes. However, this fact will increase considerably the computational efforts for the generation of the Map of Configurations without a corresponding improvement of the path. An enhancement of redundancy capability also does not seem to considerably improve the path result, although only very small increase of computational time has been experienced. Of course, great improvements will be expected with a large number of manipulator d.o.f.s.

However, the procedure does not seem to be greatly affected by small variations in the discretization size Δ of the work area and the number of d.o.f.s of the manipulator. Finally, the efficiency of the numerical procedure is greatly affected by the sizes b, h, and w of the work area. By enlarging the work area the computational time has grown considerably. This is because the procedure has been considered with more adjacent configurations in the Map generation. Thus, it is desirable to restrict the work area as much as possible, to minimize the computational time.

The example of Fig. 3.29 illustrates the free-collision path obtained with 64 configurations for an industrial PUMA 560-like robot (m = 5) with l_1 = 255 u, l_2 = 450 u, l_3 = 100 u, l_4 = 433 u l_5 = 420 u, from P_5^{START} (800, -200, 600) to P_5^{END} (800, 350, -400) assuming a C_i = (104, 232, 0, 97, 236, 450, 32, 164, 450, 451, -32, 437, 800, -200, 600). The work area with H_w (-200, -200, -400) and b = h = w = 1,000 u has been discretized by Δ = 10 u. The obstacles have been considered as a union of 25 equal circles with r_j = 100 u; to avoid interference with the robot trunk, it has been modeled as an obstacle through a further 8 circles with r_j = 150 u: in total there are v = 33 obstacles in the work area environment. The 64 configurations have been obtained from a Map of 5,687 adjacent manipulator configurations.

In the figure few configurations are reported to show the feasibility of the computed path. It is worthy of note that some adjustments and further constraints in the optimization formulation (3.2.33) have been considered to take into account the specific geometry of the robot and the real bounds on the revolute joints. However, the procedure has been worked essentially with the same features of planar cases. A further practical feasibility of the proposed path planning procedure can be deduced by observing that the obtained trajectory can be performed with an easy programming development in industrial robots, because of the imposition in Eqs (3.2.33) of straight-line displacements.

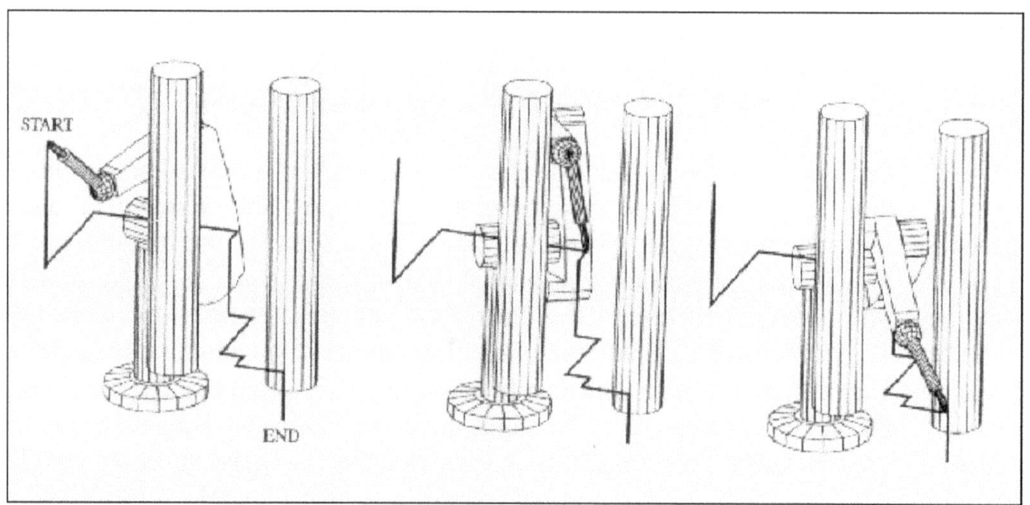

Fig. 3.29. A numerical example: free-collision path of the end-effector of a PUMA-like industrial manipulator among spherical modeled obstacles.

Thus, for the example of Fig. 3.29 with a PUMA 560-like robot, it may be very easy to obtain a program by using few instructions for linear movements of the reference point on the end-effector. In fact, it is possible to program the computed free-collision path, for example, through the MOVES motion instruction of VAL-II, which can be used to connect the extreme points of straight-line segments. The adjacent condition of the configuration sequence will ensure that during the MOVES action the robot will move along the computed path configurations.

3.3 Velocity and acceleration analysis

The motion of a manipulator can be characterized also by the motion of the links. Important motion characteristics are the velocity and acceleration of the link bodies, also for dynamic calculations. These kinematic characteristics refer to the motion of rigid bodies and they can be evaluated through linear components and angular components, which describe the translation and rotation motions, respectively.

The motion of manipulator links can be studied by referring to reference frames that are attached to the links as motion of frames relative to other frames. Thus, the motion of the manipulator links can be studied by using the approach of relative motion between two bodies.

The linear velocity describes an attribute of a point and angular velocity describes an attribute of a body. Similarly it is for linear and angular components of the acceleration. The velocity status of a rigid body is known when both linear and angular velocities are determined with respect to a reference frame by considering the point for the linear velocity also as origin for the angular velocity vector. Thus, the velocity of a body is determined when the velocity of the origin of the reference frame attached to it is calculated together with the angular velocity vector.

In Fig. 3.30 the generation of linear and angular velocities is illustrated by referring to the motion of a frame B with respect to frame A, that can be considered as fixed.

The vector ${}_A\omega_B$ indicates the angular velocity of B with respect to A, and it can be expressed in any coordinate system. By indicating ${}^C{}_A\omega_B$ the vector is expressed with Cartesian components of frame C. In order to simplify the rotation, ω_B will indicate the angular velocity of frame B with respect to and in terms of some understood frame, which is usually the base frame, and ${}^C\omega_B$ will indicate the same angular velocity but expressed with respect to frame C.

The linear velocity ${}_A\mathbf{v}_B$ is given as the time derivative of the position vector ${}_A\mathbf{p}_B$ between the origins of frames A and B.

Consequently, the velocity ${}_A\mathbf{v}_Q$ of any point Q of the body B at a position ${}_B\mathbf{p}_Q$ in frame B can be computed as described in A by summing up the contributes of the linear and angular velocities of the body with the velocity ${}_B\mathbf{v}_Q$ of point Q with respect to the frame B in the form

$${}_A\mathbf{v}_Q = {}_A\mathbf{v}_B + {}_A\boldsymbol{\omega}_B \, \mathrm{x} \, {}_A\mathrm{R}_B \, {}_B\mathbf{p}_Q + {}_A\mathrm{R}_B \, {}_B\mathbf{v}_Q \qquad (3.3.1)$$

where ${}_A\mathrm{R}_B$ represents the rotation matrix of B with respect to A.

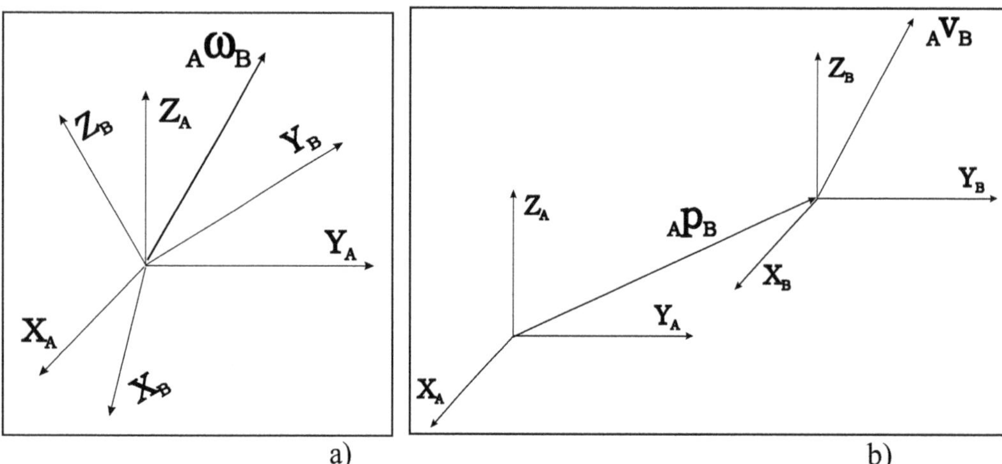

a) b)

Fig. 3.30: Velocity vectors of relative motion of rigid bodies: a) angular velocity; b) linear velocity.

The velocities of the links of a manipulator can by computed by applying Eq. (3.3.1) and considering the serial chain architecture, that gives the relative motion relative to neighbors, Fig. 3.31. Thus, when using the H–D notation, starting from the base, the velocity of link i+1 will be that of link i, plus whatever component is added by the mobility of joint i+1. In particular, the angular velocity of link i+1 is the sum of the angular velocity ${}^i\omega_i$ of links i and the rotational velocity due to the angular mobility of joint i+1. This can be expressed as

$$ {}^i\boldsymbol{\omega}_{i+1} = {}^i\boldsymbol{\omega}_i + {}_iR_{i+1}\dot{\theta}_{i+1}\,{}^{i+1}\mathbf{Z}_{i+1} \tag{3.3.2} $$

when all the vectors are written with respect to the same frame i. In Eq. (3.3.2) ${}_iR_{i+1}$ is the rotation matrix, θ_{i+1} dot is the angular velocity of joint i+1, and ${}^{i+1}Z_{i+1}$ its direction along Z_{i+1} axis. Sometimes the reference frame can be conveniently chosen as the base frame.

The description of the angular velocity of link i+1 with respect to the frame i+1 (superscript at the left of a symbol) can be obtained by premultiplying both sides of Eq. (3.3.2) by ${}_{i+1}R_i$ as

$$ {}^{i+1}\boldsymbol{\omega}_{i+1} = {}_{i+1}R_i\,{}^i\boldsymbol{\omega}_i + \dot{\theta}_{i+1}\,{}^{i+1}\mathbf{Z}_{i+1} \tag{3.3.3} $$

Similarly, the linear velocity of the origin of frame i+1 is obtained at the sum of the velocity of the origin of the frame i and a component due to the rotational velocity of link i . This can be expressed in terms of frame i as

$$ {}^i\mathbf{v}_{i+1} = {}^i\mathbf{v}_i + {}^i\boldsymbol{\omega}_i \,\mathrm{x}\, {}_i\mathbf{p}_{i+1} \tag{3.3.4} $$

and in terms of frame i+1 as

$$ {}^{i+1}\mathbf{v}_{i+1} = {}_{i+1}R_i \left({}^i\mathbf{v}_i + {}^i\boldsymbol{\omega}_i \, x \, {}_i\mathbf{p}_{i+1} \right) \tag{3.3.5} $$

in which the vector ${}_i\mathbf{p}_{i+1}$ refers to the distance vector between O_{i+1} and O_i, as shown in Fig. 3.31. It is of note that this vector can also be computed by using the link length a_i and link offset d_{i+1}.

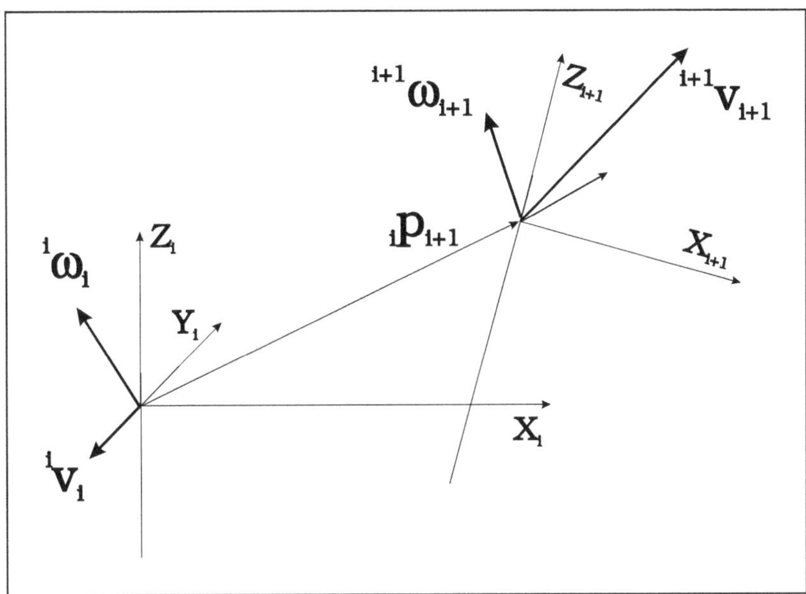

Fig. 3.31: Velocity propagation from link to link in a serial chain manipulator.

The above-mentioned expressions refer to a revolute pair at joint i+1. When joint i+1 is prismatic, one can compute the velocities similarly to obtain

$$ {}^{i+1}\boldsymbol{\omega}_{i+1} = {}_{i+1}R_i \, {}^i\boldsymbol{\omega}_i \tag{3.3.6} $$

$$ {}^{i+1}\mathbf{v}_{i+1} = {}_{i+1}R_i \left({}^i\mathbf{v}_i + {}^i\boldsymbol{\omega}_i \, x \, {}_i\mathbf{p}_{i+1} \right) + \dot{d}_{i+1} \, {}^{i+1}\mathbf{Z}_{i+1} \tag{3.3.7} $$

in which d_{i+1} dot is the velocity of the prismatic joint.

A velocity analysis of a manipulator can be carried out by using Eqs (3.3.3), (3.3.4), (3.3.6), and (3.3.7) from the base to the manipulator extremity that is from 0 to N. The velocity ${}^N\boldsymbol{\omega}_N$ and ${}^N\mathbf{v}_N$ are expressed in terms of N-th frame but a multiplication with ${}_0R_N$ gives those velocities in terms of the base coordinate reference 0.

Differentiating the fundamental velocity expressions and using the propagation approach, similar to the case of velocity analysis can carry out the acceleration analysis of a manipulator.

Linear acceleration $\mathbf{v}_A$ dot expresses the acceleration of the origin of frame A with

respect to and in terms of an understood frame; similarly for the angular acceleration ω_A dot.
Referring to the angular motion, the sum of angular velocity of B rotating relative to A with the angular velocity of C rotating relative to B is expressed with respect to frame A as

$$ {}_A\omega_C = {}_A\omega_B + {}_A R_B\, {}_B\omega_C \tag{3.3.8} $$

and its time derivative can be written as

$$ {}_A\dot{\omega}_C = {}_A\dot{\omega}_B + \frac{d}{dt}\, {}_A R_B\, {}_B\omega_C \tag{3.3.9} $$

By looking at Eq. (3.3.1) when the frame origins coincide, one can consider the expression

$$ \frac{d}{dt}\, {}_A R_B\, {}_B\mathbf{p}_Q = {}_A\omega_B \mathrm{x}\, {}_A R_B\, {}_B\mathbf{p}_Q + {}_A R_B\, {}_A\mathbf{v}_Q \tag{3.3.10} $$

that is a characteristic for the motion of rigid bodies.
One can use Eq. (3.3.10) to compute the second term of right side of Eq. (3.3.9) by substituting ${}_B\mathbf{p}_Q$ with ${}_B\omega_C$ to obtain

$$ {}_A\dot{\omega}_C = {}_A\dot{\omega}_B + {}_A R_B\, {}_B\dot{\omega}_C + {}_A\omega_B \mathrm{x}\, {}_A R_B\, {}_B\omega_C \tag{3.3.11} $$

Equation (3.3.11) is the motion composition in terms of angular acceleration and applying it to the serial chain of manipulators gives

$$ {}^{i+1}\dot{\omega}_{i+1} = \ddot{\theta}_{i+1}\, {}^{i+1}\mathbf{Z}_{i+1} + {}_{i+1}R_i\, {}^i\dot{\omega}_i + {}_{i+1}R_i\, {}^i\omega_i \mathrm{x} \dot{\theta}_{i+1}\, {}^{i+1}\mathbf{Z}_{i+1} \tag{3.3.12} $$

in which the right superscript for vectors gives the frame with respect to which the vector is described.
For linear acceleration, the differentiation of the velocity expression (3.3.1) gives the acceleration of point Q as

$$ {}_A\dot{\mathbf{v}}_Q = {}_A\dot{\mathbf{v}}_B + {}_A\dot{\omega}_B \mathrm{x}\, {}_A R_B\, {}_B\mathbf{p}_Q + {}_A\omega_B \mathrm{x} \frac{d}{dt}\left({}_A R_B\, {}_B\mathbf{p}_Q\right) + \frac{d}{dt}\left({}_A R_B\, {}_B\mathbf{v}_Q\right) \tag{3.3.13} $$

By applying the rule of Eq. (3.4.10) to terms ${}_A R_B\, {}_B\mathbf{p}_Q$ and ${}_A R_B\, {}_B\mathbf{v}_Q$ and grouping the results, the general formula can be obtained as

$$
{}_A\dot{\mathbf{v}}_Q = {}_A\dot{\mathbf{v}}_B + {}_A R_{BB}\dot{\mathbf{v}}_Q + {}_A\dot{\boldsymbol{\omega}}_B \mathrm{x}\, {}_A R_{BB}\mathbf{p}_Q + \\
+ {}_A\boldsymbol{\omega}_B \mathrm{x} \left({}_A\boldsymbol{\omega}_B \mathrm{x}\, {}_A R_{BB}\mathbf{p}_Q \right) + 2\, {}_A\boldsymbol{\omega}_B \mathrm{x}\, {}_A R_{BB}\mathbf{v}_Q \tag{3.3.14}
$$

in which one can recognize the Coriolis and centripetal components in the last terms. Equation (3.3.14) can be expressed in terms of i-th and (i+1)-th frames of a serial chain manipulator in order to give the acceleration propagation in the form

$$
{}^{i+1}\dot{\mathbf{v}}_{i+1} = {}_{i+1}R_i \left({}^i\dot{\boldsymbol{\omega}}_i \mathrm{x}\, {}_i\mathbf{p}_{i+1} + {}^i\boldsymbol{\omega}_i \mathrm{x}\, {}^i\boldsymbol{\omega}_i \mathrm{x}\, {}_i\mathbf{p}_{i+1} + {}^i\dot{\mathbf{v}}_i \right) + 2\, {}^{i+1}\boldsymbol{\omega}_{i+1} \mathrm{x} \dot{d}_{i+1}\, {}^{i+1}\mathbf{Z}_{i+1} + \ddot{d}_{i+1}\, {}^{i+1}\mathbf{Z}_{i+1} \tag{3.3.15}
$$

when the joint is revolute, the time derivatives of d vanish, and the expression is simplified.
A significant point, mainly for Dynamics calculation, is the center of mass of a link and its acceleration can be deduced by applying Eq. (3.3.15) to obtain

$$
{}^i\dot{\mathbf{v}}_{Gi} = {}^i\dot{\boldsymbol{\omega}}_i \mathrm{x}\, {}_i\mathbf{p}_{Gi} + {}^i\boldsymbol{\omega}_i \mathrm{x}\, {}^i\boldsymbol{\omega}_i \mathrm{x}\, {}_i\mathbf{p}_{Gi} + {}^i\dot{\mathbf{v}}_i \tag{3.3.16}
$$

An acceleration analysis of a manipulator can be carried out by using Eqs (3.3.12) and (3.3.15) from the base to the extremity, that is from 0 to N. Multiplying the dot vectors ${}^N\boldsymbol{\omega}_N$ and ${}^N\mathbf{v}_N$ by ${}_0R_N$ gives the acceleration of the extremity link in terms of the coordinates of the fixed base frame 0.

3.3.1 An example

The application of the outlined formulation for kinematic analysis of a manipulator is illustrated by referring to a planar 2R manipulator, since it is a basic architecture in robot structures. In Fig. 3.32 a planar 2R manipulator is modeled with its parameters. Reference frames are attached at the links and H–D parameters are indicated.
The transformation matrices can be computed as

$$
{}_0T_1 = \begin{vmatrix} c_1 & -s_1 & 0 & 0 \\ s_1 & c_1 & 0 & 0 \\ 0 & 0 & 1 & 0 \\ 0 & 0 & 0 & 1 \end{vmatrix}
$$

$$
{}_1T_2 = \begin{vmatrix} c_2 & -s_2 & 0 & a_1 \\ s_2 & c_2 & 0 & 0 \\ 0 & 0 & 1 & 0 \\ 0 & 0 & 0 & 1 \end{vmatrix} \tag{3.3.17}
$$

$$
{}_2T_3 = \begin{vmatrix} 1 & 0 & 0 & a_2 \\ 0 & 1 & 0 & 0 \\ 0 & 0 & 1 & 0 \\ 0 & 0 & 0 & 1 \end{vmatrix}
$$

The last matrix can be thought as referring to a third link of a 3R manipulator that is fixed permanently at zero degrees.
Thus, using the expressions for velocity propagation one can compute

$$
\begin{aligned}
&\boldsymbol{\omega}_1 = \left(0 \quad 0 \quad \dot{\theta}_1\right)^t \\
&{}_1\mathbf{v}_1 = \left(0 \quad 0 \quad 0\right)^t \\
&\boldsymbol{\omega}_2 = \left(0 \quad 0 \quad \dot{\theta}_1 + \dot{\theta}_2\right)^t \\
&{}_2\mathbf{v}_2 = \begin{vmatrix} c_2 & s_2 & 0 \\ -s_2 & c_2 & 0 \\ 0 & 0 & 1 \end{vmatrix} \begin{vmatrix} 0 \\ a_1\dot{\theta}_1 \\ 0 \end{vmatrix} = \left(a_1 s_2 \dot{\theta}_1 \quad a_1 c_2 \dot{\theta}_1 \quad 0\right)^t \qquad (3.3.18) \\
&\boldsymbol{\omega}_3 = \boldsymbol{\omega}_2 \\
&{}_3\mathbf{v}_3 = \left(a_1 s_2 \dot{\theta}_1 \quad a_1 c_2 \dot{\theta}_1 + a_2\left(\dot{\theta}_1 + \dot{\theta}_2\right) \quad 0\right)^t
\end{aligned}
$$

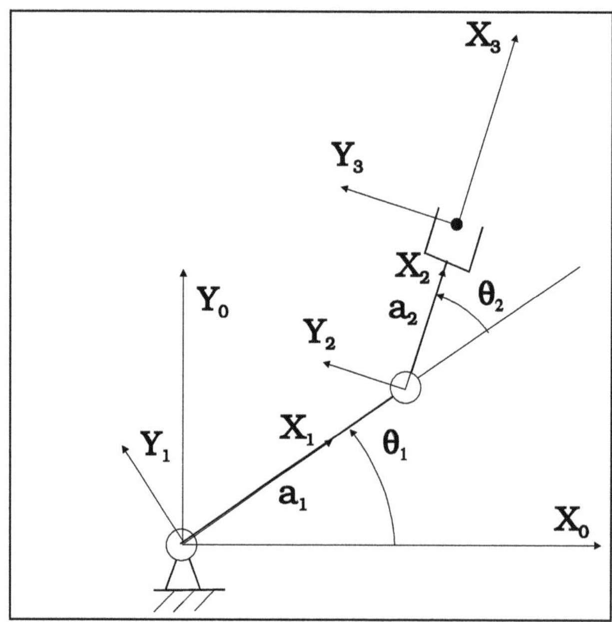

Fig. 3.32: Kinematic model of a planar 2R manipulator.

Finally, the velocity of the extremity point with respect to the fixed base frame can be computed by using rotation matrix ${}_0R_3 = {}_0R_1 + {}_1R_2 + {}_2R_3$ as

$$ {}_0\mathbf{v}_3 = \left(-a_1 s_1 \dot{\theta}_1 - a_2 s_{12}\left(\dot{\theta}_1 + \dot{\theta}_2\right) \quad a_1 c_1 \dot{\theta}_1 + a_2 c_{12}\left(\dot{\theta}_1 + \dot{\theta}_2\right) \quad 0\right)^t \tag{3.3.19} $$

Using the expressions for acceleration propagation one can compute, assuming the base accelerated at g along Y direction for gravity consideration, as

$$ \begin{aligned} &\dot{\omega}_1 = \left(0 \quad 0 \quad \ddot{\theta}_1\right)^t \\ &{}_1\dot{\mathbf{v}}_1 = \left(g s_1 \quad g c_1 \quad 0\right)^t \\ &\dot{\omega}_2 = \left(0 \quad 0 \quad \ddot{\theta}_1 + \ddot{\theta}_2\right)^t \\ &{}_2\dot{\mathbf{v}}_2 = \left(a_1 s_2 \ddot{\theta}_1 - a_1 c_2 \dot{\theta}_1^2 + g s_{12} \quad a_1 c_2 \ddot{\theta}_1 + a_1 s_2 \dot{\theta}_1^2 + g c_{12} \quad 0\right)^t \\ &\dot{\omega}_3 = \dot{\omega}_2 \end{aligned} \tag{3.3.20} $$

The analysis can be completed by computing the acceleration of the extremity point as

$$ {}_3\dot{\mathbf{v}}_3 = \left(0 \quad a_2\left(\ddot{\theta}_1 + \ddot{\theta}_2\right) \quad 0\right)^t + \left(-a_2\left(\dot{\theta}_1 + \dot{\theta}_2\right)^2 \quad 0 \quad 0\right)^t + {}^2\dot{\mathbf{v}}_2 \tag{3.3.21} $$

The above-mentioned expressions give a description of the velocity and acceleration of the manipulator and its links as function of the joint velocity and acceleration.

3.4 Jacobian and singular configurations

The Jacobian of a function f(x) is defined as

$$ J = \frac{\partial f}{\partial x} \tag{3.4.1} $$

and by using the Jacobian J the differential dy of the function f can be computed as

$$ dy = J dx \tag{3.4.2} $$

The above-mentioned expressions can be extended to multi-dimensional cases with M functions f_i of N variable x_j. By using vector notation, a vector **F** whose components are the f_i functions, can be expressed as function of a vector **X**, whose components are variables x_j, in the form

$$ \mathbf{Y} = \mathbf{F}(\mathbf{X}) \tag{3.4.2} $$

Thus, the Jacobian J is an $M \times N$ matrix that is given by

$$\mathbf{J} = \begin{vmatrix} \frac{\partial f_1}{\partial x_1} & \frac{\partial f_1}{\partial x_2} & \cdots & \cdots & \frac{\partial f_1}{\partial x_N} \\ \vdots & & & & \vdots \\ \vdots & & & & \vdots \\ \vdots & & & & \vdots \\ \frac{\partial f_M}{\partial x_1} & \cdots & \cdots & \cdots & \frac{\partial f_M}{\partial x_N} \end{vmatrix} \tag{3.4.3}$$

whose entries are the partial derivatives of the functions f.
The differential of **Y** can be expressed as

$$\mathbf{dY} = \frac{\partial \mathbf{F}}{\partial \mathbf{X}} \mathbf{dX} \tag{3.4.4}$$

and therefore in term of the Jacobian J it yields

$$\mathbf{dY} = \mathbf{JdX} \tag{3.4.5}$$

Thus, the Jacobian represents the tangent between the spaces Y and X. It is a linear transformation whose value changes with variable variation.
By dividing both sides by a differential time, Eq. (3.4.5) gives

$$\dot{\mathbf{Y}} = \mathbf{J}\dot{\mathbf{X}} \tag{3.4.6}$$

which represents a mapping of velocities in **X** to those in **Y**.
Referring to the kinematics of manipulators, vector **X** can represent the Joint Space and **Y** is the Cartesian Space. The multidimensional function **F** is the set of the expressions for the direct kinematics problem. Therefore, the Jacobian J of a manipulator maps the Joint Space into Cartesian Space in the form

$$\mathbf{V}_0 = \mathbf{J}_0 \dot{\theta} \tag{3.4.7}$$

in which $\mathbf{V}_0$ is the vector of the Cartesian velocities with respect to 0 frame and $\boldsymbol{\theta}$ dot is the vector of joint variables of the manipulator. The subscript 0 indicates the frame with respect to which the Cartesian velocities are computed and sometimes it is omitted when the frame or the base frame is evident. Vector $\mathbf{V}_0$ indicates the velocity of the manipulator extremity and generally it is given as 6×1 vector by the velocity of a tip point $\mathbf{v}_0$ and rotational velocity $\boldsymbol{\omega}_0$ of the extremity link as

$$\mathbf{V}_0 = \begin{vmatrix} \mathbf{v}_0 \\ \boldsymbol{\omega}_0 \end{vmatrix} \tag{3.4.8}$$

This expression refers to a 6 × 6 Jacobian, but a Jacobian of any dimension (including non square matrix) can be determined depending on the dimension of the Cartesian Space in which the manipulator motion is referred to. Indeed, the number of rows in a Jacobian equals the number of d.o.f.s in the Cartesian Space under examination, and the number of columns corresponds to the number of joints in the manipulator chain. For example, in the case of planar manipulators the Cartesian Space is described by 3 d.o.f.s (two translations and one angular motion) and therefore the corresponding Jacobian can be expressed by 3 rows, whereas the number of columns still depends on the number of joints in the manipulator.
The determination of the Jacobian of a manipulator can be obtained by one of the following procedures:

- by using the definition in Eq. (3.4.3) through differentiation of the function components of **F** that are formulated for the kinematics of a manipulator;
- by using the mapping in Eq. (3.4.7) when the velocities are already computed from kinematic analysis of velocity.

An additional procedure can be used as deduced from the mapping of Static actions, similar to the case for velocity mapping, as outlined in 3.5.
The Jacobian formulation is useful for computing the joint rates, when the Cartesian velocities are given, from the expression

$$\dot{\theta} = J_0^{-1}\, \mathbf{V}_0 \tag{3.4.9}$$

if matrix **J** is invertible. Equation (3.4.9) has a very practical importance but it depends on the condition of invertibility of **J**. Matrix is invertible when it is possible to compute

$$\mathbf{J}_0^{-1} = \frac{\mathbf{J}_{\mathbf{adj}}}{\det \mathbf{J}} \tag{3.4.10}$$

in which $\mathbf{J}_{adj}$ is the adjunct matrix of **J** and det **J** is the determinant of **J**. When det **J** = 0 matrix **J** is not invertible and it is named 'singular'.
Since **J** depends on the manipulator configuration, singularities of **J** occur at certain manipulator configurations that are named 'singularities of the manipulator' (in short as 'singularities'). In approaching a singularity condition $\mathbf{J}^{-1}$ goes to infinity and as Eq. (3.4.9) this corresponds that for a given finite Cartesian velocity, an infinite joint rate is computed as necessary. Since infinite joint rates are unpractical, singularities are considered as dangerous manipulator configurations and they are usually avoided in industrial robots.
When a manipulator is in a singular configuration, the infinite value for joint rates can be interpreted as corresponding to a motion that is not possible for the manipulator. This

also means that at singular configurations a manipulator has lost one or more d.o.f.s as viewed from Cartesian Space.
Singular configurations can occur at workspace boundary configurations when the manipulator is fully stretched or folded back on itself with the end-effector at its extreme position. In addition, singularities can occur within the workspace when two or more joint axes are lined up. In this last occurrence, the loss of d.o.f.s correspond also to the fact that the motion of the end-effector can also occur along a possible direction but it is not uniquely determined as function of joint motions, as shown in the example of Fig. 3.33.
Therefore, investigating the Jacobian for singular configurations gives a procedure for singularity analysis of manipulators.

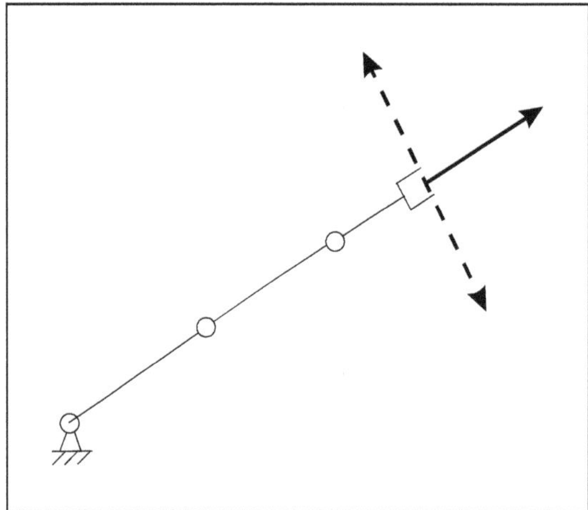

Fig. 3.33: An example of singular configuration of a planar 3R manipulator for which outward radial velocity is impossible and transversal velocity is not a unique function of joint variables.

3.4.1 An example

The case of a planar 2R manipulator can be studied by referring to the velocity analysis in the example in 3.3.1. By looking at the expression of ${}^0\mathbf{v}_3$ and ${}^3\mathbf{v}_3$ a 2×2 Jacobian matrix can be obtained as the mapping of Eq. (3.4.7) but referring to linear velocity only, in the form

$$J_0 = \begin{vmatrix} -a_1 s_1 - a_2 s_{12} & -a_2 s_{12} \\ a_1 c_1 + a_2 c_{12} & a_2 c_{12} \end{vmatrix} \tag{3.4.11}$$

and

$$J_3 = \begin{vmatrix} a_1 s_1 & 0 \\ a_1 c_1 + a_2 & a_2 \end{vmatrix} \tag{3.4.12}$$

If angular velocity is considered, a 3×2 Jacobian matrix can be obtained by including a third row with the entries equal to 1, as the expression of ${}_3\omega_3$. But in this case the Jacobian will not be a square matrix that can be useful for determining singular configurations of the manipulator.
The determinant of J_3 can be computed straightforwardly as

$$\det J_3 = a_1 a_2 s_2 \tag{3.4.13}$$

whose condition for singularity gives

$$\sin\theta_2 = 0 \tag{3.4.14}$$

which refers to the manipulator in fully stretched and folded back configurations.
In these configurations the manipulator has lost one d.o.f., as in the illustrative case of Fig. 3.33.
The computation of the Jacobian by using the derivatives of position equations would have given the same results for the 2×2 Jacobian as J_0 and J_3. But it would not have permitted the computation of the angular velocity entry since there is no orientation vector whose derivative gives the angular velocity vector.

3.5 Statics of manipulators

One of the most important features of robotic manipulators is the positioning of objects or end-effector in static configurations. In this task a manipulator acts like a structure when the joints are locked in a suitable way. Thus, it is of practical interest to evaluate and simulate the static behavior of a manipulator. In addition, design considerations can be deduced by the computations for the static equilibrium.

3.5.1 A mechanical model

In Fig. 3.34 a mechanical model is presented in the form of a free-body diagram to determine conditions for static equilibrium of a manipulator link. The free-body diagram of a rigid body is obtained by analyzing forces and torques acting on the body when it is considered as disconnected from the neighboring bodies. Thus, actions on a link body of a manipulator can be considered as determined in Fig. 3.34:

- $\mathbf{f}_i$ is the force exerted on link i by the link i-1 through the joint connection;
- $\mathbf{t}_i$ is the torque exerted on link i by the link i-1 through the joint connection;
- $\mathbf{F}_i$ is the external force exerted on link i by bodies that do not compose of the manipulator; it can include the weight of the body link;
- $\mathbf{T}_i$ is the external torque exerted on link i by bodies that do not compose of the manipulator.

All above-mentioned vectors are expressed with respect to the i-th frame on link i.
The actions $\mathbf{f}_{i+1}$ and $\mathbf{t}_{i+1}$ are the actions exerted by link i on link i+1 and they are described in the i+1-th reference.
The skew general architecture and points of application for forces and torques

characterize the geometry of the link free-body diagram of Fig. 3.34.
The general skew architecture refers also to the model for H–D notation and represents a general model of a mechanical design from which any particular case can be deduced.
The positions of the points of application for forces and torques can also be considered as structural characteristics for manipulator links. In fact, points O_i and O_{i+1} are the origins of the i-th and (i+1)-th reference frames; G_i point is the center of mass of the link. It is of note that both forces and torques will be considered as resultant vectors of actions acting on the link, even in a distributed way. In addition, torques could have a contribution regarding a moment of transportation that can be needed to consider those points as points of application for forces.
Thus, the center of mass G_i can also be considered for external actions, since in that case the weight of the link can also be taken into account. Otherwise the corresponding point of application must be identified in the link geometry.

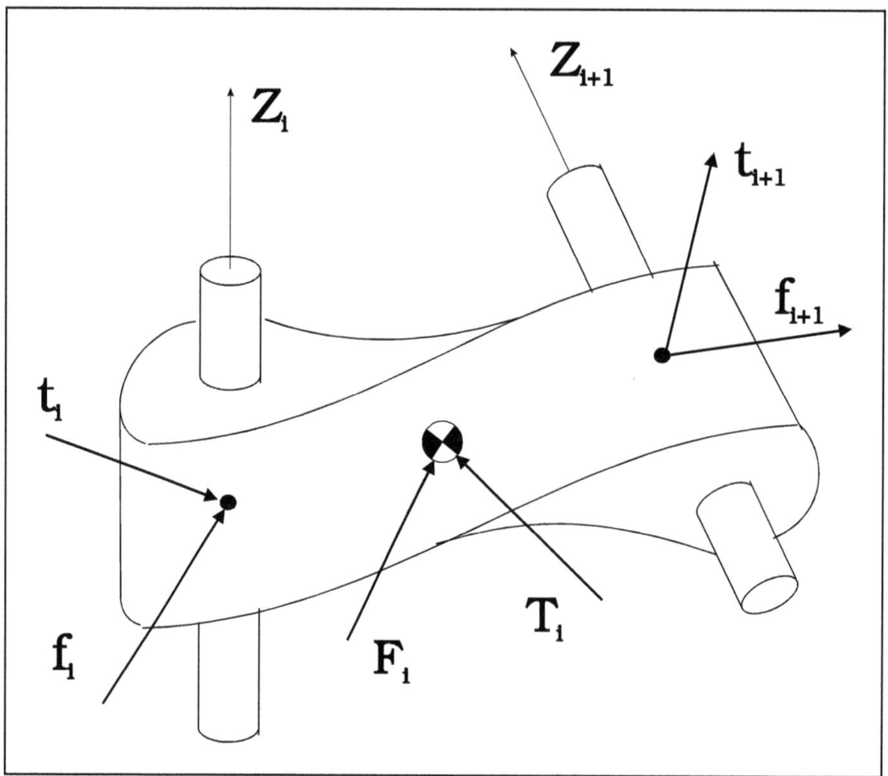

Fig. 3.34: Free-body diagram of a manipulator link.

It is of note that even the actions $\mathbf{f}_i$ and $\mathbf{t}_i$ of the neighborhood link through the joint can be considered as resultant vectors of distributed actions exerted on the joint kinematic surface. This means that a well-designed joint should have a symmetric mechanical design and operation so that O_i will be the point of application of the resultant vectors.

3.5.2 Equations of equilibrium

The mechanical model of Fig. 3.34 is useful also for deducing a formulation for the static equilibrium of a link. In fact, summing the forces and equating to zero, the equilibrium for translation is expressed as

$$\mathbf{f}_i - {}_iR_{i+1}\mathbf{f}_{i+1} + \mathbf{F}_i = 0 \tag{3.5.1}$$

in which vector $\mathbf{f}_{i+1}$ has been written with respect to the i-th frame by means of the rotation matrix ${}_iR_{i+1}$ describing the (i+1)-th frame relative to i-th frame.
The rotation equilibrium can be expressed by summing torques about the origin point O_i in the form

$$\mathbf{t}_i - {}_iR_{i+1}\mathbf{t}_{i+1} + \mathbf{T}_i - \mathbf{O}_i\mathbf{O}_{i+1} x\, {}_iR_{i+1}\mathbf{f}_{i+1} + \mathbf{O}_i\mathbf{G}_i x \mathbf{F}_i = 0 \tag{3.5.2}$$

Equations (3.5.1) and (3.5.2) express the static equilibrium of the i-th link as described with respect to the i-th frame. Usually, Eqs (3.5.1) and (3.5.2) are simplified in the form

$$\mathbf{f}_i - {}_iR_{i+1}\mathbf{f}_{i+1} = 0 \tag{3.5.3}$$

$$\mathbf{t}_i - {}_iR_{i+1}\mathbf{t}_{i+1} - \mathbf{O}_i\mathbf{O}_{i+1} x\, {}_iR_{i+1}\mathbf{f}_{i+1} = 0 \tag{3.5.4}$$

by neglecting external actions, since in general external force and torque act on the extremity link only as payload or task operation action and the link weight can be neglected as compared with the values of the actions at the joints of the link. Moreover, one can use the joint forces and torques at the extremity frame to model payload, external force, or task actions on the extremity link.
In addition, solving Eqs (3.5.1) and (3.5.2) for the i-th joint force $\mathbf{f}_i$ and torque $\mathbf{t}_i$ gives an expression for the static equilibrium of a manipulator in the form of a force propagation from link to link. This is expressed by

$$\mathbf{f}_i = {}_iR_{i+1}\mathbf{f}_{i+1} - \mathbf{F}_i \tag{3.5.5}$$

$$\mathbf{t}_i = {}_iR_{i+1}\mathbf{t}_{i+1} - \mathbf{T}_i + \mathbf{O}_i\mathbf{O}_{i+1} x\, {}_iR_{i+1}\mathbf{f}_{i+1} - \mathbf{O}_i\mathbf{G}_i x \mathbf{F}_i \tag{3.5.6}$$

Thus, starting from the manipulator extremity on which a known payload is applied, Eqs. (3.5.5) and (3.5.6) give the conditions for the static equilibrium of a manipulator.
The component of force $\mathbf{f}_i$ or torque $\mathbf{t}_i$ along Z_i-axis gives the value of the action that the i-th actuator must exert to ensure the static equilibrium. Thus, the actuator force τ_i is computed for revolute joint as

$$\tau_i = \mathbf{t}_i \cdot \mathbf{Z}_i \qquad (3.5.7)$$

and in the case the i-th joint is prismatic, it is given by

$$\tau_i = \mathbf{f}_i \cdot \mathbf{Z}_i \qquad (3.5.8)$$

in which $\mathbf{Z}_i$ indicates the unit vector of Z_i-axis.
The structure and joints of the link resist the other components of the force $\mathbf{f}_i$ and torque $\mathbf{t}_i$. In fact, a joint transmits force $\mathbf{f}_i$ and torque $\mathbf{t}_i$ properly when it preserves its geometrical and operational efficiency. Thus, the values of those components can be used to check the feasibility of the static equilibrium in a built robot by looking at the possibility by the joints and links to resist to them. In the case of manipulator design, the values of those components are used to size and shape properly the joints and links.

3.5.3 Jacobian mapping of forces

The static equilibrium of a robotic multibody system can be expressed also by using the Principle of Virtual Work.
Virtual displacements can be considered by looking at the movements that are possible because of the geometric constraints only. Thus, joints can move by $\delta\mathbf{q}$ as expressed in Joint Space and the manipulator will move the extremity by $\delta\mathbf{X}$, including translation and rotation, as expressed in Cartesian Space. Work is computed as the dot product of force and torque vectors by the corresponding displacement vectors. Therefore, assuming friction as negligible, the Principle of Virtual Work can be written for a manipulator in the form

$$\mathbf{F} \cdot \delta\mathbf{X} = \boldsymbol{\tau} \cdot \delta\mathbf{q} \qquad (3.5.9)$$

by equating the work for the virtual displacement of the manipulator extremity with the work for the corresponding virtual motion of the joints. In Eq. (3.5.9) when considering a general 6 d.o.f. manipulator in the space, $\mathbf{F}$ is the 6×1 vector with force and torque acting on the manipulator extremity; $\delta\mathbf{X}$ is the 6×1 vector of infinitesimal Cartesian displacement of manipulator extremity; $\boldsymbol{\tau}$ is the 6×1 vector of torques acting on the joints of the manipulator; $\delta\mathbf{q}$ is the 6×1 vector of the infinitesimal joint displacements.
Equation (3.5.9) can be expressed in matrix form as

$$\mathbf{F}^t \delta\mathbf{X} = \boldsymbol{\tau}^t \delta\mathbf{q} \qquad (3.5.10)$$

where t is the transpose operator.
In addition, displacement vectors are not independent but they are related to each other through the Jacobian

$$\delta\mathbf{X} = \mathrm{J}\delta\mathbf{q} \qquad (3.5.11)$$

Therefore, Eq. (3.5.9) can be rewritten in the form

$$\mathbf{F}^t J \delta \mathbf{q} = \boldsymbol{\tau}^t \delta \mathbf{q} \tag{3.5.12}$$

which must hold for any $\delta\mathbf{q}$ to give, after transposing both sides,

$$\boldsymbol{\tau} = J^t \mathbf{F} \tag{3.5.13}$$

Equation (3.5.13) express that at static equilibrium the Jacobian transpose maps Cartesian forces acting on the manipulator extremity into the corresponding joint torques. Expression (3.5.13) is usually referred to with those vectors written with respect to the reference frame at the manipulator extremity. When referred to the base frame 0 Eq. (3.5.13) takes the form

$$\boldsymbol{\tau} = {}_0J^t \, {}_0\mathbf{F} \tag{3.5.14}$$

and similarly it can be written when it is referred to other frames, by expressing both J and $\mathbf{F}$ in those frames. Equation (3.5.13) and (5.5.14) also express an alternative way to compute the Jacobian of a manipulator.
In addition, Eq. (3.5.13) gives an important interpretation of a singular configuration. In fact, when J is singular and J^t goes to infinity there are certain directions in which very large joint torques are needed for exerting small actions by manipulator extremity. For example, with outstretched arm, one will make great efforts to push along with the hand to the outside direction of the arm.

3.5.4 An example

The case of planar 2R manipulator in 3.3.1 is considered when it applies a force $\mathbf{F}_3$ with components F_x and F_y by its end-effector.
Using the force propagation through Eqs (3.5.5) and (3.5.6), one can compute

$$\begin{aligned}
&\mathbf{f}_2 = \begin{pmatrix} F_x & F_y & 0 \end{pmatrix}^t \\
&\mathbf{t}_2 = a_2 \mathbf{X}_2 \times \mathbf{f}_2 = \begin{pmatrix} 0 & 0 & a_2 F_y \end{pmatrix}^t \\
&\mathbf{f}_1 = \begin{pmatrix} c_2 F_x - s_2 F_y & s_2 F_x + c_2 F_y & 0 \end{pmatrix}^t \\
&\mathbf{t}_1 = {}_1R_2 \mathbf{t}_2 + a_1 \mathbf{X}_1 \times \mathbf{f}_1 = \begin{pmatrix} 0 & 0 & a_1 s_2 F_x + a_1 c_2 F_y + a_2 F_y \end{pmatrix}^t
\end{aligned} \tag{3.5.15}$$

Consequently, the actuator torques can be determined as

$$\begin{aligned}
&\tau_1 = a_1 s_2 F_x + a_1 c_2 F_y + a_2 F_y \\
&\tau_2 = a_2 F_y
\end{aligned} \tag{3.5.16}$$

The expressions for actuator torques can be grouped in a matrix form as

$$\boldsymbol{\tau} = \begin{vmatrix} a_1 s_2 & a_1 c_2 + a_2 \\ 0 & a_2 \end{vmatrix} \begin{vmatrix} F_x \\ F_y \end{vmatrix} = J^t \mathbf{F} \tag{3.5.16}$$

in which one can recognize the transposition of the Jacobian that has been calculated in the example 3.4.1.

3.6 Dynamics of manipulators

The dynamical behavior of a robotic multibody system can be studied by formulating the relations between the motion of the system and actions acting on it.
Indeed, dynamic equations can be formulated by looking at two different problems, namely 'direct dynamics' and 'inverse dynamics'. The main features of these dynamic problems are illustrated in Fig. 3.35.
In direct dynamics the actions on the system are known and the problem is to find the resulting motion. This can be expressed in a general form as

$$\ddot{q} = \mathbf{f}(\boldsymbol{\tau}; \mathbf{L}; \mathbf{m}) \tag{3.6.1}$$

in which **q** double dot is the vector for joint acceleration; $\boldsymbol{\tau}$ is the vector for joint torques; **L** describes the architecture geometry in terms of H–D parameters; and **m** indicates the vector of the inertial parameters of the system. The function vector **f** indicates the differential nature of dynamic equation.
In inverse dynamics the motion of the system is prescribed and the problem is to find the corresponding actuation forces. This can be expressed in a general form as

$$\boldsymbol{\tau} = \mathbf{g}(\ddot{q}; \mathbf{L}; \mathbf{m}) \tag{3.6.2}$$

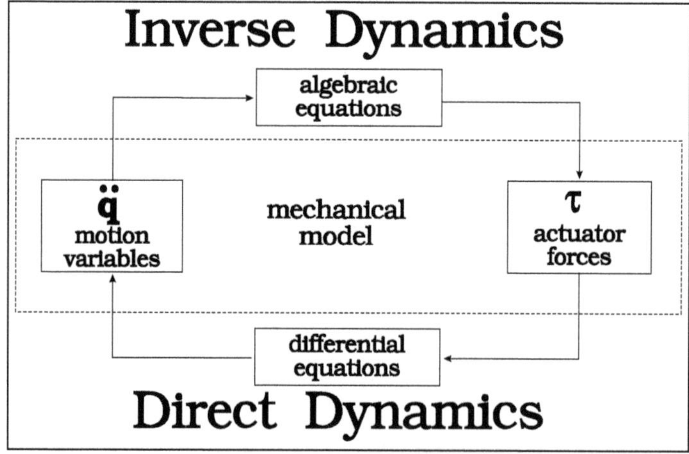

Fig.3.35: A general view of dynamics problems for robotic manipulators.

in which **g** function vector indicates the algebraic nature of those dynamic equations.
Both problems have practical interest when they are solved with efficient computational algorithms. Direct problem is fundamental for simulating the robot operation before building or operating them. Inverse problem is useful for regulating the robot operation during a given manipulation task.
In addition, the solutions of both problems can be used and are used in control algorithms for high performance robot motion.

3.6.1 Mechanical model and inertia characteristics

The model of Fig.3.34 can be extended to the case of dynamics by adding the consideration of inertia forces and torques, as shown in Fig. 3.36. The inertia forces and torques are expressed as

$$\mathbf{F}_i^{in} = -m_i \mathbf{a}_{Gi} \tag{3.6.3}$$

$$\mathbf{T}_i^{in} = -I_i^{Gi} \dot{\boldsymbol{\omega}}_i - \boldsymbol{\omega}_i \, x \, I_i^{Gi} \boldsymbol{\omega}_i \tag{3.6.4}$$

in which inertia characteristics are evaluated in terms of mass m_i and inertia tensor I_i^{Gi} of the i-th link. Indeed, inertia characteristics are determined when both mass distribution and kinematic status of a link are known.

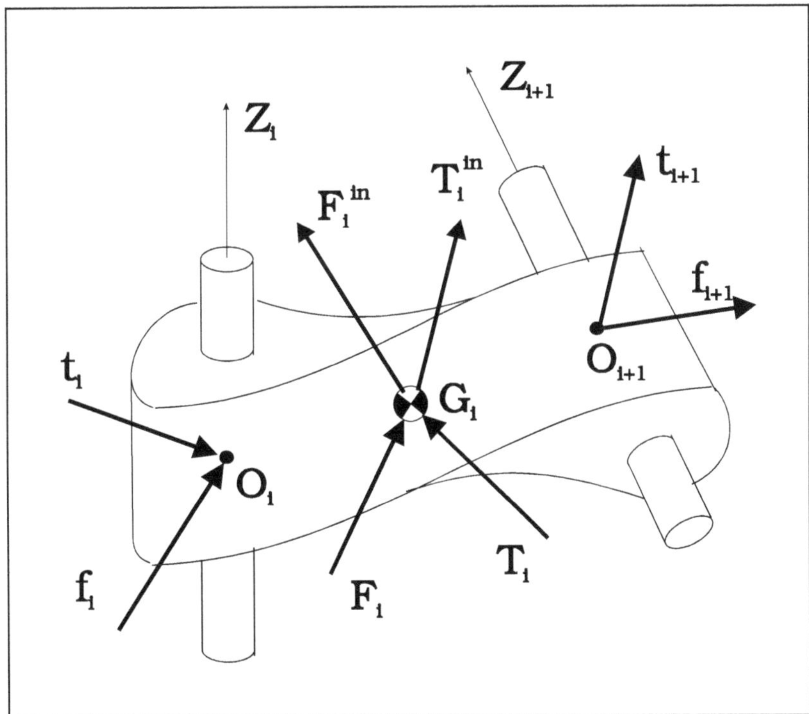

Fig. 3.36: The free-body diagram of a manipulator link for dynamics.

Mass distribution is characterized by the mass of a link and inertia tensor.
The mass m_i of a link can be evaluated by computing

$$m_i = \int_{Vi} \rho dxdydz \tag{3.6.5}$$

in which ρ is the material density and the integral is calculated over the volume Vi of the link body. The expression (3.6.5) requires knowing an analytic function both for the volume and material density. But it is not always possible to have such functions and therefore alternative procedures are used to compute the mass through:

- numerical approximation;
- experimental determination.

Usually the material density is assumed as a constant as obtained from a mechanical design of a homogeneous link structure.
Numerical approximation of link volume is obtained by approximating the integral operation over differential volumes by means of a discrete sum of known finite volumes with simple shapes. Experimental determination can be obtained by a direct evaluation of the weight of the link or through indirect evaluation of the mass by using a pendulum test. This can be performed by oscillating the link body in free mode with small angles about a joint and by measuring the frequency f of the oscillations to evaluate

$$f = \frac{1}{2\pi}\sqrt{\frac{mgh}{J}} \tag{3.6.6}$$

in which m is the mass of the body, g is the gravity acceleration, h is the distance between the joint and center of mass, and J is the moment of inertia of the body with respect to the axis of the joint.
It is of note that mass evaluation of a link also includes the determination of the position of the center of mass that is given by the Cartesian coordinates in the i-th frame as

$$x_{Gi} = \frac{\int_{Vi} \rho x dV}{\int_{Vi} \rho dV} \qquad y_{Gi} = \frac{\int_{Vi} \rho y dV}{\int_{Vi} \rho dV} \qquad z_{Gi} = \frac{\int_{Vi} \rho z dV}{\int_{Vi} \rho dV} \tag{3.6.7}$$

The inertia tensor is a matrix grouping the inertia characteristics for rotation movements in the form

$$I_i^{Gi} = \begin{vmatrix} I_{xxi} & -I_{xyi} & -I_{xzi} \\ -I_{xyi} & I_{yyi} & -I_{yzi} \\ -I_{xzi} & -I_{yzi} & I_{zzi} \end{vmatrix} \tag{3.6.8}$$

in which the scalar elements are given by

$$\begin{aligned} I_{xxi} &= \int_{Vi} \rho\left(y^2 + z^2\right) dxdydz \\ I_{yyi} &= \int_{Vi} \rho\left(x^2 + z^2\right) dxdydz \\ I_{zzi} &= \int_{Vi} \rho\left(x^2 + y^2\right) dxdydz \\ I_{xyi} &= \int_{Vi} \rho xydxdydz \\ I_{xzi} &= \int_{Vi} \rho xzdxdydz \\ I_{yzi} &= \int_{Vi} \rho yzdxdydz \end{aligned} \tag{3.6.9}$$

where the integrals are computed over the volume Vi of the link body by using the coordinates of the points.
The elements I_{xxi}, I_{yyi}, and I_{zzi} are named as moments of inertia of the body and they are characterized by square distance of the points in the computing formula.
The elements I_{xyi}, I_{xzi}, and I_{yzi} are named as products of inertia of the body and they are characterized by mixed distance coefficients in the computing formula.
The elements of the inertia tensor are dependent on the position and orientation of the frame, with respect to which they are defined. But it is possible to determine a reference frame for which the products of inertia are zero and the corresponding moments of inertia are called 'principal moments of inertia'. The axes of this reference frame are named 'principal axes of inertia of the body'. They can be found also as eigenvalues and eigenvectors of the inertia tensor.
The position change of a reference frame for the inertia tensor can be computed by using the Huygens Theorem (also known as the Parallel Axis Theorem), which gives

$$\begin{aligned} I_{zzi}^{A} &= I_{zzi}^{Gi} + m_i\left(x_{Gi}^2 + y_{Gi}^2\right) \\ I_{xyi}^{A} &= I_{xyi}^{Gi} + m_i x_{Gi} y_{Gi} \end{aligned} \tag{3.6.10}$$

when the reference with origin in A is an arbitrarily translated frame with the coordinates x_{Gi}, y_{Gi}, z_{Gi} locating the center of mass G_i with respect to A.
The orientation change of reference frame in computing the inertia tensor is expressed by using the rotation matrix.
Similarly to mass of a body, the elements of the inertia tensor can be difficult to be evaluated, mainly for built manipulators. Therefore, numerical approximation can be obtained by again using a discrete sum of known volumes with simple shapes in combination with application of the Huygens Theorem.
Experimental determination can also be carried out by using the pendulum test to evaluate the moment of inertia about the axis of the used joint through the formula of Eq. (3.6.6).

3.6.2 Newton–Euler equations

The Newton–Euler formulation can use the model of Fig. 3.36 to deduce the dynamic equations for a link. Thus, one can write the Newton–Euler equations to express the dynamic equilibrium with respect to the i-th frame in the form of the D'Alembert Principle by summing forces acting on the body link in the form

$$ {}_i\mathbf{F}_i^{in} + \mathbf{f}_i - {}_iR_{i+1}\mathbf{f}_{i+1} + \mathbf{F}_i = 0 \tag{3.6.11}$$

and summing torques about the center of mass in the form

$$ {}_i\mathbf{T}_i^{in} + \mathbf{t}_i - {}_iR_{i+1}\mathbf{t}_{i+1} + \mathbf{T}_i - \mathbf{O}_i\mathbf{O}_{i+1} x\, {}_iR_{i+1}\mathbf{f}_{i+1} + \mathbf{O}_i\mathbf{G}_i x \mathbf{F}_i = 0 \tag{3.6.12}$$

in which the inertia force and torque of Eqs (3.6.3) and (3.6.4) are expressed with respect to the i-th frame. Equations (3.6.11) and (3.6.12) express the instantaneous dynamic equilibrium of a body link. Usually, Eqs (3.6.11) and (3.6.12) are simplified as

$$ {}_i\mathbf{F}_i^{in} + \mathbf{f}_i - {}_iR_{i+1}\mathbf{f}_{i+1} = 0 \tag{3.6.13}$$

$$ {}_i\mathbf{T}_i^{in} + \mathbf{t}_i - {}_iR_{i+1}\mathbf{t}_{i+1} - \mathbf{O}_i\mathbf{O}_{i+1} x\, {}_iR_{i+1}\mathbf{f}_{i+1} = 0 \tag{3.6.14}$$

when external actions are not applied to the links.
As in the case of static equilibrium, solving Eqs (3.6.13) and (3.6.14) gives an expression for dynamic equilibrium of a manipulator in the form of force propagation from extremity link to the base frame. This is expressed by

$$ \mathbf{f}_i = {}_iR_{i+1}\mathbf{f}_{i+1} - {}_i\mathbf{F}_i^{in} \tag{3.6.15}$$

$$ \mathbf{t}_i = {}_iR_{i+1}\mathbf{t}_{i+1} + \mathbf{O}_i\mathbf{O}_{i+1} x\, {}_iR_{i+1}\mathbf{f}_{i+1} - {}_i\mathbf{T}_i^{in} \tag{3.6.16}$$

The actions on the extremity link can be given from the manipulating task of the robot, including the case for a movement of unloaded manipulation.
Equations (3.6.15) and (3.6.16), like the previous ones, can be computed after the kinematics has been solved to compute inertia actions. Thus, the dynamics of a manipulator can be computed in two steps, namely an outward computation for the kinematics and an inward computation for the dynamic equations, as synthetically summarized in Fig. 3.37.
An iterative numerical algorithm can be outlined for serial manipulators with revolute and prismatic pairs by summarizing the kinematic analysis in 3.3 and force propagation of Eqs (3.6.15) and (3.6.16) in the following:
- Outward kinematic iteration with i from 0 to (N-1), to compute:

$$^{i+1}\boldsymbol{\omega}_{i+1} = {}_{i+1}R_i \, {}^{i}\boldsymbol{\omega}_i + \dot{\theta}_{i+1} \, {}^{i+1}\mathbf{Z}_{i+1}$$

$$^{i+1}\dot{\boldsymbol{\omega}}_{i+1} = {}_{i+1}R_i \, {}^{i}\dot{\boldsymbol{\omega}}_i + \ddot{\theta}_{i+1} \, {}^{i+1}\mathbf{Z}_{i+1} + {}_{i+1}R_i \, {}^{i}\boldsymbol{\omega}_i x \dot{\theta}_{i+1} \, {}^{i+1}\mathbf{Z}_{i+1}$$

$$^{i+1}\dot{\mathbf{v}}_{i+1} = {}_{i+1}R_i \left({}^{i}\dot{\boldsymbol{\omega}}_i x_i \mathbf{p}_{i+1} + {}^{i}\boldsymbol{\omega}_i x \, {}^{i}\boldsymbol{\omega}_i x_i \mathbf{p}_{i+1} + {}^{i}\dot{\mathbf{v}}_i \right) + 2 \, {}^{i+1}\boldsymbol{\omega}_{i+1} x \dot{d}_{i+1} \, {}^{i+1}\mathbf{Z}_{i+1} + \ddot{d}_{i+1} \, {}^{i+1}\mathbf{Z}_{i+1}$$

$$^{i}\dot{\mathbf{v}}_{Gi} = {}^{i}\dot{\boldsymbol{\omega}}_i x_i \mathbf{p}_{Gi} + {}^{i}\boldsymbol{\omega}_i x \, {}^{i}\boldsymbol{\omega}_i x_i \mathbf{p}_{Gi} + {}^{i}\dot{\mathbf{v}}_i \qquad (3.3.17)$$

$$\mathbf{F}_i^{in} = -m_i \mathbf{a}_{Gi}$$

$$\mathbf{T}_i^{in} = -I_i^{Gi} \dot{\boldsymbol{\omega}}_i - \boldsymbol{\omega}_i x I_i^{Gi} \boldsymbol{\omega}_i$$

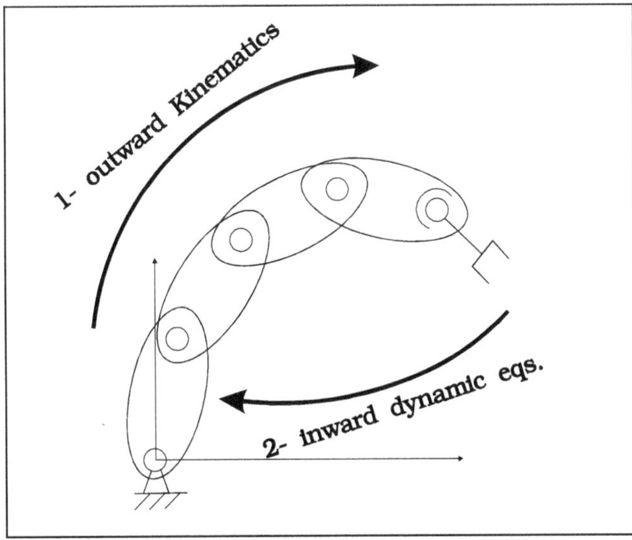

Fig. 3.37: Recursive computation flows for the dynamics of a manipulator.

- Inward iteration for dynamic equations with i from N to 1, to compute:

$$\mathbf{f}_i = {}_iR_{i+1}\mathbf{f}_{i+1} - {}_i\mathbf{F}_i^{in}$$
$$\mathbf{t}_i = {}_iR_{i+1}\mathbf{t}_{i+1} + \mathbf{O}_i\mathbf{O}_{i+1} \times {}_iR_{i+1}\mathbf{f}_{i+1} - {}_i\mathbf{T}_i^{in} \tag{3.6.18}$$

In this computational algorithm the weight of links can be simply taken into account by setting the base acceleration equal to the gravity acceleration g. In fact this base acceleration will give the same effect on the links as the weight force would have. But no additional computational efforts are spent for the formulation of the dynamics of a manipulator.
The above-mentioned recursive formulation for dynamics analysis can be synthesized in matrix form to give a general structure of the mathematical model for the equation of motion in the form

$$\tau = M(\theta)\ddot{\theta} + C(\theta, \dot{\theta}) + G(\theta) \tag{3.6.19}$$

in which $\boldsymbol{\tau}$ is a $N \times 1$ vector of actuator forces for a manipulator with N links; M is the $N \times N$ inertia matrix of the manipulator; C is the $N \times 1$ vector of the centrifugal and Coriolis forces; G is the $N \times 1$ vector of the gravity terms. Vector θ and its derivatives are related to the kinematic status of the joints of the manipulator. Equation (3.6.19) refers to the State Space since its coefficient C is function of both position and velocity of joints. Writing Eq. (3.6.19) for the i-th actuator by using the recursive algorithm gives the expressions of the entries of the coefficient matrices in the form

$$M_{ij} = \sum_{k=j}^{N}\left[({}_0R_kZ_i)^t I_k^{Gk}({}_kR_0Z_j)\right] + \sum_{k=j}^{N}\left[m_kZ_k \times (O_0G_k - O_0O_k)\right]\cdot\left[Z_k \times (O_0G_k - O_0O_k)\right]$$

$$\begin{aligned} C_i = {} & \sum_{r=1}^{N} m_r \sum_{k=1}^{r-1}\sum_{s=1}^{k}\dot{\theta}_sZ_s \times \sum_{t=1}^{k}\dot{\theta}_tZ_t \times O_kO_{k+1} + \sum_{s=2}^{k}\sum_{t=1}^{s-1}\dot{\theta}_tZ_t \times \dot{\theta}_sZ_s \times O_kO_{k+1} \\ & \cdot Z_i \times (O_0G_r - O_0O_i) + \sum_{r=i}^{N} m_r \sum_{s=1}^{r}\dot{\theta}_sZ_s \times \sum_{t=1}^{r}\dot{\theta}_tZ_t \times O_rG_r + \sum_{s=2}^{r}\sum_{t=1}^{s-1}\dot{\theta}_tZ_t \times \dot{\theta}_sZ_s \times O_rG_r \\ & \cdot Z_i \times (O_rG_r - O_0O_i) + \sum_{r=i}^{N}({}_rR_0Z_i)^t I_r^{Gr}\sum_{j=1}^{r}(\dot{\theta}_j\,{}_rR_0Z_j) \times \sum_{k=j+1}^{r}(\dot{\theta}_k\,{}_rR_0Z_k) \\ & + \left[{}_rR_0Z_i \times \sum_{s=1}^{r}(\dot{\theta}_s\,{}_rR_0Z_s)\right]^t I_r^{Gr}\sum_{t=1}^{r}(\dot{\theta}_t\,{}_rR_0Z_t) \end{aligned} \tag{3.6.20}$$

$$G_i = -\mathbf{g} \cdot \left[\mathbf{Z}_i \times \sum_{j=1}^{N} m_j \left(\mathbf{O_0G}_j - \mathbf{O_0O}_j \right) \right]$$

From the expressions of M_{ij}, C_i, and G_i of Eq. (3.6.20) the computational convenience and interpretation clarity of the recursive algorithm of Eqs. (3.6.17) and (3.6.18) is evident. In particular, the expression for C_i is formulated as function of the mass and inertia tensor with differentiated forms showing the complication of the coupling effects of the centrifugal and Coriolis forces.
Alternatively, the matrix coefficients of Eq. (3.6.19) can be determined by direct comparison with the closed-form results of the recursive algorithm of Eqs (3.6.17) and (3.6.18).
It is of note that the matrix expression (3.6.19) is related to the equation of motion only and does not refer to the possibility of computing all the reaction forces and torques at the joints as the Newton–Euler recursive algorithm does.
In addition, when the dependence of coefficient vector C from velocity is expressed the dynamic equation of motion can be written in the form

$$\tau = M(\theta)\ddot{\theta} + C_1(\theta)\left[\dot{\theta}\dot{\theta}\right] + C_2(\theta)\left[\dot{\theta}^2\right] + G(\theta) \tag{3.6.21}$$

as referring to the Joint Space since all the equation coefficients are functions of the manipulator configuration vector θ. The matrix C has been partitioned into C_1 and C_2 for Coriolis and centrifugal terms, respectively.
Another important view for the dynamic equation is the one referring to Cartesian Space. In particular, by referring to the Cartesian variable Eq. (3.6.19) can be written in the form

$$\mathbf{F} = M_X(\theta)\ddot{\mathbf{X}} + C_X\left(\theta, \dot{\theta}\right) + G_X(\theta) \tag{3.6.22}$$

where $\mathbf{F}$ is 6×1 vector of the force and torque acting on the end-effector, and $\mathbf{X}$ is the vector of Cartesian coordinates representing position and orientation of the end-effector. The matrices M_X, C_X, and G_X are similar to those in Eq. (3.6.19) in Joint Space but they are described in the Cartesian Space.
In fact, they can be obtained by looking at the change of reference space. Thus, the kinetic energy of a manipulator can be expressed as referring to Joint Space or to Cartesian Space and equating these expressions as

$$\frac{1}{2}\dot{\mathbf{X}}^t M_X \dot{\mathbf{X}} = \frac{1}{2}\dot{\theta}^t M_X \dot{\theta} \tag{3.6.23}$$

gives

$$M_X(\theta) = J^{-t} M(\theta) J^{-1} \tag{3.6.24}$$

when the Jacobian relationship is considered as

$$\dot{\mathbf{X}} = J\dot{\theta} \tag{3.6.25}$$

Similarly, the other matrices can be obtained one from another space by using the Jacobian in the form

$$C_X(\theta) = J^{-t}\left(C - M(\theta)\, J^{-1}\dot{J}\right)$$
$$G_X(\theta) = J^{-t} G \tag{3.6.26}$$

From this description of the dynamics the influence of the Jacobian is evident and the effect of a singularity can be interpreted as an increase to infinity for the Cartesian inertia matrix M_X, and Coriolis and centrifugal matrix C_X. Consequently, in the direction of the singularity the end-effector should exert large action F to obtain small kinematic motion by feeling large inertia effects.

3.6.2.1 An example

In this section an analysis on the specific system of a general spatial 2R manipulator, whose mechanical and kinematical models are shown in Fig. 3.38, is carried out to show the direct application of Newton–Euler equations with computer simulation and mechanical design purposes. It is also useful to show from a teaching viewpoint the dynamical phenomena and the computational difficulties in deriving a closed-form formulation with direct understanding purposes.

Referring to the scheme of Fig. 3.38 for a two-link spatial manipulator, links 1 and 2 can be identified with the rigid bodies with G_1 and G_2 as centers of mass. Assuming no forces are acting on the link 2, when the payload has been included in its mass characteristics, the dynamic equations can be expressed in the following form

- link 1 of the chain:

$$\mathbf{f}_1 = {}_1\mathbf{R}_2\, \mathbf{f}_2 - \mathbf{F}_1^{in}$$
$$\mathbf{t}_1 = {}_1R_2\, \mathbf{t}_2 + \mathbf{O}_1\mathbf{G}_1 \times \mathbf{F}_i^{in} + \mathbf{O}_1\mathbf{O}_2 \times {}_1R_2\, \mathbf{f}_2 - \mathbf{T}_1^{in} \tag{3.6.27}$$

- link 2 of the chain:

$$\mathbf{f}_2 = -\mathbf{F}_2^{in}$$
$$t_2 = \mathbf{O}_2\mathbf{G}_2 \times \mathbf{F}_2^{in} - \mathbf{T}_2^{in} \tag{3.6.28}$$

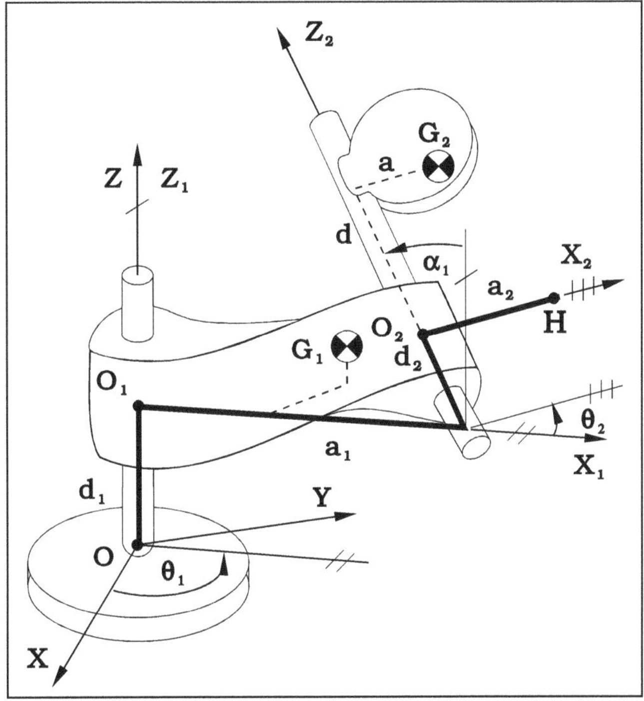

Fig. 3.38: A scheme for a spatial 2R manipulator and its mechanical parameters.

Evaluation of inertia actions on the links requires a previous kinematic analysis.
The kinematics of the chain is computed through the expression of the acceleration $\mathbf{a}_{Gi}$ of the mass center, of the link angular velocity $\boldsymbol{\omega}_i$,and acceleration $\dot{\boldsymbol{\omega}}_i$ as outlined in 3.3. In particular, it yields

$$\mathbf{a}_{Gi} = \dot{\omega}_i \text{ x } \mathbf{O}_i\mathbf{G}_i + \omega_i \text{ x } (\omega_i \text{ x } \mathbf{O}_i\mathbf{G}_i) + \mathbf{a}_i \tag{3.6.29}$$

where the involved angular velocities and accelerations can be calculated as functions of the joint parameters θ_1 and θ_2 in the form

$$\begin{aligned}
\omega_1 &= \dot{\theta}_1 \mathbf{Z}_1 \\
\omega_2 &= {}_2R_1 \omega_1 + \dot{\theta}_2 \mathbf{Z}_2 \\
\dot{\omega}_1 &= \ddot{\theta}_1 \mathbf{Z}_1 \\
\dot{\omega}_2 &= {}_2R_1 \dot{\omega}_1 + {}_2R_1 \omega_1 \text{x} \dot{\theta}_2 \mathbf{Z}_2 + \ddot{\theta}_2 \mathbf{Z}_2
\end{aligned} \tag{3.6.30}$$

The acceleration of the coordinate origins can be given as

$$\begin{aligned}
\mathbf{a}_1 &= {}_0R_1 \, \mathbf{a}_0 \\
\mathbf{a}_2 &= {}_2R_1 \left(a_1 + \dot{\omega}_1 \text{ x } \mathbf{O}_1\mathbf{O}_2 + \omega_1 \text{ x } \omega_1 \text{ x } \mathbf{O}_1\mathbf{O}_2\right)
\end{aligned} \tag{3.6.31}$$

The above-mentioned formulation refers to the standard procedure that is outlined in 3.3. It is worthy of note that the acceleration $\mathbf{a}_0$ on the manipulator base can be considered due to both the gravity field g and the base motion acceleration.
By using Eqs (3.6.27)–(3.6.31) the expression of the equations of motion can be deduced by extracting the Z components τ_1 and τ_2 of $\mathbf{t}_1$ and $\mathbf{t}_2$, respectively.
It is interesting to consider the vectors in Eq. (3.6.30) through their components in the form

$$\dot{\boldsymbol{\omega}}_2 \equiv \left(-\dot{\theta}_1\dot{\theta}_2 s_1 \quad -\ddot{\theta}_1 s_1 \quad \ddot{\theta}_1 c_1 + \ddot{\theta}_2\right)$$

$$\mathbf{a}_2 \equiv \left(k_1 c_2 - k_2 s_2 \quad k_1 c_1 s_2 + k_2 c_1 c_2 - s_1 g \quad k_1 s_1 s_2 + k_2 s_1 c_2 + c_1 g\right)$$

$$\mathbf{a}_{G1} \equiv \left(x_{G1}\dot{\theta}_1^2 - y_{G1}\ddot{\theta}_1 \quad y_{G1}\dot{\theta}_1^2 + x_{G1}\ddot{\theta}_1 \quad g\right) \tag{3.6.32}$$

$$\mathbf{a}_{G2} \equiv \left(-ds_1\ddot{\theta}_1 + a\left(\dot{\theta}_1^2 + \dot{\theta}_2^2 + 2c_1\dot{\theta}_1\dot{\theta}_2\right) \quad a\left(c_1\ddot{\theta}_1 + \ddot{\theta}_2\right) + ds_1\dot{\theta}_1\left(c_1\dot{\theta}_1 + 2\dot{\theta}_2\right) \quad as_1\ddot{\theta}_1 + ds_1^2\dot{\theta}_1^2\right) + \mathbf{a}_2$$

where

$$k_1 = a_1\dot{\theta}_1^2 - d_2 s_1\ddot{\theta}_1$$

$$k_2 = a_1\ddot{\theta}_1 + d_2 s_1\dot{\theta}_1^2 \tag{3.6.33}$$

and $c_1 = \cos\alpha_1$, $s_1 = \sin\alpha_1$, $c_2 = \cos\theta_2$, and $s_2 = \sin\theta_2$.
These expressions describe a highly coupled motion between the joints.
The inertial actions strongly stress this interaction between the joint motions when they are computed in explicit form even as the Z components only. In fact, t_{2z} has all the terms according to Eqs (3.6.3) and (3.6.4), since both $\boldsymbol{\omega}_2$ and $\mathbf{a}_2$ have no-zero components. Additionally, the inertial tensor I_2^{G2} cannot be the principal axis tensor because of the general geometry, since it might further take into account the inertia of the grasped payload according to the formula

$$I_2^{G2} = I_{L2} + I_L \tag{3.6.34}$$

where I_{L2} is the link 2 inertial tensor and I_L is the inertial tensor of the payload. Since I_2^{G2} is referred to a frame with origin in G_2 and parallel to link 2 frame, both I_{L2} and I_L are in general not principal axes tensors, and the products of inertia I_{2xy}, I_{2yz}, I_{2xz} must be considered.
Consequently, a general form of the equations of motion for a two-link spatial manipulator can be deduced by solving with respect to the acceleration of θ_1 and θ_2 the

torque equations (3.6.27) and (3.6.28) in the form:
- link 2

$$A_2\ddot{\theta}_2 = B_2\ddot{\theta}_1 + C_2\dot{\theta}_1^2 + D_2\dot{\theta}_1^2\dot{\theta}_2^2 + V_2 \quad (3.6.35)$$

with

$$\begin{aligned} A_2 &= I_{2zz} - m_2 a_2 \\ B_2 &= I_{2yz}s_1 + I_{2yz}c_1 + m_2 a c_1 (a + d_2 s_1 s_2 + a_1 c_2) \\ C_2 &= I_{2xy}s_1^2 - I_{2xz}s_1 c_1 + m_2 a (d s_1 c_1 + a_1 + d_2 s_1 c_1 c_2) \\ D_2 &= 2 m_2 a d s_1 \\ V_2 &= -t_2 - m_2 a s_1 g \end{aligned} \quad (3.6.36)$$

- link 1

$$A_1\ddot{\theta}_1 = B_1\ddot{\theta}_2 + C_2\dot{\theta}_2^2 + D_2\dot{\theta}_1\dot{\theta}_2 + E_1\dot{\theta}_1^2 + V_1 \quad (3.6.37)$$

with

$$\begin{aligned} A_1 &= - I_{1xy} - I_{1yz} - I_{1zz} - I_{2yy}\, s_1^2 - I_{2yz}\, s_1\, c_1 - B_2\, c_1 - m_1\,(x_{G1}^2 + y_{G1}^2) \\ B_1 &= I_{2yz}s_1 - A_2 c_1 \\ C_1 &= I_{2xz}s_1 \\ D_1 &= I_{2yz}s_1 - A_2 c_1 \\ E_1 &= -I_{2xy}s_1^2 c_1 + I_{2xz}c_1^2 + C_2 c_1 \\ V_1 &= -t_1 + G_2 c_1 + P_2 m_2 \\ P_2 &= (d_2\, s_1 - a_1\, s_2 - d_2\, s_1\, c_2)[a(\dot{\theta}_1^2 + \dot{\theta}_2^2 + 2c_1\, \dot{\theta}_1\dot{\theta}_2) - d\, s_1\, \ddot{\theta}_1 + k_1\, c_2 - k_2\, s_2] + \\ &+ (a_1\, c_1\, c_2 - d_2\, s_1\, c_1\, s_2 - a\, s_1)\begin{bmatrix} a\,(c_1\, \ddot{\theta}_1 + \ddot{\theta}_2) + d\, s_1\, \dot{\theta}_1\, (c_1\, \dot{\theta}_1 + 2\,\dot{\theta}_2) + k_1\, c_1\, s_2 \\ + k_2\, c_1\, c_2 - s_1\, g \end{bmatrix} \end{aligned} \quad (3.6.38)$$

The analytical complexity of Eqs (3.6.35) and (3.6.38), through the coefficient expressions (3.6.36) and (3.6.38), can be ascribed to the inertial actions. Particularly, the link 1 motion seems to be more complicated due to the m_2 mass effect, which has been emphasized in the formulation by means of the definition of P_2. Moreover, m_2 might be expressed as

$$m_2 = m_{L2} + m_L \quad (3.6.39)$$

to take into account the link 2 mass m_{L2} and the payload mass m_L. When link 2 does not carry any load, but interacts with the environment so that force **F** and torque **T** arise, a fictitious load mass and moment of inertia can be considered for instantaneous equivalence in the form

$$m_L = \frac{\mathbf{F} \cdot \mathbf{a}_{G2}}{a_{G2}^2}$$

$$I_L = \frac{\mathbf{T} \cdot \dot{\omega}_2}{\dot{\omega}_2^2} \tag{3.6.40}$$

Therefore, the payload or the environment interaction may influence strongly and directly the dynamical behavior of both joint motions through m_2 and I_2 evaluation by means of Eqs (3.6.34), (3.6.39), and (3.6.40).
The deduced motion equations show a great amount of involved calculations and stress the fact that unavoidable coupling effects are due mostly to the products of inertia of the links. But, acting on the manipulator geometrical parameters the mass effects can be limited or at least forecast, as it is expressed by P_2 definition, without modifying the generality of the geometry. There are great difficulties, on the contrary, in evaluating and reducing the inertial tensors of general structures since the values of products of inertia depend on the link skew geometry when the reference frame is assigned as the link frame.
Looking at Eq. (3.6.35), the link 2 motion is not strongly affected by the results of the motion equations of link 1 as the contrary happens through Eq. (3.6.38). Nevertheless, it might be considered that the link 1 motion directly influences the kinematics of link 2, as the Eqs.(3.6.32) formulation stresses.
A first used practical simplification is achieved by adopting anthropomorphous design solutions, with $s_1 = 0$ or $c_1 = 0$. In this case a great amount of terms in the equations goes to zero, but moreover the structural geometry becomes particular so that the products of inertia vanish and principal axes inertia tensors are obtained.

3.6.3 Lagrange formulation

Dynamics of robotic multibody manipulators can also be formulated by using energy aspects of the motion and the most used approach is the Lagrangian formulation to obtain the equations of motion of the system.
This approach consists in using the Lagrange equation

$$\frac{d}{dt}\frac{\partial L}{\partial \dot{q}_i} - \frac{\partial L}{\partial q_i} = \tau_i \tag{3.6.41}$$

as referring to the i-th link body.
In Eq. (3.6.41) the Lagrangian function L is defined as the sum of the kinetic energy T and potential energy U of the body system as

$$L = T + U \tag{3.6.42}$$

The variable q_i is the so-called 'Lagrangian coordinate' for the motion of the link body and can be identified with the joint variable in the H–D notation as θ_i for revolute joints and d_i for prismatic pairs. The actuator torque τ_i is the non-conservative Lagrangian force that supplies energy for the link motion.
Equation (3.6.41) can be written when it is assumed that other non-conservative forces, such as friction actions, are not applied or negligible. A conservative force can be formulated through its potential function U that in the case of the weight is expressed as the potential U of gravitational field. It is of note that Eq. (3.6.41) gives one dynamic equation for each link body (or better for each d.o.f.) and can be used to formulate equation of motion only. Indeed, the Lagrangian formulation is very useful for simulating the manipulator motion or regulating the actuator force. The main complication can be recognized in the formulation of the Lagrangian function L. Both kinetic and potential energies in Eq. (3.6.41) are expressed with respect to the base frame.
Thus, potential energy of gravity field can be formulated as a sum for the N links of the manipulator as

$$U = \sum_{i=1}^{N} U_i \tag{3.6.43}$$

where

$$U_i = -m_i g \; {}_0h_{Gi} \tag{3.6.44}$$

in which ${}_0h_{Gi}$ is the component of the position vector ${}_0\mathbf{r}_{Gi}$ of the center of mass G_i along the direction of the gravity acceleration with respect to the base frame. It can be computed as

$${}_0\mathbf{r}_{Gi} = {}_0T_i \; {}_i\mathbf{r}_{Gi} \tag{3.6.45}$$

by using the position vector ${}_0\mathbf{r}_{Gi}$ given with respect to the i-th frame.
The kinetic energy is given by

$$T = \sum_{i=1}^{N} T_i \tag{3.6.46}$$

where

$$T_i = \frac{1}{2}\left(m_i \; {}_0v_{Gi}^2 + I_i^{Gi}\omega_i^2\right) \tag{3.6.47}$$

is the kinetic energy of the i-th link body under the motion that is described by translation velocity ${}_0\mathbf{v}_{Gi}$ and angular velocity $\boldsymbol{\omega}_i$ with respect to the base frame.
Then, the generalized force τ_i for the i-th link can be computed by performing the computations that are involved in Eq. (3.6.41) with respect to the corresponding generalized Lagrangian coordinate q_i. The obtained differential expression will be the equation of motion for the manipulator as a function of the variable coordinate q_i. Thus, one can obtain one equation of motion for each variable coordinate only. Of course, those equations of motion can be coupled through the presence of the value of the joint variables and their derivatives in more than one equation.
A step-by-step computational procedure can be outlined by computing:

- the kinematics of the given manipulator in order to obtain position and velocity vectors of the link bodies;
- the kinetic energy through Eq. (3.6.47) and potential energy through Eq. (3.6.44);
- the Lagrangian function of Eq. (3.6.42);
- the derivatives of the Lagrangian function with respect to each Lagrangian coordinate and its time derivative as requested in Eq. (3.6.41);
- the expression of the equation of motion through Eq. (3.6.41).

The above-mentioned Lagrangian formulation can be synthesized in a matrix form to give a general structure of the mathematical model, when a manipulator has several d.o.f.s and several equations of motion will be computed as coupled expressions.
The matrix form for the Lagrangian formulation can be written as

$$\tau = M(\theta)\ddot{\theta} + C(\theta, \dot{\theta}) + G(\theta) \tag{3.6.48}$$

in which $\boldsymbol{\tau}$ is an $N \times 1$ vector of actuator forces for a manipulator with N links; M is the $N \times N$ inertia matrix of the manipulator; C is the $N \times 1$ vector of the centrifugal and Coriolis forces; G is the $N \times 1$ vector of the gravity terms. Vector θ and its derivatives are related to the kinematic status of the joints of the manipulator.
Of course the structure is similar to that of Eq. (3.6.19) that has been obtained in the Newton–Euler formulation, but the computations are different as shown in the following. Writing Eq. (3.6.48) for the i-th actuator the equation of motion for the i-th link can be expressed as

$$\tau_i = \sum_{k=1}^{N} D_{ik}\ddot{\theta}_k + \sum_{k=1}^{N}\sum_{r=1}^{N} H_{ikr}\ddot{\theta}_k\ddot{\theta}_r + G_i \tag{3.6.49}$$

for which

$$D_{ik} = \sum_{j=\max(i,k)}^{N} \mathrm{Tr}\left[\frac{\partial\, {}_0T_j}{\partial\theta_k}\, J_j^{Gj} \left(\frac{\partial\, {}_0T_j}{\partial\theta_i}\right)^t\right] \tag{3.6.50}$$

$$H_{ikr} = \sum_{j=\max(i,k,r)}^{N} Tr\left[\frac{\partial^2 {}_0T_j}{\partial\theta_k \partial\theta_r} J_j^{Gj} \left(\frac{\partial {}_0T_j}{\partial\theta_i}\right)^t\right] \tag{3.6.50}$$

$$G_i = -\sum_{j=1}^{N} m_j \frac{\partial {}_0T_j}{\partial\theta_i} r_{Gi}$$

in which Tr is the trace operator for matrices, max(i,k) gives the maximum value between i and k, and the so-called 'global inertia matrix', including mass effect, is expressed as

$$J_j^{Gj} = \begin{vmatrix} Ixx_j^{Gj} & Ixy_j^{Gj} & Ixz_j^{Gj} & m_j x_{Gj} \\ Iyx_j^{Gj} & Iyy_j^{Gj} & Iyz_j^{Gj} & m_j y_{Gj} \\ Izx_j^{Gj} & Izy_j^{Gj} & Izz_j^{Gj} & m_j z_{Gj} \\ m_j x_{Gj} & m_j y_{Gj} & m_j z_{Gj} & m_j \end{vmatrix} \tag{3.6.51}$$

The above-mentioned expressions involve derivatives of the transformation matrix with respect to joint variables.
The matrix entry D_{ik} is related to inertia effects due to the acceleration in joint motion and gives the coupling effects of inertia actions between the i-th and k-th joints. Since inertia matrix is symmetric and the trace operator gives the same values for a matrix and its transpose, entries of D fulfil the condition $D_{ik} = D_{ki}$.
The entry H_{ikm} is related to inertia action due to the velocity of joint motions. In particular, when k=m, H_{ikk} gives an expression for centrifugal force that is generated by angular velocity of joint k and felt at joint i. When k differs from m, H_{ikm} gives an expression for the Coriolis force that is generated by the velocity in motion of joint k and felt at joint i.
The entry G_i represents the action of the gravity field on the mass of the link.
It is of note that the Lagrangian formulation gives the equations of motion only and does not permit evaluation of other reaction force and torque at the joints. But the Lagrangian formulation also considerably reduces the computational efforts as compared with the Newton–Euler algorithm.

3.6.3. An example

Figure 3.39 shows a scheme for a telescopic RP manipulator. The manipulator is represented in the plane of the motion and the Lagrangian coordinates can be chosen as the joint angle θ_1 and the stroke a_1 of the prismatic pair. H–D notation is shown in Fig. 3.39 and the mass distribution has been described by the point masses m_R and m_L that are the masses for the link associated with revolute and prismatic joints, respectively. In

particular, m_R is the mass of the links including the fixed part of the prismatic joint mass and it is moved by the revolute joint. The mass m_L is the mass that is moved by the prismatic pair.
The kinetic energy associated with the motion of m_R is given by

$$T_R = \frac{1}{2} m_R \, a_1^2 \, \dot{\theta}_1^2 \tag{3.6.52}$$

and potential energy can be expressed as

$$U_R = -m_R \, g \, a_1 \sin\theta_1 \tag{3.6.53}$$

in which the stroke a_1 is assumed variable from the minimum value a_{1min}.
Similarly, the kinetic energy associated with the motion of mass m_L is given by

$$T_L = \frac{1}{2} m_L \left[\dot{a}_1^2 + (a_1 + a_2)^2 \dot{\theta}_1^2 \right] \tag{3.6.54}$$

and the potential energy is expressed

$$U_L = -m_L \, g \, (a_1 + a_2) \sin\theta_1 \tag{3.6.55}$$

Thus, computing the Lagrangian function and its derivatives as outlined in the step-by-step procedure yields to the expression of the equations of the motion as functions of θ_1 and a_1.
Potential energy can be neglected by assuming $g = 0$, when the plane X_0Y_0 of the motion is orthogonal to the gravity field direction.

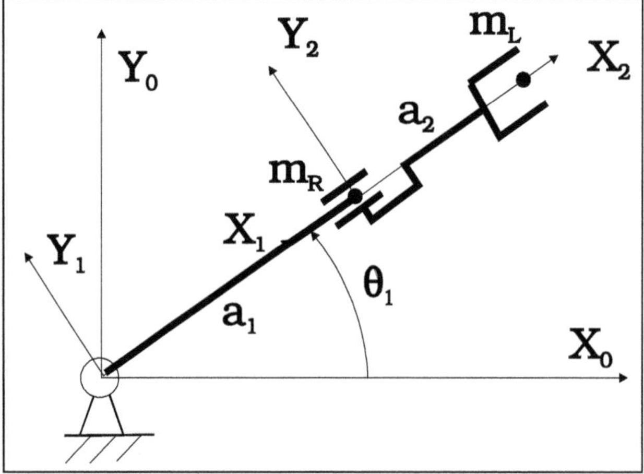

Fig. 3.39: A scheme of an RP manipulator with concentrated masses.

Therefore, the computed quantities are

$$
\begin{aligned}
&\frac{\partial L}{\partial \theta_1} = -\left[m_R a_1 \cos\theta_1 + m_L (a_1 + a_2)\right] g \cos\theta_1 \\
&\frac{\partial L}{\partial a_1} = \left[m_L (a_1 + a_2) + m_R a_1\right] \dot{\theta}_1^2 - (m_L + m_R) g \sin\theta_1 \\
&\frac{\partial L}{\partial \dot{\theta}_1} = \dot{\theta}_1 \left[m_R a_1^2 + m_L (a_1 + a_2)^2\right] \\
&\frac{\partial L}{\partial \dot{a}_1} = -m_L \dot{a}_1 \\
&\frac{d}{dt}\frac{\partial L}{\partial \dot{\theta}_1} = \ddot{\theta}_1 \left[m_R a_1^2 + m_L (a_1 + a_2)^2\right] \\
&\frac{d}{dt}\frac{\partial L}{\partial \dot{a}_1} = -m_L \ddot{a}_1
\end{aligned}
\qquad (3.6.56)
$$

Consequently, the generalized force τ for θ_1 motion can be computed as

$$
\tau = \ddot{\theta}_1 \left[m_R a_1^2 + m_L (a_1 + a_2)^2\right] - \left[m_R a_1 \cos\theta_1 + m_L (a_1 + a_2)\right] g \cos\theta_1 \qquad (3.6.57)
$$

and the generalized force for a_1 motion can be calculated as

$$
f = m_L \ddot{a}_1 - \left[m_L (a_1 + a_2) + m_R a_1\right] \dot{\theta}_1^2 - (m_L + m_R) g \sin\theta_1 \;{}_1^2 \qquad (3.6.58)
$$

It is of note that the dynamic coupling effects shown in Eqs (3.6.57) and (3.6.58) occur when the joints move simultaneously and when the motion is performed in a gravity field.

3.7 Stiffness of manipulators

Stiffness evaluation is concerned with the static response of a manipulator under the action of external force **F** and torque **T**. This response can be analyzed through a determination of compliant displacements and/or stiffness matrix.

Referring to a Cartesian frame fixed on a manipulator base, a general 6×1 vector $\Delta\mathbf{S}$ (Δx, Δy, Δz, $\Delta\phi$, $\delta\theta$, $\Delta\psi$) can be defined for the compliant displacement whose components are the variations Δx, Δy, Δz of Cartesian coordinates of the origin of a moving frame on the manipulator extremity and the variations of $\Delta\phi$, $\delta\theta$, $\Delta\psi$ of angular coordinates of the moving frame.

A stiffness matrix expresses the relationship between the compliant displacement occurring at the moving frame, that is fixed on the manipulator extremity and the acting external wrench **W** whose components are **F** and **T**.

Thus, the stiffness matrix can be defined from the expression

$$\mathbf{W} = \mathrm{K}\ \Delta\mathbf{S} \tag{3.7.1}$$

when compliant displacement occurs as related with linear elastic deformations of the manipulator components. Therefore, a general so-called 'Cartesian stiffness matrix' is defined by Eq. (3.7.1) as a 6×6 matrix.
The stiffness matrix, that is expressed from Eq. (3.7.1), can also be considered as related to the wrench for unit vector of displacement in the range of linear deformation of manipulator components.
Stiffness evaluation can be carried out by a stiffness analysis with numerical simulations and/or experimental tests to determine the stiffness matrix as a characteristic of the manipulator architecture, but for a given configuration.
In fact, since the vector $\Delta\mathbf{S}$ of compliant displacement depends on the manipulator configuration, the stiffness matrix is configuration dependent.
In addition, a reduced number of d.o.f.s can refer to a stiffness matrix with reduced dimension, likewise the Jacobian. Thus, depending on the size of the space in which the actions and displacements can occur, a stiffness matrix can be formulated with less rows and columns.
In the following a general case will be considered to outline the fundamentals for a suitable formulation that can be useful for numerical analysis and experimental determination of the stiffness matrix of a robotic manipulator.

3.7.1 A mechanical model

The stiffness of a robotic multibody system is given by the stiffness of the components of the system that is described by means of a suitable model of elastic response of those components. A main source of compliance for a manipulator can be considered the links, joints, and actuators with their mechanical transmissions. A suitable model can refer to lumped stiffness parameters for each component that can be identified by means of suitable linear and torsion springs. An example of such a model is shown in Fig. 3.40 in which the main compliance source of a PUMA-like robot are identified and represented as linear springs for translation deformations and as torsion springs for angular deformations. A linear spring will take into account mainly axial deformation of a link body and the corresponding spring coefficient k is the lumped parameter that describes the elastic characteristic of the body. Transversal deformations can be considered by using a cantilever model to compute the compliant displacement of an extremity point that can be modeled through a suitable torsion spring. Thus, a torsion spring is identified to consider both angular deformations and transversal compliant displacements of a link body of a manipulator.
A model for stiffness evaluation will be defined by analyzing each manipulator component to identify the spring coefficients and by determining the manipulator configuration with kinematic parameters affecting the compliant response of a manipulator.
In the example of Fig. 3.40 a PUMA-like robot is analyzed for stiffness evaluation by

identifying the lumped springs in a kinematic diagram. Mass distribution can be considered by determining the center of mass for links and actuators.

Thus, a mechanical model for stiffness evaluation will consist of:

- a kinematic diagram;
- an identification of lumped springs and their relative locations in the kinematic diagram;
- a distribution of masses with indication of the positions of the concentrated masses in the link kinematic diagram.

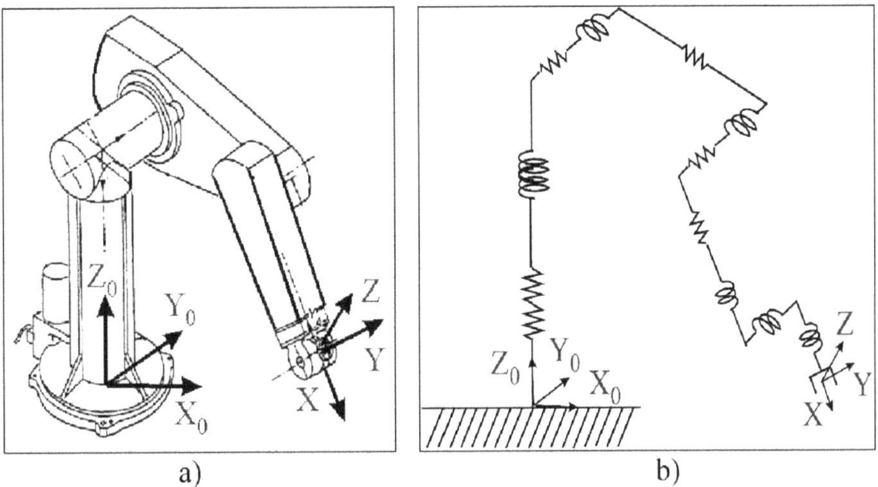

Fig. 3.40: A model for stiffness evaluation of a PUMA-like robot: a) the mechanical design; b) a stiffness model with lumped spring parameters.

3.7.2 A formulation for stiffness analysis

Several approaches can be considered in deriving a formulation for the stiffness matrix of a manipulator.

We shall consider two approaches, namely a so-called 'Jacobian-dependent formulation' and a so-called 'component matrix formulation', in order to outline the problems of deducing a formulation and its limits.

In the first approach, main focus is addressed to the configuration dependence of the stiffness matrix and in the second one the functionality of basic aspects is described by the determination of matrices that will contribute separately to the formulation of the stiffness matrix. These separate contributions refer to the kinematic configuration of a manipulator, the lumped spring parameters of manipulator components, and the displacement kinematics. Indeed, one can find these contributions in any formulation for stiffness matrix, since they are the main source of stiffness response of a manipulator.

In the Jacobian-dependent formulation the stiffness matrix is expressed by using a Jacobian of a manipulator so that the configuration dependence is strongly indicated.

The relationship between the vector $\boldsymbol{\theta}$ of joint coordinates and the vector $\mathbf{S}$ of position and orientation coordinates of a manipulator extremity can be written as

$$\boldsymbol{\theta} = \mathrm{G}\,\mathbf{S} \tag{3.7.2}$$

in which G is a matrix representing the inverse kinematics formulation for a manipulator. A Jacobian J of Eq. (3.7.2) can be defined as

$$\mathrm{J} = \frac{\partial \theta}{\partial x} \tag{3.7.3}$$

so that the compliant displacement $\Delta\mathbf{S}$ can be obtained from

$$\Delta\boldsymbol{\theta} = \mathrm{J}\,\Delta\mathbf{S} \tag{3.7.4}$$

The angular deformation $\Delta\theta$ for joint angles can be evaluated as function of the actuator stiffness from the expression

$$\mathbf{F} = \mathrm{K}_\theta\,\Delta\boldsymbol{\theta} \tag{3.7.5}$$

in which **F** is the vector of forces acting on the links, and K_θ is a diagonal matrix whose non zero entries are the lumped stiffness parameters of the joints, including the actuators. Vector **F** can be computed from the external wrench **W** acting on the manipulator extremity from the expression

$$\mathbf{W} = \mathrm{J}^t\,\mathbf{F} \tag{3.7.6}$$

Thus, the relationship between **W** and $\Delta\mathbf{S}$ can be obtained from Eqs (3.7.2)–(3.7.6) in the form

$$\mathbf{W} = \mathrm{J}^t\,\mathrm{K}_\theta\,\mathrm{J}\,\Delta\mathbf{S} = \mathrm{K}\,\Delta\mathbf{S} \tag{3.7.7}$$

which gives the expression for the stiffness matrix

$$\mathrm{K} = \mathrm{J}^t\,\mathrm{K}_\theta\,\mathrm{J} \tag{3.7.8}$$

Equation (3.7.8) gives a stiffness evaluation of a manipulator that requires the computation of inverse kinematics for determining J. This computation can be very complex, although it is very significant since it determines the effect of singularity on compliant response.
In addition, Eq. (3.7.8) is limited to the case in which the link deformation is negligible, since Eq. (3.7.5) describes the effect of joint and actuator deformations only.
The component matrix formulation is based on a numerical procedure that determines

step-by-step the matrices whose composition gives the stiffness matrix of a manipulator. Thus, the stiffness matrix of a manipulator can be formulated as

$$K = C_F \, K_p \, C_K \tag{3.7.9}$$

in which C_F is a matrix for the force and torque transmission from the manipulator extremity to the joints; K_p is a matrix including the lumped spring parameters; and C_K is a matrix describing the compliant displacement of the manipulator extremity as function of the deformations of the manipulator components.
In particular, C_F transmission matrix can be computed by using a static analysis of a manipulator when external force and torque are considered in order to determine the reaction forces at the joints. Thus, C_F matrix will describe the relationship between an external wrench **W** and the vector **R** of reaction forces at the joints in the form

$$\mathbf{W} = C_F \, \mathbf{R} \tag{3.7.10}$$

and its entries are functions of the dimensional parameters and kinematic variables of manipulator configuration.
The stiffness parameter matrix K_p can be determined as grouping the lumped spring parameters that are identified in a suitable model of a manipulator. Thus, its entries are the lumped spring parameters and the matrix gives the possibility to compute the vector $\Delta\mathbf{v}$ of the corresponding deformations in the manipulator components from the expression

$$\mathbf{R} = K_p \, \Delta\mathbf{v} \tag{3.7.11}$$

The compliant displacement matrix C_F describes the relationship between the deformation vector $\Delta\mathbf{v}$ and the vector $\Delta\mathbf{S}$ of the corresponding compliant displacement of the manipulator extremity in the form

$$\Delta\mathbf{v} = C_K \, \Delta\mathbf{S} \tag{3.7.12}$$

Therefore, combining Eqs (3.7.10) to (3.7.12) one can obtain the expression of Eq. (3.7.9) to compute Eq. (3.7.1) in order to identify the stiffness matrix K of a manipulator in a prescribed configuration for a given wrench.

3.7.3 A numerical example

In Fig. 3.41 a telescopic RP manipulator is modeled for stiffness analysis by identifying the lumped spring parameters K1, K_2, K_T in the kinematic diagram.
The parameter K_1 refers to the stiffness of link a_1 including the radial compliance of the revolute joint and corresponding actuator.
The parameter K_2 describes the stiffness of the linear actuator and the link of length s.

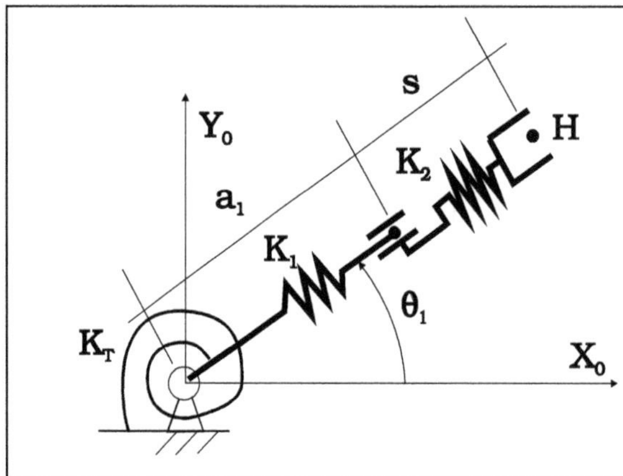

Fig. 3.41: A scheme of a planar RP manipulator and its stiffness-lumped parameters.

The parameter K_T gives the angular compliance characteristic of the revolute joint and corresponding actuator. It can also model the bending deformation of the links when the compliant displacement of link extremity is considered as due to a link rotation because of an angular deformation of the joint.
The stiffness matrix of the telescopic RP manipulator is a 3 × 3 matrix when a planar compliant motion is considered as function of elastic deformations Δa_1, Δs, and $\Delta\theta_1$ due to the spring parameters K_1, K_2, K_T, respectively.
The stiffness matrix can be computed by using the matrix component formulation through a numerical procedure that has been outlined through Eqs (3.7.10)–(3.7.12).
In particular, for the case of Fig. 3.41, by means of a static analysis one can obtain

$$C_F = \begin{vmatrix} c_1 & c_1 & 0 \\ s_1 & s_1 & 0 \\ (a_1+s)s_1 & (a_1+s)c_1 & 1 \end{vmatrix} \tag{3.7.13}$$

The stiffness parameter matrix is computed referring to the model of Fig. 3.41 as a diagonal matrix whose no zero entries are the lumped spring parameters of the manipulator in the form

$$K_p = \begin{vmatrix} K_1 & 0 & 0 \\ 0 & K_2 & 0 \\ 0 & 0 & K_T \end{vmatrix} \tag{3.7.14}$$

The kinematic analysis gives the equations of the direct kinematics that can be used to compute the variations of the extremity coordinates that are the components of $\mathbf{\Delta v}$

vector so that it yields

$$C_K = \begin{vmatrix} c_1 & c_1 & -(a_1+s)s_1 \\ s_1 & s_1 & (a_1+s)c_1 \\ 0 & 0 & 1 \end{vmatrix} \tag{3.7.15}$$

3.8 Performance criteria for manipulators

Performances of robots are evaluated through synthetic measures of main characteristics of their design and operation. As illustrated in 1.3 those performance criteria give an indication of the capability of a robot and they can be used to choose, check, design, or evaluate the feasibility of a robot for a given application or specific task.

In this section the most significant used performance criteria are discussed by looking at their meaning and formulation for numerical computation.

Performances of a robot are evaluated through the magnitude of errors of the performance criteria. Experimental tests and numerical simulation can evaluate those errors.

Experimental tests are carried out to verify performances and operation characteristics with standardized experiments that are generally ruled by ISO standards and codes. They are performed to identify magnitude of errors with respect to an expected behavior that is coded in the programming unit of a robot. Experimental tests are also used for calibration procedures that allow the regulation of the operation and programming of a robot with limited known errors.

Numerical simulation can be useful to determine the effects of modeled parameters and operations on performance criteria. They can also give calibration procedures for experimental tests and they give computational values for structural errors that can be simulated.

A calibration procedure is useful to determine the model parameters for identifying suitable corrections to vanish or limit errors for performance criteria and operation efficiency. In particular, kinematic calibration refers to an experimental determination of H–D parameters of a manipulator chain.

Although numerical simulation can be convenient and useful, evaluation of errors and corresponding calibrations must be performed on a built system. Indeed, they are carried out for each single robot since the reality of the design and operation of each single robot is strongly affected by a source of errors that are hazardous and specific for each single system and its operational status.

Performance errors are an evaluation of the difference between the behavior of a real system and its model. Robotic manipulators are modeled with a deterministic analysis of their design and operation, neglecting phenomena that are considered of secondary importance and therefore are assumed negligible. In addition, models can also be affected by errors for its parameters. The hazardous character of errors is due to phenomena that are variable as function of many parameters of the system and environment in which it operates, even with unpredictable functionality. A typical example of hazardous phenomena is the friction forces that can be considered of influence for manipulator operation and efficiency. Friction forces depend on many variables that cannot be taken into account in a suitably efficient computational model for manipulators, such as for example the material structure or surface profile of a body. Therefore, in the case of evaluation of those hazardous errors a range of errors is usually considered whose bounds are identified by experimental tests in statistical procedures, as outlined in the following sections.

Summarizing the above-mentioned synthetic considerations, one can formulate errors for robot manipulators by referring to a mathematical model for kinematics and dynamics analysis to express the actuation torque in the form

$$\boldsymbol{\tau} = M(\theta)\ddot{\theta} + \Delta M(\theta)\ddot{\theta} + C(\theta,\dot{\theta}) + \Delta C(\theta,\dot{\theta}) + G(\theta) + \Delta G(\theta) + \ldots \tag{3.8.1}$$

in which M, C, and G are given by a deterministic model and ΔM, ΔC, and ΔG are the uncertainties or errors of the parameters of the model; and the final dots indicate the neglected phenomena and actions that in a real system require additional torque action to achieve the desired motion of the manipulator.

The uncertainties and errors ΔM, ΔC, and ΔG can be evaluated as errors in the determination of the structural parameters of a manipulator, i.e. inertia characteristics, link sizes, and position of centers of gravity. But they can also be due to unexpected variation of the model and its parameters as function of the environmental conditions (such as temperature, humidity, and air pressure) and task characteristics (such as configuration of link chain).

Thus, calibration procedures can be useful to determine the bounds of the performance errors in standardized experimental tests that hopefully can describe the majority of situations. But the above-mentioned considerations also make clear the necessity for a regulation and control of the manipulator action during its operation in order to keep to a minimum the difference between a desired manipulation and the real manipulator commanded movement.

For general manipulation purposes main performance errors are usually considered regarding:

- accuracy and repeatability, as related to fundamental motion parameters;
- dynamics and payload capability, as related to task goals;
- compliance response, as related to static behavior.

Additionally, a specific task will require consideration and evaluation of other specific performance errors. They can be identified and determined specifically for each case.

In the following we shall examine the main characteristics and their evaluation since they can be found as general references for robot characterization as listed in 1.3.

3.8.1 Accuracy and repeatability

Accuracy and repeatability errors are important measures of fundamental characteristics of the manipulation capability of a manipulator. In fact, they are always mentioned in technical data of robots, as illustrated in the examples of industrial robots in 1.3. They characterize static performance of a robotic mechanical system in terms of positioning

precision.
Both repeatability and accuracy are usually coded as referring to position capability. But similar approaches can also be developed for a characterization of orientation capability of a manipulator or specifically a wrist. The codes generally refer to industrial robots whose manipulators are open chain architectures, but they can also be extended to parallel manipulators, although some difference in the specific procedures must be examined.
Several definitions of accuracy and repeatability can be used depending also on what specific aspect is considered. For example, one can examine position and orientation capabilities of a manipulator as due to a specific joint or a specific movement of the manipulator.
Therefore, assuming $\mathbf{h}_d$ as a desired position vector and $\mathbf{h}_i$ as the i-th measured reached position, the i-th positional accuracy error e_{ai} can be evaluated as the deviation distance between the two vectors in the form

$$e_{ai} = \left| \mathbf{h}_d - \mathbf{h}_i \right| \tag{3.8.2}$$

This error is a local error and it refers to the i-th measure. But due to the variety and variability of error sources, usually the accuracy error is better evaluated on a statistical basis by repeating several times (some codes suggest 50 times) the reach attempt to the same desired point. Thus, the local accuracy error can be evaluated by using the mean value e_{aM} as

$$e_{aM} = \frac{\sum_{i=1}^{n} e_{ai}}{n} \tag{3.8.3}$$

or the standard deviation e_{aS} as

$$e_{aS} = \sqrt{\frac{\sum_{i=1}^{n} (e_{ai} - e_{aM})}{n-1}} \tag{3.8.4}$$

when the measures are repeated over n reaches.
The local accuracy error refers to a specific point in the workspace and therefore a complete accuracy characterization would need an evaluation of many points and a dense measure mapping in workspace domain. In general, for industrial robots a global accuracy error is given as referring to a value that is the worst evaluation over an indicated workspace area or even total workspace.
It is out of note that the evaluation of accuracy errors need to be specified in terms of conditions at which the robot has operated, and in particular referring to a warming-up with suitable pre-determined movements. Moreover, the accuracy errors also depend on

the path that the manipulator performs to reach the desired position and therefore the movement trajectory must be specified in the procedure for the experimental tests.
In addition, the uncertainty of accuracy errors is also affected by the resolution of robot movements. The resolution of a joint can be described as the smallest joint motion that is felt by the control system. By extension, the resolution of a robot movement is given by the smallest movement that the control system can feel and achieve with the resolution of all the joints and actuators.
Repeatability error can be defined as a measure of the difference between position vectors for reached manipulator positions and the mean of these positions, when the robot is asked to reach a desired position several times in a repeated operation cycle.
Therefore, one can evaluate the repeatability error on a statistical basis by using the difference e_{ri} of the i-th measure from the mean value of the reached positions as

$$e_{ri} = | \mathbf{h}_M - \mathbf{h}_i | \tag{3.8.5}$$

when the mean value e_{rM} is computed as

$$e_{rM} = \frac{\sum_{i=1}^{n} e_{ri}}{n} \tag{3.8.6}$$

The repeatability error can also be evaluated by means of the standard deviation value e_{rS} as

$$e_{rS} = \sqrt{\frac{\sum_{i=1}^{n} (e_{ri} - e_{rM})}{n-1}} \tag{3.8.7}$$

Similar to accuracy errors, repeatability errors can be given for a local or global evaluation. In addition, it depends on the chosen motion to and from the desired position. Even the number of attempts may affect the numerical evaluation, since the distribution of the measure results can be more or less accurate, but other performances, such as heating and friction can modify the robot positioning over the time of the repeated movements. Thus, one can use short-term or long-term repeatability errors when the operation of a robot is evaluated for periods of time, e.g., hours or days, respectively.
An interpretation of the relationship between accuracy and repeatability errors is illustrated in Fig. 3.50 referring to a general view from the Theory of Measures. The accuracy error can be interpreted as a measure of the difference between a desired position and the mean position of the statistical distribution of the measures. The repeatability error can be related to the dispersion size of the repeated measures and it can be represented with a sphere of radius e_{rS} inside which the measures are contained.

The smaller the sphere, the better repeatability the manipulator shows. This corresponds to having a thicker distribution curve in Fig. 3.42 so that there is a high probability that with one measure only one can obtain a measure within the repeatability error bands.

3.8.2 Dynamic characteristics

Dynamic characteristics are related to motion capability of a robot ensuring repeatability and accuracy performances after a manipulator motion.
Thus, dynamic characteristics are generally described by the values of: - maximum velocities; - maximum accelerations; - cycle times; - payload data.

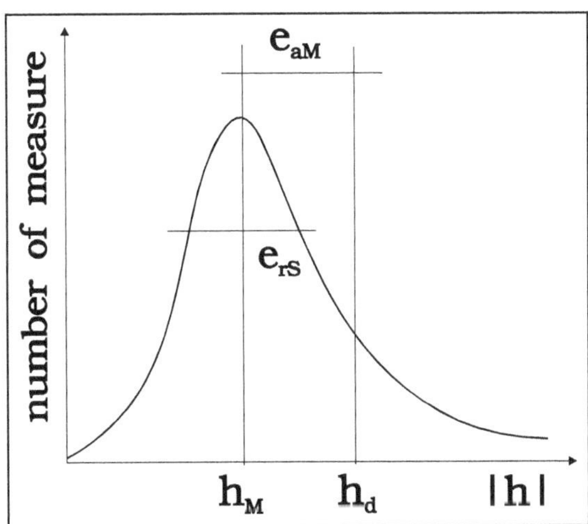

Fig. 3.42: A statistical interpretation of accuracy and repeatability errors.

Maximum velocities and accelerations are indicated with respect to a single joint or a motion axis. In addition, maximum values can be determined also for a specific movement of a manipulator. Indeed, general values can be given in technical data, as shown in 1.3, as those referring to a general tested operation of a manipulator. These maximum values for velocities and accelerations depend on the actuator capabilities, but also the manipulator design and its mass distribution can strongly affect the mechanical efficiency and force transmission for the kinematic properties of the motion operations.
In fact, referring to the case of a mechanical transmission, an actuator torque will move its inertia J_ψ to ψ acceleration and the link inertia J_L to θ acceleration. Assuming a gear transmission from the actuator to the link with transmission ratio tr, the actuator torque can be considered as partitioned into a torque T_ψ for its shaft motion and a torque (tr T_L) for the link motion. Since T_L is the second axis of the transmission system the torque is transmitted with tr transmission ratio. In addition, the efficiency of the transmission can be described by the ratio η between the output power and input power in the transmission and it is named 'mechanical efficiency'. It can be considered to evaluate the real output torque for the link motion as (tr TL / η). Thus, the actuator torque will operate to give

$$\tau = T_\psi + \frac{tr}{\eta} T_L = J_\psi \ddot{\psi} + \frac{tr}{\eta} J_L \ddot{\theta} \tag{3.8.8}$$

It can be interpreted as an actuator acting on an equivalent inertia felt by the actuator as

$$\tau = J_{eq} \ddot{\psi} = \left(J_\psi + \frac{tr^2}{\eta} J_L \right) \ddot{\psi} \tag{3.8.9}$$

Cycle time is a time measure of practical repetitive task capability that a manipulator can effect with its declared performances. Usually the cycle is defined in terms of Cartesian movements and can be described for a specific task. The time is a measure of the general efficiency and it is considered an important evaluation criterion for the productivity of any machinery.
Payload characteristics can be identified by the inertia or equivalent inertia that a manipulator can carry and manipulate by ensuring a safe manipulation that is characterized also by other fundamental manipulator characteristics. Thus, generally a mass or a weight of nominal payload is indicated, since in most of the manipulative applications the size of the manipulated body is small when compared with the manipulator size. This is the case also due to the size of grippers, that are characterized by the weight only.
However, more generally the inertia tensor of the payload body is also significant . Thus, in the most recent robot indications are given moments of inertia about several axes as the maximum values that can be handled by the manipulator. These data can be given on a local basis as function of manipulator configurations, as in the similar case of cranes, or in terms of a global evaluation. Sometimes, payload limits are expressed in terms of torques that can be exerted by a wrist on the extremity flange.
It is of note that those dynamic data characteristics are given by referring to operations that are tested by looking at other robot performances. In the case of those limits being overpassed the robot integrity cannot be damaged but the manipulation performance cannot be ensured as in the declared technical data of a manipulator.

3.8.3 Compliance response

Compliance characteristics of a manipulator strongly affect both positioning and dynamics characteristics of a manipulator. This is the reason because specific attention can be required in evaluating compliance or stiffness response of a manipulator. Tests for compliance characterization can be conveniently carried out by looking at the static behavior. In particular, evaluations of compliant displacements of the end-effector are obtained under given load conditions at different manipulator configurations. Generally, translation displacements are considered with respect to the base frame. Tests are repeated to obtain significant statistical results.
Stiffness evaluation can be obtained from results of a compliant analysis.
Experimental tests are carried out by measuring the compliant displacements along

specific directions. Then, the evaluation can be expressed by giving the magnitudes of the displacements or the equivalent spring coefficients.
Compliant response evaluation depends also on the overshooting characteristic of a manipulator. The overshoot can be defined as the largest distance of over-travel passing a target position. This is due both to elastic characteristics of a manipulator and dynamic performance of the braking motion. The overshoot can be characterized by the above-mentioned largest distance and the time settling the manipulator at the desired configuration within a given band of uncertainty. Compliance response is usually evaluate also after given motions but considering the settling time and overshoot.

3.9 Fundamentals of mechanics of parallel manipulators

A parallel robot can be defined as a closed-loop mechanism whose moving platform is connected to the base by several independent kinematic chains. A 6-d.o.f.s (degrees of freedom) fully parallel manipulator, also called ‘a hexapod’, is a parallel manipulator with 6 legs. The moving platform is controlled by 6 actuators that are placed at some of the joints of the serial kinematic chains constituting the legs.
Parallel manipulators have inherent advantages with respect to serial ones: they can be faster, stiffer, and more precise. They have a high ratio between payload and weight of the structure. This is because the load is distributed among all the legs, and, for some architectures, the legs are only subjected to axial loads. Parallel manipulators can be more precise since they are stiffer because of the parallel architecture of the leg assembly. These robots are faster since in many designs they have the actuation mounted on the base.
On the other hand, parallel robots suffer some drawbacks. They have a limited and complex-shaped workspace. Moreover, rotation and position capabilities of 6-d.o.f.s parallel manipulators are highly coupled, which makes complicated their kinematics, control, and calibration procedure. Another important drawback is related to the presence of singularities inside the workspace. Kinematics is generally difficult to solve, and in fact, computing direct kinematics is a complex problem allowing a great number of multiple solutions.
In the last two decades, the above-mentioned characteristics have addressed the attention of many robot designers and industrial users with the aim of developing robotic parallel manipulators for industrial and non-industrial applications.
One of the main goals of using parallel manipulators is to obtain better performance with respect to serial ones. But some companies failed in their projects to build and commercialize parallel manipulators with high features since they do not know the limits and drawbacks of parallel manipulators.
Furthermore, customers that want to use a parallel manipulator in a robotized cell should have the necessary know-how, since it is well-known that parallel manipulators are more difficult to control and calibrate, if compared with serial ones. Thus, it is very important to consider theoretical aspects when designing a parallel architecture, but without being too distant from practical feasibility.
The paradigm of parallel manipulators is the so-called ‘Gough–Stewart Platform’ as a general 6-d.o.f.s manipulator with the kinematic diagram shown in Fig. 3.43.

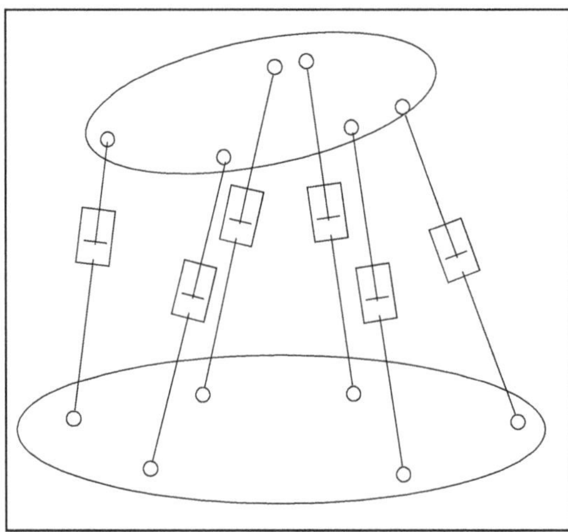

Fig. 3.43: A kinematic diagram for Gough–Stewart parallel manipulators.

The manipulator was first introduced in 1949 as a tire testing mechanism by Gough and later in 1965 as flight-simulator by Stewart. It took almost fifteen years from Stewart's flight simulator to see a parallel mechanism being first combined in an assembly cell. In 1964 Klaus Cappel patented a machine to be used as a motion simulator. Both Cappel and Gough presented a hexapod with a symmetrical octahedral structure.

Only recently machine industry has discovered the potential applications of parallel manipulators even with a reduced number of d.o.f.s. These mechanisms can have simple mechanical structure, simple control system, high-speed performance, low manufacturing and operations cost, and low price. Hence they can be applied widely not only in production lines, performing tasks where full mobility (6-d.o.f.s) is not required due to the nature of the task. For instance, 3-d.o.f.s spatial parallel mechanisms have been presented for telescope applications; flight simulation; beam aiming applications; earthquake simulators; and pointing devices.

At the beginning, the trend of the study on parallel manipulators was to solve problems related to 6-d.o.f.s parallel architectures, such as hexapods, and to develop new architectures mainly with 6-d.o.f.s. Recently, the trends are to better design and consequently use existent architectures, mainly with the reduced number of d.o.f.s.

Today, the motivation behind the great interest in the design of parallel manipulators is related to application areas, such as machine tool centers, robot-assisted surgery, robot surveillance, telescope guidance, and simulation and operation of precise motion.

The first successful parallel manipulator in industry is the Delta robot, which was designed and patented by Prof. Reymond Clavel in Switzerland, Today it is commercialized and used for pick and place applications, as the parallel counterpart of the SCARA robot.

Several parallel manipulators are available in the market as industrial robots for practical applications. In the following a few examples are presented by discussing their design and operation main characteristics with the aim of illustrating the current successful applications but the great potentiality of parallel manipulators.

A hexapod machining center has its roots in the Gough-–Stewart Platform, successfully used as a flight simulator to train pilots. As a multi-axis machine, it is unconventional. A single rigid position is ensured for the mechanism and its position changes as the leg lengths are varied. A general configuration of the hexapod uses 12 attachment points from an octahedral framework, with the spindle pointing down toward the workpiece.
In Fig. 3.44 a hexapod 6-d.o.f.s parallel manipulator is shown in the version that is commercialized by CMW for precision machining of mechanical parts. It has a maximum speed along X and Y axes of 50 m/min and along Z axis of 20 m/min. Acceleration is up to 1 g. The spindle motor has a power of 40 kW. Its workspace volume is approximately a half-sphere having a diameter of 700 mm.
Today Delta architecture can be considered as a traditional parallel manipulator design and it is used in several robots that are built and commercialized by several companies. A very successful example is the ABB version shown in Fig. 3.45.

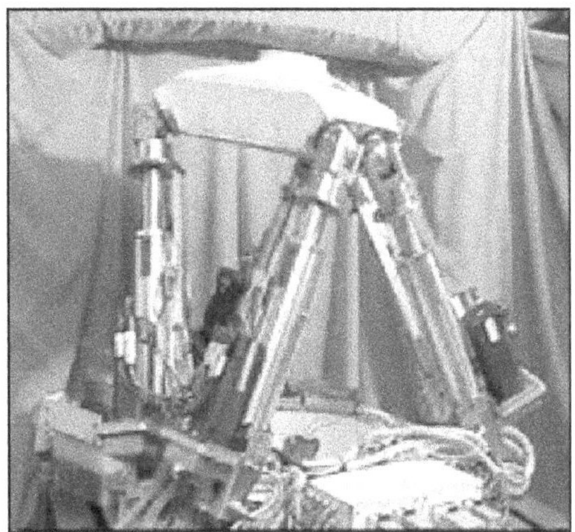

Fig. 3.44: A robot parallel manipulator hexapod commercialized by CMW 300. (Photo taken from CMW webpage).

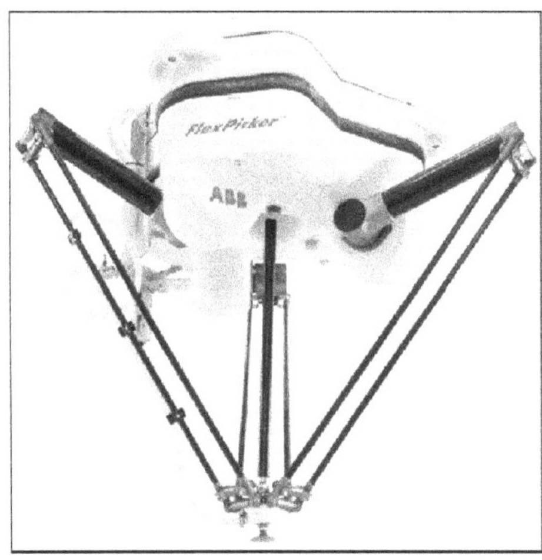

Fig. 3.45: A Delta robot commercialized by ABB. (Photo taken from ABB webpage).

The basic idea of the Delta parallel robot design is the use of the so-called ‘Pi joints’. These joints are constituted of a parallelogram, which allows the output link to remain at a fixed orientation with respect to the input link. Indeed, the use of 3 such parallelograms completely restrains the orientation of the mobile platform that remains only with 3 purely translational d.o.f.s.

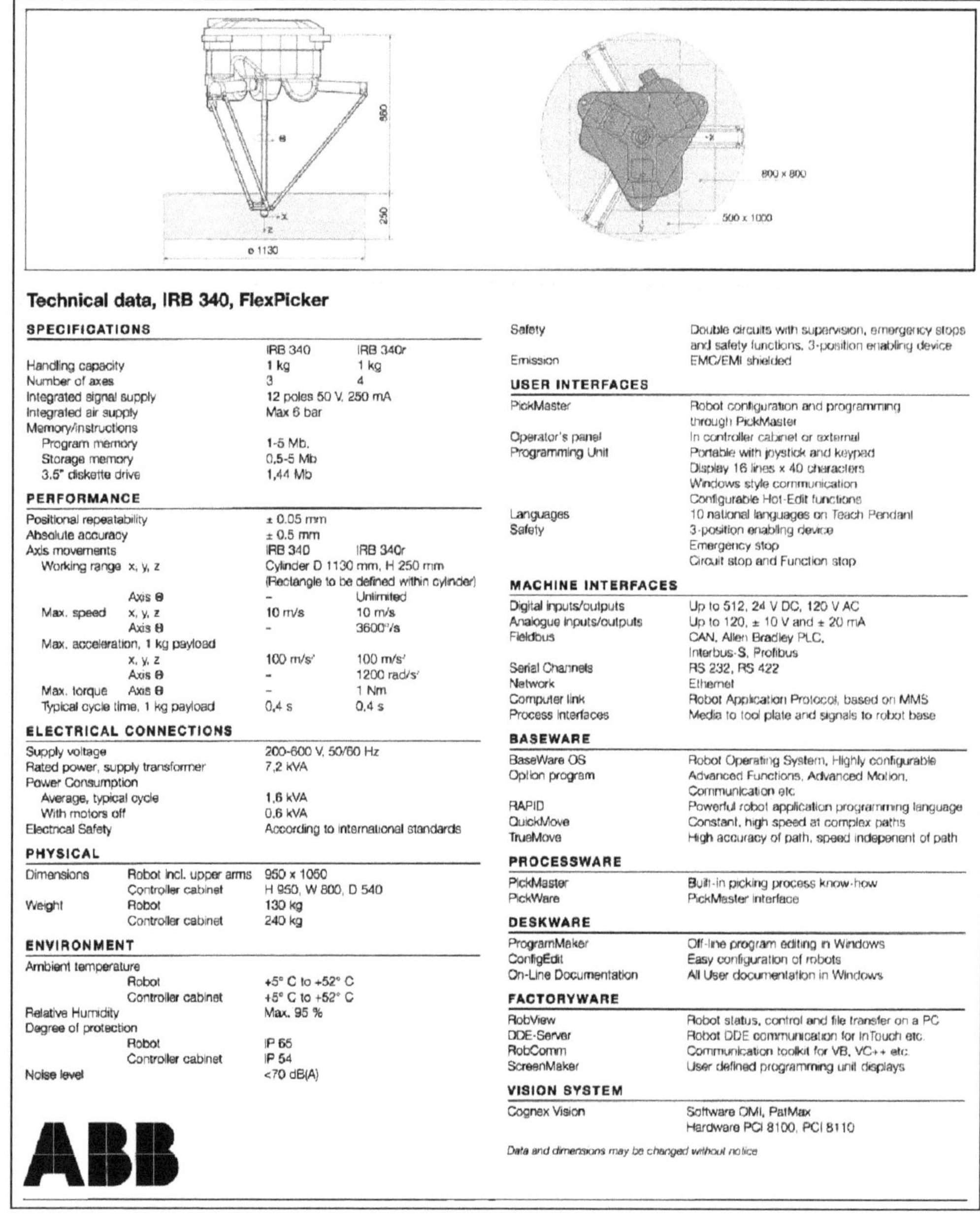

Technical data, IRB 340, FlexPicker

SPECIFICATIONS

	IRB 340	IRB 340r
Handling capacity	1 kg	1 kg
Number of axes	3	4
Integrated signal supply	12 poles 50 V, 250 mA	
Integrated air supply	Max 6 bar	
Memory/instructions		
Program memory	1-5 Mb.	
Storage memory	0,5-5 Mb	
3.5" diskette drive	1,44 Mb	

PERFORMANCE

	IRB 340	IRB 340r
Positional repeatability	± 0.05 mm	
Absolute accuracy	± 0.5 mm	
Axis movements	IRB 340	IRB 340r
Working range x, y, z	Cylinder D 1130 mm, H 250 mm (Rectangle to be defined within cylinder)	
Axis θ	–	Unlimited
Max. speed x, y, z	10 m/s	10 m/s
Axis θ	-	3600°/s
Max. acceleration, 1 kg payload		
x, y, z	100 m/s²	100 m/s²
Axis θ	-	1200 rad/s²
Max. torque Axis θ	–	1 Nm
Typical cycle time, 1 kg payload	0,4 s	0,4 s

ELECTRICAL CONNECTIONS

Supply voltage	200-600 V, 50/60 Hz
Rated power, supply transformer	7,2 kVA
Power Consumption	
Average, typical cycle	1,6 kVA
With motors off	0,6 kVA
Electrical Safety	According to international standards

PHYSICAL

Dimensions	Robot incl. upper arms	950 x 1050
	Controller cabinet	H 950, W 800, D 540
Weight	Robot	130 kg
	Controller cabinet	240 kg

ENVIRONMENT

Ambient temperature		
	Robot	+5° C to +52° C
	Controller cabinet	+5° C to +52° C
Relative Humidity		Max. 95 %
Degree of protection		
	Robot	IP 65
	Controller cabinet	IP 54
Noise level		<70 dB(A)

ABB

Safety	Double circuits with supervision, emergency stops and safety functions, 3-position enabling device
Emission	EMC/EMI shielded

USER INTERFACES

PickMaster	Robot configuration and programming through PickMaster
Operator's panel	In controller cabinet or external
Programming Unit	Portable with joystick and keypad Display 16 lines x 40 characters Windows style communication Configurable Hot-Edit functions
Languages	10 national languages on Teach Pendant
Safety	3-position enabling device Emergency stop Circuit stop and Function stop

MACHINE INTERFACES

Digital inputs/outputs	Up to 512, 24 V DC, 120 V AC
Analogue inputs/outputs	Up to 120, ± 10 V and ± 20 mA
Fieldbus	CAN, Allen Bradley PLC, Interbus-S, Profibus
Serial Channels	RS 232, RS 422
Network	Ethernet
Computer link	Robot Application Protocol, based on MMS
Process interfaces	Media to tool plate and signals to robot base

BASEWARE

BaseWare OS	Robot Operating System, Highly configurable
Option program	Advanced Functions, Advanced Motion, Communication etc
RAPID	Powerful robot application programming language
QuickMove	Constant, high speed at complex paths
TrueMove	High accuracy of path, speed indepenent of path

PROCESSWARE

PickMaster	Built-in picking process know-how
PickWare	PickMaster interface

DESKWARE

ProgramMaker	Off-line program editing in Windows
ConfigEdit	Easy configuration of robots
On-Line Documentation	All User documentation in Windows

FACTORYWARE

RobView	Robot status, control and file transfer on a PC
DDE-Server	Robot DDE communication for InTouch etc.
RobComm	Communication toolkit for VB, VC++ etc.
ScreenMaker	User defined programming unit displays

VISION SYSTEM

Cognex Vision	Software OMI, PatMax Hardware PCI 8100, PCI 8110

Data and dimensions may be changed without notice

Fig. 3.46: A data sheet for an ABB parallel manipulator.

The input links of the 3 parallelograms are mounted on rotating levers by revolute joints. The use of base-mounted actuators and low-mass links allows the mobile platform to achieve accelerations of up to 50 g in experimental environments and 12 g in industrial applications. This makes the Delta robot a perfect candidate for pick and place operations of light objects (from 10 g to 1 kg). Ideally, its workspace is the intersection of three right circular tori. The Delta robots available on the market operate typically in a cylindrical workspace, which is 1 m in diameter and 0.2 m high.
In Fig. 3.46 a technical data sheet is reported for an industrial ABB Delta robot. Characteristics of the parallel manipulator are indicated likewise in the case of traditional serial robotic manipulators in 1.3 with the aim of proposing the parallel architecture even for traditional applications. In fact the data sheet describes the robot characteristics according to the standard codes that are summarized in Table 1.4.
In Fig. 3.47 a parallel manipulator, named 'Hexabot', is shown in a commercialized design for working table applications. The workspace has a diameter of about 305 mm on the plane X-Y and a height of 178 mm. Hexabot has an accuracy of $\pm$ 25 μm and a repeatability of 10 μm for a payload of 91 kg. The Hexel company sells the Hexabot for applications such as: flexible fixturing for welding, polishing, deburringg grinding; transfer lines; precision positioning; high speed pick and place for packaging, loading/unloading; programmable welding; motion simulation for animation and exhibitions; testing equipment.

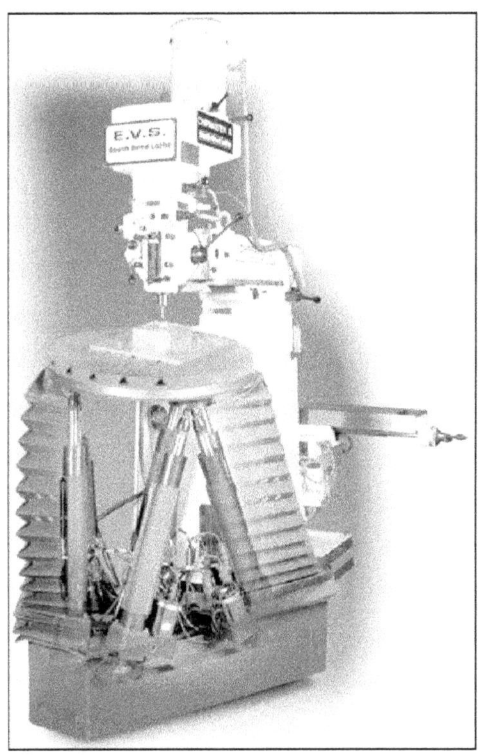

Fig. 3.47: The Hexabot commercialized by Hexel. (Photo taken from Hexel webpage).

In Fig. 3.48 a commercialized 3-d.o.f.s parallel manipulator, named 'Tricept', is shown as a robotic arm carrying a suitable wrist. The SMT Tricept 850 that is shown in Fig. 3.48 is a 5-axis machine with a declared positioning accuracy of ±50μm and repeatability of ±10μm within the workspace. Its maximum velocity is 90 m/min and its acceleration is 2 g.

The F-200iB robot in Fig. 3.49 is a 6-d.o.f.s servo-driven parallel manipulator commercialized by FANUC. It has been designed for manufacturing and automotive assembly. Fanuc F-200iB has a repeatability of ± 0.1 mm. Currently it is used for several applications, primarily spot welding. Maximum declared velocities are 1,500 mm/sec on X and Y directions and 300 mm/sec on Z direction.

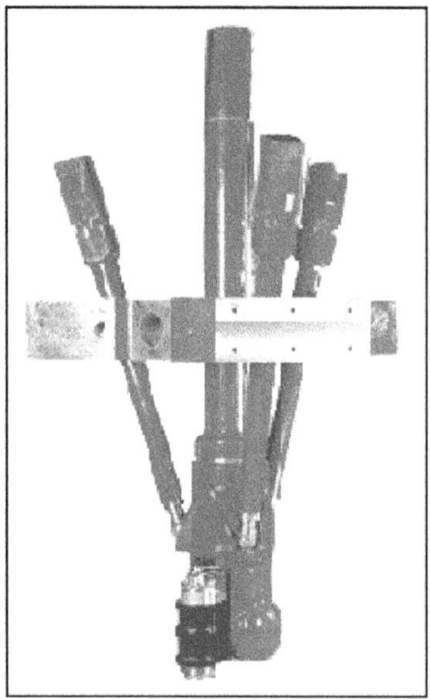

Fig. 3.48: The Tricept 850 commercialized by SMT. (Photo taken from SMT webpage).

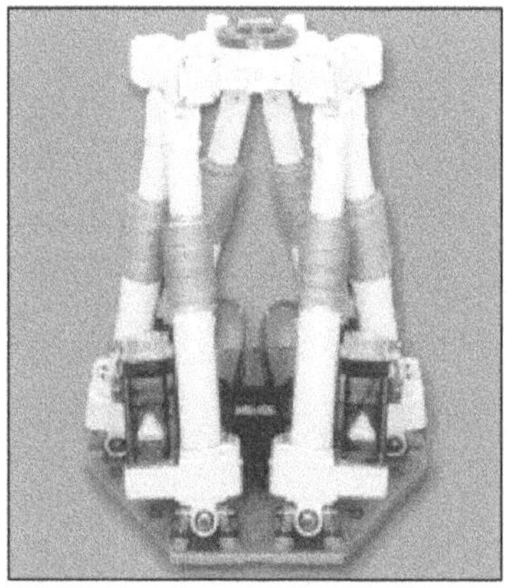

Fig. 3.49: The F-200iB commercialized by FANUC. (Photo taken from FANUC webpage).

The Eclipse parallel manipulator in Fig. 3.50 has been conceived at Seoul National University. It has been designed to overcome the problem of the limited tilting angle of the Gough-–Stewart Platform. It has a tilting angle of 90 degrees. The prototype has 6 d.o.f.s with the maximum speed of 1.2 m/min and the maximum acceleration of 0.1 g. It is used for 5-axis machining.
A prototype of year 2000, shown in Figure 3.50, has been developed to test the Eclipse for rapid machining. The Eclipse parallel manipulator is an illustrative example of the variety of kinematic chain that can be conceived with the parallel architecture, as shown in the kinematic diagram of Fig. 3.50 a). The Eclipse consists of three vertical columns, each of which slides independently on the circular guides. The movement of each column along the circular guide is achieved by a servomotor and a pinion and ring gear transmission. Each column has a carriage, which moves vertically along the linear guide of the column. A fixed length rod is attached to each of the carriages through a pin joint. Each of the 2 R joints is actuated by a servomotor–harmonic drive mechanism. The other end of the fixed length rod is attached to the tool spindle plate via a ball-socket joint. The prototype also has a vertical workpiece spindle unit in the middle span of the machine for fixing an object. For turning and grinding processes, the vertical workpiece spindle can be rotated to achieve spinning of the workpiece. In this case the tool spindle can be fixed to maintain a horizontal posture for turning processes; this is equivalent to a vertical turning process. In general, it can also be rotated. For the cases of milling, drilling, boring, and tapping processes, the vertical workpiece spindle unit is fixed, and only the tool spindle is rotated. It is worthy of note that the tool spindle can move freely from the vertical posture to the horizontal posture and vice versa. Hence, it can realize five-face milling processes within a single set-up of the workpiece.

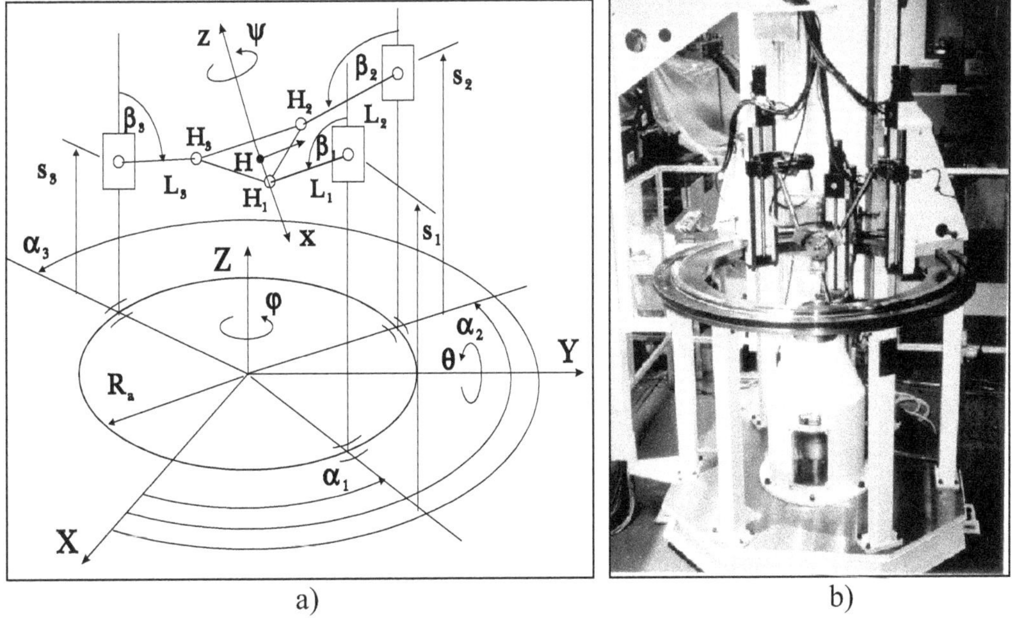

Fig. 3.50: The Eclipse robot developed at Seoul National University: a) a kinematic scheme; b) a first prototype, (Courtesy of Seoul National University).

The class of parallel manipulators with extensible legs has in general the disadvantage that the legs, which are movable, consist of heavy and bulky actuators. Thus, these robots may have problems with high-speed applications. The hexaglide architecture is a hexapod parallel manipulator in which the actuation is placed on the base on linear guides. The main advantage of this kind of mechanism is its dynamics. Another benefit is related to the legs, which are made of thin rods, in order to reduce problems of leg interference.

An hexaglide parallel manipulator is shown in Fig. 3.51 in the version that has been developed at ETHZ, Swiss Federal Institute of Technology, in Zurich. The actuation system is Direct Linear Drives. The workspace of the prototype is 800 × 700 × 400mm. The axes of the prismatic joints along which the centers of the universal joints can be translated are referred to as rail axes. The legs are connected to the moving platform through spherical joints.

Figure 3.52 shows a 3-d.o.f.s parallel architecture named 'PORTYS' that has been developed at IWF, Institute of Machine Tools and Production Technology, at the Technical University of Braunschweig.

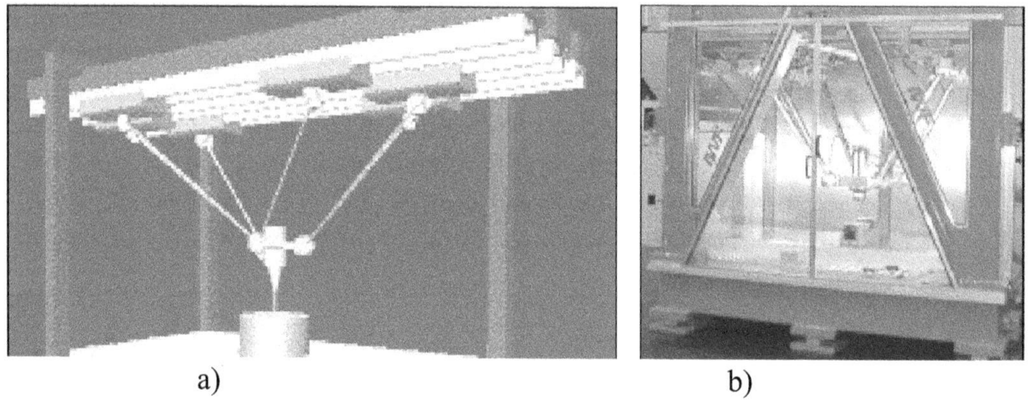

a) b)

Fig. 3.51: The Hexaglide robot developed at ETHZ in Zurich:): a) a kinematic scheme; b) a prototype. (Photos taken from ETHZ webpage

Fig. 3.52: The PORTYS robot developed at IWF in Braunschweig Technical University: (Photo taken from IWF webpage)

Two linear synchronous drives move the structure in X and Y directions. Its main application is pick and place with the following characteristics: payload of 5 kg, maximum velocity of 4 m/s, maximum acceleration of 3 g, repeatability is below 10 μm, and workspace is about 300×630 mm.
Besides the many existent prototypes, great attention is still addressed to modeling, analyzing, and formulating performances of parallel manipulators. Fundamental characteristics can be recognized in:

- complex direct kinematics
- limited complex-shaped workspace
- high precision and stiff performances
- volume occupancy
- large payload
- high speed and acceleration
- singularities

The inverse kinematics problem for parallel manipulators, which is essential for trajectory planning, is quite straightforward. The inverse kinematics describes the mapping from the generalized coordinates describing the position and orientation of the moving platform to the joint coordinates. Furthermore, there is a unique solution and each joint variable may be computed being given the desired pose of the robot.
One can consider as generalized coordinates XYZ the position of the center point P of the moving platform with respect to a fixed frame (Xo Yo Zo), Fig. 3.53, and Euler angles defining the orientation of the moving platform with respect to a fixed frame. A matrix R defines the orthogonal 3×3 rotation matrix defined by the Euler angles, which describes the orientation of the frame attached to the moving platform with respect to the fixed frame, Fig. 3.53. Let A_i and B_i be the attachment points at the base and moving platform, respectively, and d_i the leg lengths. Let $\mathbf{a}_i$ and $\mathbf{b}_i$ be the position vectors of points A_i and B_i in the fixed and moving coordinate frames, respectively. Thus, the inverse kinematics problem can be solved by using

$$A_iB_i = \mathbf{p} + R\mathbf{b}_i - \mathbf{a}_i \tag{3.9.1}$$

to extract the joint variables from leg lengths. The length of the i-th leg can be obtained by taking the dot product of the vector A_iB_i with itself, for i:1,...,6 in the form

$$\mathbf{d}_i^2 = [\mathbf{p} + R\mathbf{b}_i - \mathbf{a}_i]^t [\mathbf{p} + R\mathbf{b}_i - \mathbf{a}_i] \tag{3.9.2}$$

The direct kinematics problem describes the mapping from the joint coordinates to the generalized coordinates. The problem is quite difficult since it involves the solution of a system of nonlinear coupled algebraic equations (3.9.1), and has many solutions that refer to assembly modes.
For the general case with planar base and platform the direct kinematics problem may have up to 40 solutions. A 20-th degree polynomial can be derived whose solutions lead to 40 mutually symmetric assembly modes.

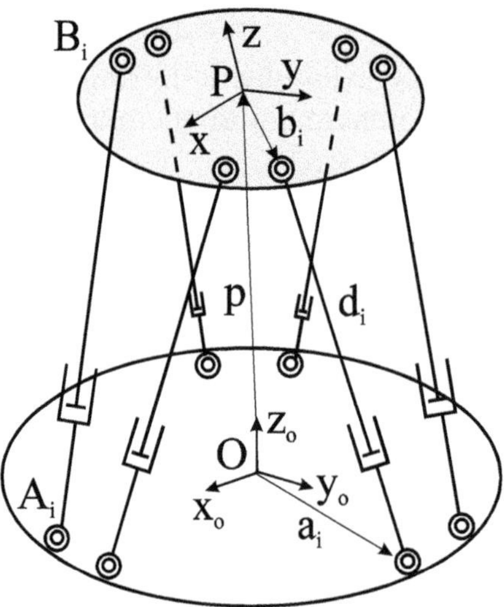

Fig. 3.53: Parameters for the inverse kinematic problem of parallel manipulators.

The position and orientation of the movable plate can be computed by using the geometry of the mechanical design of plates and legs, separately. In particular, the plate designs provide the possibility to deduce specific formulation as function of the typical solutions. Generally, the plate designs are characterized by having the joint connections in the plane plate and typical design configurations are illustrated in Fig. 3.54.

In particular, the general configuration 6S of Fig. 3.54 a) of the Gough–Stewart Platform refers to the case of six spherical joints that are installed on both plates on the vertices of the hexagon. The case of Fig. 3.54 b), that is named '3S plate design', is characterized by the geometry of an equilateral triangle in which vertices are located the spherical joints. This design plate and its modifications is usually used in parallel manipulators with a reduced number of d.o.f.s. Similar is the plate design of Fig. 3.54 c), which is an illustrative example of the many combinations that can be obtained from the design plate of Fig. 3.54 b) by substituting the spherical joints with revolute joints or by a combination of these. Indeed, even the general plate design of Fig. 3.54 a) can give a great variety of manipulator architectures when spherical joints are substituted by revolute and prismatic joints or a combination of these.

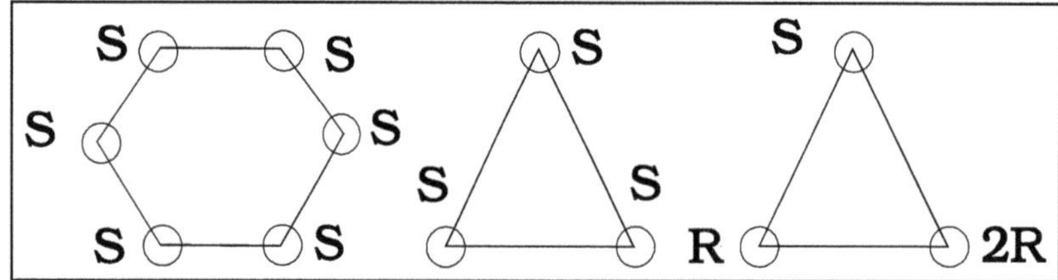

Fig. 3.54: Plate designs for attachments in the plates of parallel manipulators: a)general 6S design in the Gough–Stewart Platform; b) 3S triangular design; c) S-2R-R plate design.

In the case of Fig. 3.54 the spherical joint connections in the plates are located to give a symmetrical design 3S with an equilateral triangle with the joints A_j and B_j (j = 1, 2, 3). Thus, referring to Fig. 3.53, the position vector of the center point P with respect to O $X_0Y_0Z_0$ can be expressed as

$$\mathbf{OP} = \frac{1}{3}\left(\mathbf{OB}_1 + \mathbf{OB}_2 + \mathbf{OB}_3\right) \tag{3.9.3}$$

with

$$\mathbf{OB}_j = R_{\gamma j}\, \mathbf{A}_j\mathbf{B}_j + \mathbf{OA}_j \qquad (j = 1, 2, 3) \tag{3.9.4}$$

where $R_{\gamma j}$ expresses the coordinate transformation from frame $A_jX_jY_jZ_j$ to the fixed frame O $X_0Y_0Z_0$. It can be written as a function of a structural rotation angle γ_j among X_0 and X_j in the form

$$R_{\gamma j} = \begin{vmatrix} \cos\gamma_j & -\sin\gamma_j & 0 \\ \sin\gamma_j & \cos\gamma_j & 0 \\ 0 & 0 & 1 \end{vmatrix} \tag{3.9.5}$$

Frame $A_jX_jY_jZ_j$ has been determined with X_j axis lying on the plane of the fixed plate pointing outward from O; Z_j axis is orthogonal to the plate plane. In addition, vector $\mathbf{A}_j\mathbf{O}$ = r_B $\mathbf{e}_j$ is the distance vector of A_j from O and $\mathbf{e}_j$ with components ($\cos\gamma_j$, $\sin\gamma_j$, 0) is its unit vector. The position vector $\mathbf{A}_j\mathbf{B}_j$ is given with respect to $A_jX_jY_jZ_j$ and x_j , y_j , z_j are its coordinates.
The orientation of the movable plate can be determined by the angles θ, ψ, and ϕ and it can be synthetically expressed through an orientation matrix R_P which can be obtained from the rotation matrices Rot(Z_0, θ), Rot(Y', φ), Rot(Z_0, ψ), as outlined in 3.1, in the form

$$R_P = \begin{vmatrix} -s\theta\, s\psi + s\varphi\, c\theta\, c\psi & -s\theta\, c\psi - s\varphi\, c\theta\, s\psi & c\theta\, c\varphi \\ c\theta\, s\psi + s\varphi\, s\theta\, c\psi & +c\theta\, c\psi - s\varphi\, s\theta\, s\psi & s\theta\, c\varphi \\ -c\varphi\, c\psi & c\varphi\, s\psi & s\varphi \end{vmatrix} \tag{3.9.6}$$

where sθ is for sinθ, cθ is for cosθ, and so on. The Euler angles are generally used for applications as a working table and movable platform device, but any other angles, such as pitch, yaw, and roll can be used to describe the orientation of the movable plate as the orientation of the frame P XYZ with respect to O $X_0Y_0Z_0$.
The calculation of the quantities x, y, z, θ, ψ, and φ is stated as a direct problem of kinematics for parallel manipulators. Particularly, the components x_j ,y_j,and z_j of $\mathbf{A}_j\mathbf{B}_j$

with j = 1, 2, 3 can be solved from the nine Cartesian components of the expressions

$$\mathbf{OB}_j = \mathbf{OP} + r_P R_P \mathbf{e}_j \qquad (j = 1, 2, 3) \tag{3.9.7}$$

where $r_P = r_B$ is the size of the plates and is a constant distance of B_j from P; and $\mathbf{e}_j$ with components $\cos\gamma_j$, $\sin\gamma_j$, 0, is the unit vector of $\mathbf{B}_j\mathbf{P}$ as measured on the movable plate at the initial configuration with respect to the base frame O $X_0Y_0Z_0$. Thus, it is possible to derive through some algebraic manipulations the six independent expressions of the position and orientation in the form

$$x = \frac{y_3 - y_2}{\sqrt{3}} - \frac{r_P}{2}(1 - \sin\varphi)\cos(\psi - \theta)$$

$$y = y_1 - r_P(\sin\psi \cos\theta + \cos\psi \sin\varphi \sin\theta)$$

$$z = \frac{z_1 + z_2 + z_3}{3}$$

$$\psi = \tan^{-1}\left[\sqrt{3}\frac{z_3 - z_2}{2z_1 - z_2 - z_3}\right] \tag{3.9.8}$$

$$\varphi = \cos^{-1}\left[\pm\frac{2}{3r_P}\sqrt{z_1^2 + z_2^2 + z_3^2 - z_1 z_2 - z_2 z_3 - z_1 z_3}\right] \quad ,(z \geq z_1 \Rightarrow +; z < z_1 \Rightarrow -)$$

$$\theta = \sin^{-1}\left[2\frac{y_1 + y_2 + y_3}{3r_P(1 + \sin\varphi)}\right] - \psi$$

When $z_1 = z_2 = z_3$ occur, ψ and θ are undetermined because of their definition, but the sum can be given by the expression for θ.

The first of Eqs (3.9.8) is obtained from an algebraic manipulation of Eqs (3.9.3), (3.9.4), and (3.9.7). Particularly, the expression for x from Eq. (3.9.3) can be written in terms of $x_j \cos\gamma_j$ - $y_j \sin\gamma_j$ by using the coordinate transformation expressed by Eq. (3.9.5). Then the sum $(x_2 + x_3)$ can be obtained by equating the y expression of Eqs (3.9.4) and (3.9.7) and it can be used in the formula for x. The coordinate x_1 of B_1 can be deduced from Eq. (3.9.7) by considering that the x component of $\mathbf{OB}_1$ is equal to $(x + r_B)$. Finally, the expression of x is obtained as in Eqs (3.9.8) by using Eq. (3.9.6) for the involved R_P elements.

The second of Eqs (3.9.8) can be solved from the relation which can be written by equating the y component of Eqs (3.9.4) and (3.9.7) for B_1.

The third of Eqs (3.9.8) can be straightforwardly obtained from the geometry of the articulation points on the movable plate by expressing the z coordinate of Eq. (3.9.3), taking into account that there is no difference for z coordinates between $A_jX_jY_jZ_j$ and O $X_0Y_0Z_0$ frames.
The formula of the fourth of Eqs (3.9.8) can be deduced by using Eqs (3.9.7) to express z_1, z_2, and z_3 . Then, computing (z_3 - z_2) and (z_3 + z_2), it is possible to obtain tan φ by making a suitable ratio of the two expressions.
By using again (z_3 - z_2) and (z_3 + z_2), from Eqs (3.9.7) the fifth of Eqs (3.9.8) can be formulated by squaring and summing them to eliminate ψ and obtain ($\cos^2 \varphi$). The sign ambiguity, which is due to the square root, can be solved by considering the geometry and the model of the device of Fig. 3.54 b) that gives φ greater than $\pi/2$ when B_1 is higher than P , and φ less than $\pi/2$ when B_1 is below P .
The sixth of Eqs (3.9.8) can be obtained starting from Eq. (3.9.3) and expressing the y component of $\mathbf{OB}_j$ through Eq. (3.9.7). The right-hand expression contains x_2 and x_3 that can be written in a suitable form by using the previous computed expression for x and a formula for x_3 , which can be obtained by equating the x component of Eq. (3.9.4) and (3.9.7). In addition, x_1 and y_1 can be expressed through the specific formulation of Eq. (3.9.7) by taking into account that X_0 axis coincides with X_1 as well as Y_0 being parallel to Y_1 . Finally, by expressing the involved R_P elements by Eq. (3.9.6), θ is obtained in the form of the sixth of Eqs (3.9.8).
It is worthy of note once more that the formulation, which is expressed by Eqs (3.9.3), (3.9.4), and (3.9.7), is specific for the plate design of Fig. 3.54 b). Indeed, this algebraic formulation for the parallel manipulators is given by the three articulation points in an equilateral triangle design.
Thus, Eqs (3.9.8) are functions of the input motion of the legs through the Cartesian coordinates y_j, z_j of the articulation points on the moving plate only. Therefore, the kinematics of the leg motion can be studied and designed separately, with the specific task og giving those coordinates. Different formulation can be obtained for different plate designs and leg kinematics.
Many algorithms have been proposed for workspace analysis of parallel manipulators but they are usually very cumbersome and numerically not efficient for repetitive computations in design algorithms.
Since the complete workspace of a 6-d.o.f.s parallel manipulator cannot be represented in a 6D space, different subsets of it are usually considered for a natural intuitive representation. In particular, the constant-orientation workspace, or position workspace, the reachable workspace, and the dexterous workspace can be defined. All of them can be represented in 3D Cartesian space. The position workspace can be defined as the region in the Cartesian space that can be reached by a reference point on the moving platform with a constant orientation. On the other hand, orientation workspace is defined as the set of all attainable orientations of the mobile platform about a fixed point and it can be represented in Cartesian, spherical, or cylindrical frames. Orientation workspace is probably the most difficult characteristic of a parallel manipulator to determine and represent for a natural view.
For evaluating the orientation capabilities of a parallel manipulator Euler angles can be

used. The main advantage of this choice is that they are easy to represent and intuitively interpreted as referred to a wrist motion. The drawbacks are related to the existence of singularity at which the current orientation of the moving platform cannot be represented. A suitable procedure can be proposed by specifically formulating the direct kinematics for specific chains and a general evaluation of workspace performance can be carried out as outlined in the following.
The specific direct kinematics solves the determination of position and orientation of the moving plate as function of the input variables of legs. Numerical algorithms can be deduced or specific formulation can be expressed, Eqs (3.9.8) for the case of Fig. 3.54 b).
A general numerical evaluation of the workspace can be deduced by formulating a suitable binary representation of a cross-section, such as for serial manipulators in 3.1. A cross-section can be obtained with a suitable scan of the computed reachable positions and orientations, once the direct kinematic problem has been solved to give the positions and orientations as functions of the kinematic input variables of the manipulator legs. This is also the case for serial manipulators, where a binary matrix P_{ij} can be defined for a cross-section of the workspace as follows: if the (i, j) grid pixel includes a reachable point, then $P_{ij} = 1$; otherwise $P_{ij} = 0$, as shown in Fig. 3.55.
For example, one can consider a cross-section at a given value of Z-Coordinate, then a point in the grid is indicated as P_{ij}, with i along X axis and j along the Y axis, namely,

$$i = \left[\frac{x + \Delta x}{x} \right]$$

$$j = \left[\frac{y + \Delta y}{y} \right] \qquad (3.9.9)$$

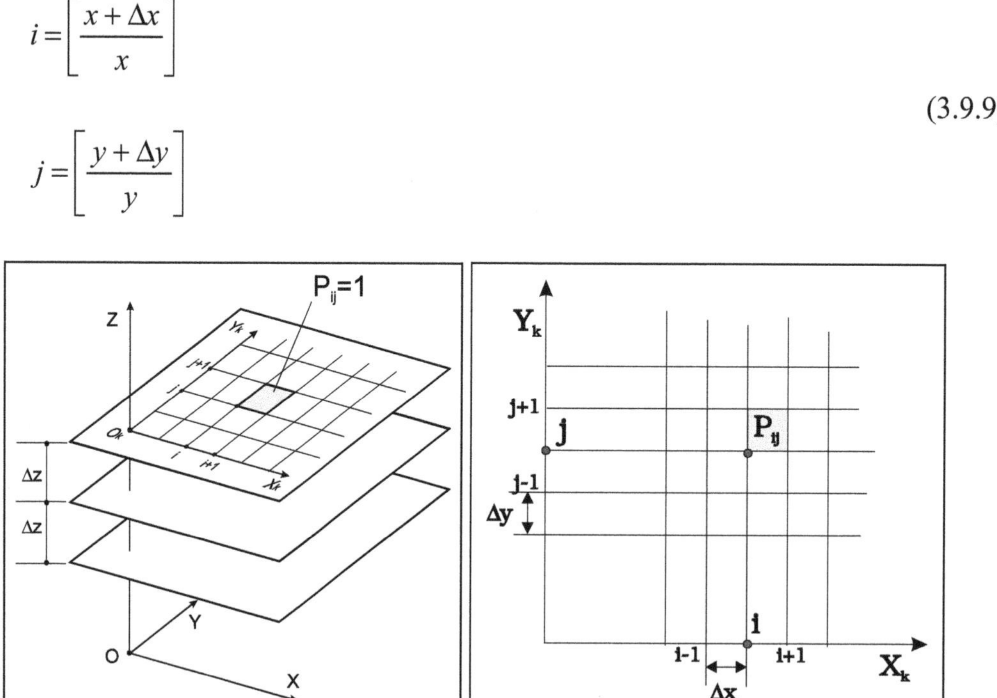

Fig. 3.55: A general scheme for binary representation and evaluation of manipulator workspace.

where i and j are computed as integer numbers. Therefore, the binary mapping for the workspace cross-section can be given as

$$P_{ij} = \begin{cases} 0 \text{ if } P_{ij} \notin \mathrm{W(H)} \\ 1 \text{ if } P_{ij} \in \mathrm{W(H)} \end{cases} \tag{3.9.10}$$

where W(H) indicates workspace region; ∈ stands for 'belonging to' and ∉ is for '"not belonging to'.
In addition, the proposed formulation is useful for a numerical evaluation of the position workspace by computing the cross-sections areas A_z as

$$A_z = \sum_{i=1}^{i_{max}} \sum_{j=1}^{j_{max}} \left(P_{ij}\ \Delta x\ \Delta y\right) \tag{3.9.11}$$

and finally, the workspace volume V can be computed as

$$V = \sum_{z=z_{min}}^{z=z_{max}} A_z\ \Delta z \tag{3.9.12}$$

Similarly, a numerical evaluation of orientation workspace can be carried out by using the formulation of Eqs (3.9.9) to (3.9.12) in order to compute the corresponding performance measures cross-sections areas A_φ , and orientation workspace volume V_φ, when a 3D representation of the orientation capability is obtained by using three angular coordinates as Cartesian coordinates.
Of course, one can use Eqs (3.9.9) to (3.9.12) in order to evaluate any cross-section by properly adapting the formulation to the scanning cross-section plane and intervals.
Stiffness and accuracy of a robotic architecture are strongly related to each other since positioning and orientating errors are due to compliant displacements and construction and assembling errors. The last errors can be evaluated by a kinematic analysis (calibration) by considering uncertainties in the kinematic parameters due to tolerances of construction and assembling of the robotic manipulator chain.
The stiffness properties of a parallel manipulator can be defined through a 6 × 6 matrix that is called 'stiffness matrix K', likewise for the case of serial manipulators in 3.7. This matrix gives the relation between the vector of the compliance displacements $\mathbf{\Delta S}$ = (Δx, Δy, Δz, $\Delta\varphi$, $\Delta\psi$, $\Delta\theta$) occurring at the movable plate when a static wrench $\mathbf{W}$ = (F_X, F_Y, F_Z, T_X, T_Y, T_Z) acts upon it, and $\mathbf{W}$ itself in the form

$$\mathbf{W} = \mathrm{K}\ \mathbf{\Delta S} \tag{3.9.13}$$

The stiffness matrix can be numerically computed by defining a suitable model of the

manipulator, which takes into account the lumped stiffness parameters of links and motors. Figure 3.56 shows a model with lumped spring parameters for a Gough–Stewart parallel manipulator. In this model each spring coefficient k_i (i=1,...,6) refers to the sum of the lumped stiffness parameters of the motor and leg structure along the axial direction of the i-leg link. The coefficients K_{Tif}, and K_{Tim} (i=1,...,6) are the torsion stiffness parameter of the joint on fixed plate or mobile plate for each leg, respectively.
It is well-known that the stiffness matrix is configuration dependent. Therefore, it must be computed as a function of input parameters, which are the strokes of linear actuators or angles of revolute actuators.

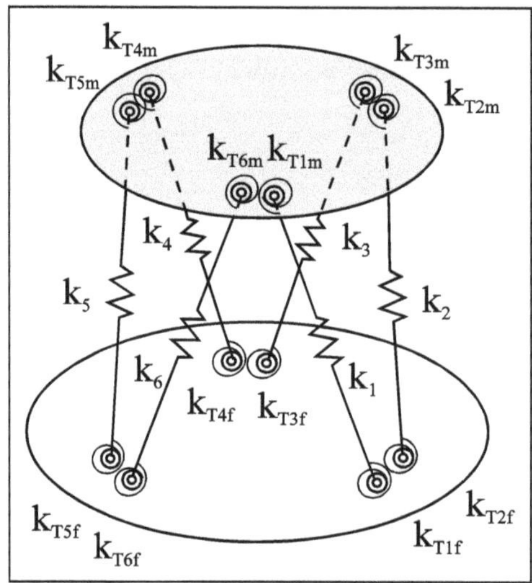

Fig. 3.56: A model for stiffness analysis of parallel manipulators.

A 6×6 stiffness matrix can be derived through the composition of suitable matrices, as in the case of serial manipulators as outlined in 3.7.
The first matrix C_F gives all the wrenches $\mathbf{W}_L$, acting on legs when a wrench $\mathbf{W}$ acts on the movable plate according to the expression

$$\mathbf{W} = C_F \ \mathbf{W}_L \tag{3.9.14}$$

with the matrix C_F representing the force transmission capability of the parallel mechanism.
The second matrix K_p gives the possibility to compute the vector $\Delta\mathbf{v}$ of all the deformations of the legs when each wrench $\mathbf{W}_{Li}$ given by $\mathbf{W}$ acts on the legs according to

$$\mathbf{W}_L = K_p \ \Delta\mathbf{v} \tag{3.9.15}$$

with the matrix Kp grouping the spring coefficients of the deformable components of a

parallel manipulator.
The third matrix C_K gives the vector $\mathbf{\Delta S}$ of compliant displacements of the movable plate due to the displacements of the legs, as expressed in the form

$$\Delta \mathbf{v} = C_K \ \mathbf{\Delta S} \tag{3.9.16}$$

This last matrix can be obtained by analyzing the kinematics of the Gough–Stewart parallel manipulator and considering the variations of kinematic variables due to the deformations and compliant displacements in the legs.
Therefore, the stiffness matrix K can be computed as

$$K = C_F \ K_p \ C_K \tag{3.9.17}$$

Equations (3.9.14) to (3.9.17) for parallel manipulators are similar to Eqs (3.7.9) to (3.7.12) for serial manipulators, but numerical evaluation can be computationally more complicated and laborious.
The stiffness matrix can also be used to compute accuracy performances. In fact, the vector of compliance displacements $\mathbf{\Delta S}$ can be computed by using Eq. (3.9.13) once the matrix K is determined when a static wrench acting on the movable plate is given.
One of the most practical drawbacks of parallel manipulators can be considered the overall size of the mechanical design that occupies considerable volume without any possibility to use it for other equipment. The volume occupancy cannot be avoided since the legs cannot be recovered when they are not used, likewise for the case of serial manipulators, and they act simultaneously to move and keep the platform at certain kinematic status (i.e. firm and stiff position, speedy operation). But this drawback is somehow overpassed by the fact that the simultaneous action of the legs also gives the possibility to carry considerable payloads with high performances in terms of repeatability and precision, high velocity, and acceleration.
The volume occupancy of a mechanical design of a parallel manipulator can be evaluated straightforwardly by looking at the size of the base and mobile plates that are defined by the locations of the joints. The volume can be computed as related to a truncated cone having the base and mobile plates as top and bottom covers.
The large payload can be computed by examining the parallel manipulator as a reticular structure and considering the stiffness properties to ensure a desired positioning accuracy.
Similarly, high velocity and high acceleration can be achieved during the operation of parallel manipulators because of their reticular architecture. Computation of these characteristics attains kinematic analysis and simulation of parallel manipulators with numerical procedures that refers to direct kinematics.
Similarly for the case of serial manipulators, velocity and acceleration analysis of parallel manipulators can be carried out by differentiating the formulation for the position problem, or by computing the Jacobian matrix, or by solving the loop-closure equations that can be formulated for vector circuits in the parallel architecture.

The link interference problem could be solved through an algorithm that allows checking for the workspace in interference-free regions and then to determine in which region the initial assembly modes can be located to obtain the interference-free workspace of the robot. This workspace analysis represents a difficult task even for robots with very simple kinematic chains.
Workspace analysis is an application of direct kinematics through procedures that are based on a search of all attainable positions and orientations of the movable plate, as outlined previously.
Design requirements and operation feasibility can also be focused conveniently on a free singularity condition. In fact, it is desirable to ensure a given workspace volume within which the platform can be movable, controllable, and far enough from singularities.
The instantaneous relationship between the velocity in the Cartesian Space and active joint velocity can be expressed for parallel manipulators as

$$\mathbf{A}\dot{\boldsymbol{\vartheta}} = \mathbf{B}\mathbf{t} \tag{3.9.18}$$

where A and B are two Jacobian matrices of the manipulator; $\dot{\boldsymbol{\vartheta}}$ is the vector of joint rates, and $\mathbf{t}$ is the twist array containing the linear velocity vector $\mathbf{v}$ and the angular velocity vector $\boldsymbol{\omega}$. This expression can be used for a suitable numerical analysis of singularities that can also be useful in experimental tests.
Usually condition of singular configurations can be represented by surfaces in the Configuration Space and they can be obtained by vanishing the determinant of the two Jacobian matrices A and B. In particular, matrix A gives the inverse kinematics singularities; and B gives the direct kinematics singularities. Direct kinematics singularities are inside the workspace and in such configurations a manipulator loses its rigidity, becoming locally movable, even if the actuate joints are locked.
The concept of singularity has been extensively studied and several classification methods have been defined. Singularities for parallel manipulators can be classified into three main groups. The first type of singularity occurs when the manipulator reaches internal or external boundaries of its workspace and the output link loses one or more d.o.f.s. The second type of singularity is related to those configurations in which the output link is locally movable even if all the actuated joints are locked. This is called 'self-motion'. The third type is related to linkage parameters and occurs when both the first and second type of singularities is involved.
Singularities can also be differentiated as configuration singularities, architecture singularities, and formulation singularities. The first type of singularity is related to particular configurations of the manipulator. Architecture singularities are caused by certain architectures; they do not depend on the specific configuration of the manipulator, and they are inherent to the kinematic design of a manipulator. Formulation singularities are due to the adopted model and formulation for numerical analysis and they can be avoided simply by changing formulation method.
In summary, in parallel manipulators singularities arise whenever A, B, or both, become

singular. Thus, a distinction can be made among three types of singularities, by considering Eq. (3.9.18), namely:
- the first type of singularity occurs when A becomes singular but B is invertible, being

$$\det A = 0 \quad \text{and} \quad \det B \neq 0 \tag{3.9.19}$$

- the second type of singularity occurs only in closed kinematic chains and arises when B becomes singular but A is invertible, i.e.

$$\det A \neq 0 \quad \text{and} \quad \det B = 0 \tag{3.9.20}$$

- the third type of singularity occurs when A and B are simultaneously singular, while none of the rows of B vanish. Under a singularity of this type, configurations arise for which the movable plate can undergo finite motions even if the actuators are locked or, equivalently, it cannot resist forces or moments into one or more directions over a finite portion of the workspace, even if all actuators are locked. A finite motion can be very small but even very large to be considered sometimes as an extra d.o.f. for specific manipulator configuration.
In general, in parallel manipulators each leg is connected to the moving platform through articulation points B_i by means of spherical joints, as shown in Fig. 3.54 a). The determinants of A and B are a function of the shape and size of the moving platform, magnitude, and direction of the vectors $\mathbf{d}_i$ of each leg, and unit vectors $\mathbf{e}_i$ of vectors $\mathbf{b}_i$ from the center point of the moving platform to the connecting joints. This can be formulated as

$$\mathbf{B} = \begin{bmatrix} \mathbf{d}_1 & (\mathrm{R}\mathbf{b}_1)\times\mathbf{d}_1 \\ \ldots & \ldots \\ \mathbf{d}_6 & (\mathrm{R}\mathbf{b}_6)\times\mathbf{d}_6 \end{bmatrix} \tag{3.9.21}$$

$$\mathbf{A} = \begin{bmatrix} \mathbf{d}_1 \times \mathrm{R}\, r_P\, \mathbf{e}_1 & 0 \\ \ldots & \ldots \\ 0 & \mathbf{d}_6 \times \mathrm{R}\, r_P\, \mathbf{e}_6 \end{bmatrix} \tag{3.9.22}$$

in which R is the rotation matrix of the moving platform with respect to the base frame. Matrices A and B can be given in a specific form for any specific parallel architecture through a suitable analysis.
For most parallel manipulators the Jacobian matrix is posed as dependent and non-isotropic. Consequently, performances such as rigidity, velocities, and forces, which can be expressed as functions of the Jacobian are pose-dependent. The above-mentioned considerations make the design of a parallel manipulator a complex task.
The high nonlinearity of the formulation of basic characteristics of parallel manipulators

makes a formulation for design purposes a complex mathematical problem.
The strong influence of fundamental characteristics to each other will require considering those formulations in one design model and nowadays this can be achieved with numerical efficiency by using optimization techniques.
Thus, an optimum design of parallel manipulators can take into account requirements and optimization criteria concerned with basic performances of workspace, stiffness, and singularity-free condition, since they are recognized as basic characteristics for designing and applying parallel manipulators. Therefore, an optimum design can be formulated as an optimization problem with a multi-objective function that takes into account the above-mentioned variety of performance criteria. A multi-objective function can be used also to consider design requirements and constraints when they are of relevant significance for design and application solution.
For the optimum design of parallel manipulators a multi-objective optimization problem can be formulated as the minimization of a function vector **f** in the form

$$\min \mathbf{f}(\mathbf{X}) = \left[\mathrm{f}_W(\mathbf{X}), \mathrm{f}_S(\mathbf{X}), \mathrm{f}_K(\mathbf{X})\right] \tag{3.9.23}$$

Subjected to

$$\begin{aligned} &\mathbf{g}\,(\mathbf{X}) < \mathbf{0} \\ &\mathbf{h}\,(\mathbf{X}) = \mathbf{0} \end{aligned} \tag{3.9.24}$$

where **X** is the vector of design variables; each component f_W, f_S, f_K of the objective function is an expression of a design optimum criterion for workspace, singularity condition, and stiffness, respectively; each component g_k (k=1,...,m) describes an inequality design constraint; and each component h_l (l=1,..., n) describes an equality design constraint. Constraints can be formulated through the functions **g** and **h** to express design requirements but also limitations for the design variables and objective functions.
Referring to workspace, singularity condition, and stiffness of parallel manipulators one can express one or more measures for each of them in the components of **f**.
In general, one can evaluate workspace by means of its volume for both position and orientation capabilities; stiffness can be considered by means of the corresponding measure of compliant displacements or a stiffness matrix measure; singularity condition can be formulated by means of the Jacobian matrices measures or related indices.
Regarding the design variables, in addition to the dimensional parameters of parallel manipulator architectures, a designer must take into account stiffness properties of the links of the architecture and even the mobility range of the joints, since both stiffness and singularity properties are also configuration-dependent.
There are a number of alternative methods to solve numerically a multi-objective optimization problem that is formulated with Eqs (3.9.23) and (3.9.24).
Nevertheless, in a simplified approach the multi-objective problem can be conveniently solved by using a single objective function F(**X**) and standard constrained optimization methods when the number N of objective components is limited. For example, a

weighted sum strategy can be used conveniently. Therefore, the optimal design problem of Eqs (3.9.23) and (3.9.24) can be formulated as minimizing a weighted sum F(**X**) of the objective functions in the form

$$\min F(\mathbf{X}) = \sum_{\mathbf{i}=1}^{\mathbf{N}} w_i f_i(\mathbf{X}) \tag{3.9.25}$$

in which w_i are weighting factors, which may correspond or be chosen referring to the relative importance of the objectives. Thus, a designer can control the design process and gives alternative inputs depending on the chosen values for the weighting factors.
By using the above-mentioned considerations and numerical evaluations of basic performance of workspace, stiffness, and singularity condition, objective functions can be formulated as

$$\begin{aligned}
f_{W1}(X) &= \left| 1 - \frac{V_{pos}}{V_{posg}} \right| \\
f_{W2}(X) &= \left| 1 - \frac{V_{or}}{V_{org}} \right| \\
f_{K1}(X) &= \left(1 - \frac{|\Delta x|}{|\Delta x_g|} \right) + \left(1 - \frac{|\Delta y|}{|\Delta y_g|} \right) + \left(1 - \frac{|\Delta z|}{|\Delta z_g|} \right) \\
f_{K2}(X) &= \left(1 - \frac{|\Delta \varphi|}{|\Delta \varphi_g|} \right) + \left(1 - \frac{|\Delta \psi|}{|\Delta \psi_g|} \right) + \left(1 - \frac{|\Delta \theta|}{|\Delta \theta_g|} \right) \\
f_S(X) &= \min(-\det \mathbf{B}_{nor})
\end{aligned} \tag{3.9.26}$$

in which | | represents the absolute value of a quantity, subscript g indicates given values, subscript pos is for position, subscript or is for orientation, and subscript nor indicates a normalized dimensionless form of matrix B.
In Eq. (3.9.26) the objective functions have been formulated by using procedures of analysis for fundamental characteristics, as outlined previously. In particular, a volume evaluation of the workspace has been considered in terms of position value Vpos and orientation value Vor. Stiffness properties have been characterized by computing linear and angular components of compliant displacements. Singularity-free condition has been imposed through a numerical evaluation of Jacobian matrix B.
The above-mentioned characteristics can also have a practical interpretation for assembling applications that can be characterized through suitable values of workspace, precision error, velocity performance as given by values of workspace volume,

compliant displacements, and velocity mapping.

3.9.1 A numerical example for CaPaMan (Cassino Parallel Manipulator)

This example illustrates the fundamental concepts for analysis and design of parallel manipulators as applied to CaPaMan (Cassino Parallel Manipulator), a specific 3-d.o.f.s parallel manipulator that has been developed at LARM Laboratory of Robotics and Mechatronics in Cassino.

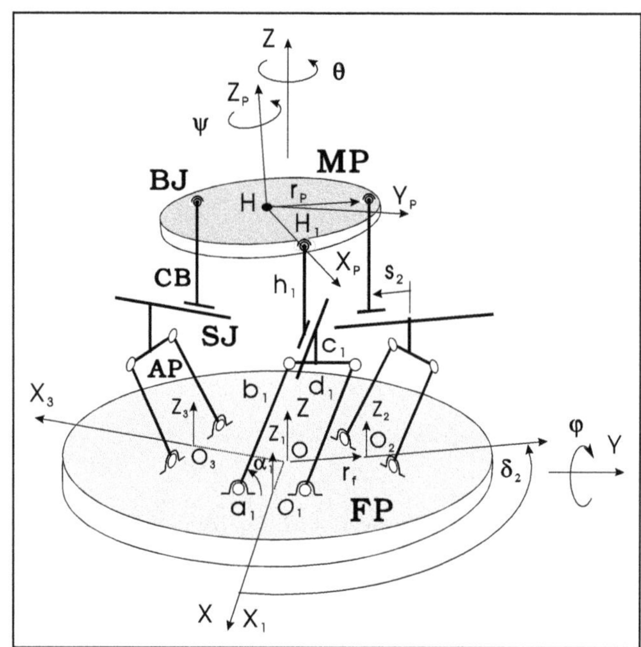

Fig. 3.57: Kinematic chain and design parameters of CaPaMan.

Fig. 3.58: A built prototype of CaPaMan at LARM in Cassino.

The kinematic chain of CaPaMan is illustrated in Fig. 3.57 and a built prototype that is available at LARM is reported in Fig. 3.58.
Figure 3.57 shows the kinematic chain of the parallel mechanism, which is composed of a movable plate MP that is connected to a fixed plate FP by means of three-leg mechanisms. Each leg mechanism is composed of an articulated parallelogram AP whose coupler carries a prismatic joint SJ, a connecting bar CB which transmits the motion from AP to MP through SJ, and a spherical joint BJ which is installed on MP. Thus, CB may translate along the prismatic guide of SJ keeping its vertical posture while the BJ allows the MP to rotate in the space.
Each AP plane is rotated $\pi/3$ with respect to its neighbor so that it lies along the vertices of equilateral triangle geometry to give symmetry properties to the mechanism.
Particularly, links of a k leg mechanism (k = 1, 2, 3) are identified through: a_k , which is the length of the frame link; b_k , which is the length of the input crank; c_k , which is the length of the coupler link; d_k , which is the length of the follower crank; h_k , which is the length of the connecting bar.
The kinematic variables are: α_k , which is the input crank angle; s_k , which is the stroke of the prismatic joint.
Finally, the size of MP and FP are given by r_p and r_f, respectively, where H is the center point of MP, O is the center point of FP, H_k is the center point of the k-th BJ, and O_k is the middle point of the frame link a_k.
Platform MP is driven by the three leg mechanisms through the corresponding articulation points H_1, H_2, H_3. A motor on the input crankshaft can actuate each leg mechanism so that the device is a 3-d.o.f.s spatial mechanism.
In addition, the kinematic architecture has been useful to propose an alternative low-cost solution for the leg design, as shown in Fig. 3.59. The versions CaPaMan 2 in Fig. 3.59 a) and CaPaMan 3 in Fig. 3.59 b) have been obtained from the kinematic design of CaPaMan in Fig. 3.57 by substituting the prismatic joint with a suitable revolute joint and the four-bar linkage with a slider-crank mechanism, respectively.

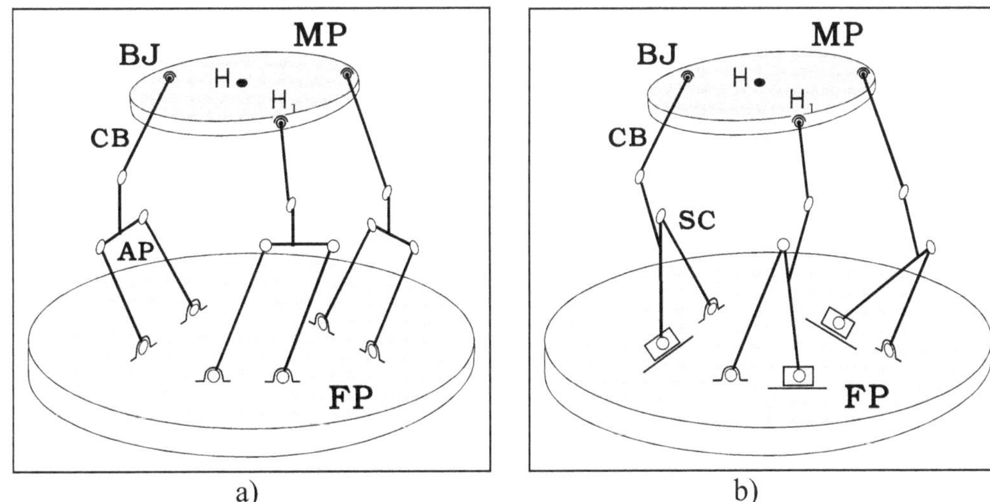

a) b)

Fig. 3.59: Kinematic design for: a) CaPaMan 2; b) CaPaMan 3.

In order to describe the motion of MP with respect to FP, a world frame OXYZ has been assumed as fixed to FP and a moving frame $HX_PY_PZ_P$ has been fixed to MP, as shown in Fig. 3.57. Particularly, OXYZ has been fixed with Z axis orthogonal to the FP plane, X axis as coincident with the line joining O to O_1 , and Y axis to give a Cartesian frame. The moving frame $HX_PY_PZ_P$ has been fixed in an analogous way to the movable plane MP with Z_P orthogonal to the MP plane, X_P axis as coincident to the line joining H to H_1 and Y_P to give a Cartesian frame.
A prototype of Fig. 3.57 has been built as in Fig. 3.58 at LARM, the Laboratory of Robotics and Mechatronics in Cassino, with the dimensional size of Table 3.1.
The linkage geometry of the parallelograms with commercial prismatic joints on the couplers can be conveniently exploited for an easy modular mechanical design and assembly, which may allow easy changes in the leg dimensions.
The kinematic feasibility of CaPaMan has been studied and characterized in terms of an analytical formulation for direct and inverse kinematics, which has been obtained as closed-form expressions because of the geometry of the parallel assembly and the parallelograms in the legs.

Table 3.1 – Sizes and motion parameters of the built prototype for CaPaMan, Fig. 3.58.

$a_k = c_k$ (mm)	$b_k = d_k$ (mm)	h_k (mm)	$r_P = r_f$ (mm)	α_k (deg)	s_k (mm)
200	80	100	250	45 ; 135	-50 ; 50

The platform design of type 3S in Fig. 3.54 b) gives the possibility to deduce a closed-form formulation of Eqs (3.9.8) so that the motion of the movable plate is determined when the motion of the three attachment points H_k is known as function of the input motion of the legs. Equations (3.9.8) for CaPaMan have been expressed as a function of H_k coordinates y_k , z_k (k=1, 2, 3), which can be given by the geometry of the legs from the input variable α_k as

$$y_k = b_k \cos \alpha_k$$
$$z_k = b_k \sin \alpha_k + h_k \tag{3.9.27}$$

The x_k coordinate coincides with the sliding displacement s_k in the corresponding prismatic joint and it can be computed as

$$x_k = \frac{x}{\cos \gamma_k} + r_P \left(R_{11} + R_{12} \tan \gamma_k\right) + y_k \tan \gamma_k - r_P \tag{3.9.28}$$

where R_{11} and R_{12} are elements of the rotation matrix between XYZ and xyz given from the expression of the rotation matrix of Eq. (3.9.6)
Furthermore, the feasibility of motion performances requires that the displacement

range for x_k must be within the mechanical limits s_{kmax} of the prismatic joint. This can be easily verified by checking that

$$-s_{k\,max} \le x_k \le s_{k\,max} \tag{3.9.29}$$

when the prismatic guide is installed symmetrically with respect to the coupler link.
The input motion can be given by cubic polynomial according to considerations in 3.2.2 as

$$\alpha_k = \alpha_{k0} + \frac{3\,\Delta\alpha_k}{t_{kf}^2}\,t^2 - \frac{2\,\Delta\alpha_k}{t_{kf}^3}\,t^3 \tag{3.9.30}$$

where α_{k0} and $\Delta\alpha_k$ are the initial value and the working interval for the input angle of the k-th leg, respectively; t is the time variable and t_{kf} the time interval within which the motion is prescribed. The cubic polynomial (3.9.30) can be usefully used for simulation and control purposes since motor units for robotic applications are usually controlled both in position and velocity.
Thus, Eqs (3.9.8) to (3.9.27) can be used to compute the direct position capability of CaPaMan when dimensional sizes and motion parameters are given. In particular by using the data of Table 3.1 for the built prototype, Figs 3.60 to 3.62 have been obtained from a computer analysis with the scanning process and binary representation of Eqs (3.9.9) to (3.9.12).
Figure 3.60 shows the position workspace of CaPaMan with its characteristic geometry whose peculiarities can be recognized as a bulk umbrella-shaped volume with an hexagonal middle cross-section and six separate lower vertices. The orientation workspace is reported in Fig. 3.61 as a result of the numerical scanning process. It is expressed in terms of the Euler angles that are used as in the Cartesian representation of Fig. 3.61, likewise the Fig. 3.60 for position workspace. The CaPaMan orientation workspace can be characterized as a bulk volume of parallelepiped shape with several voids that however do not exclude continuous orientating path for the movable plate.
In Fig. 3.62 an alternative practical representation of the orientation workspace has been computed as related to the position workspace capability. It shows the orientation capability for each reachable position in practical maps.
The direct analysis is also useful for motion planning as function of the input variables of the legs. Because of the closed-form formulation of direct kinematics, indeed, CaPaMan can be easily programmed for motion planning by using the expressions of Eq. (3.9.27) for input variables.
Illustrative practical examples are reported in Figs 3.63 and 3.64. In particular, Fig. 3.63 regards a practical application for which MP is moved for a vertical path of H. The trajectory has been obtained by a simultaneous input motion of the leg parallelograms. On the contrary, by using different input motions for the leg parallelograms a complex motion for MP can be performed, as in the case of Fig. 3.64.

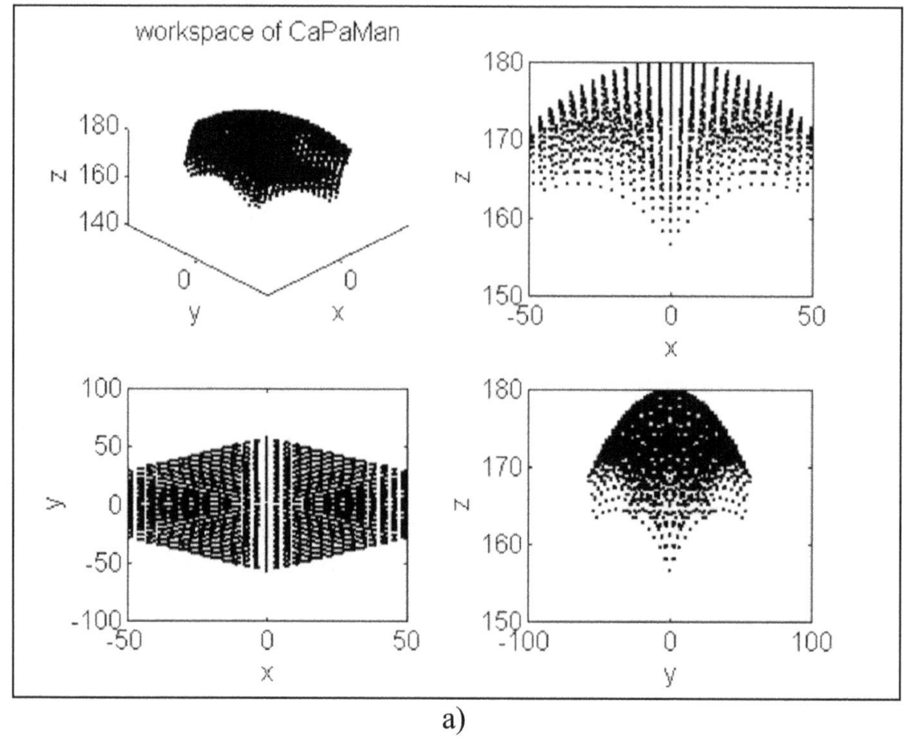

a)

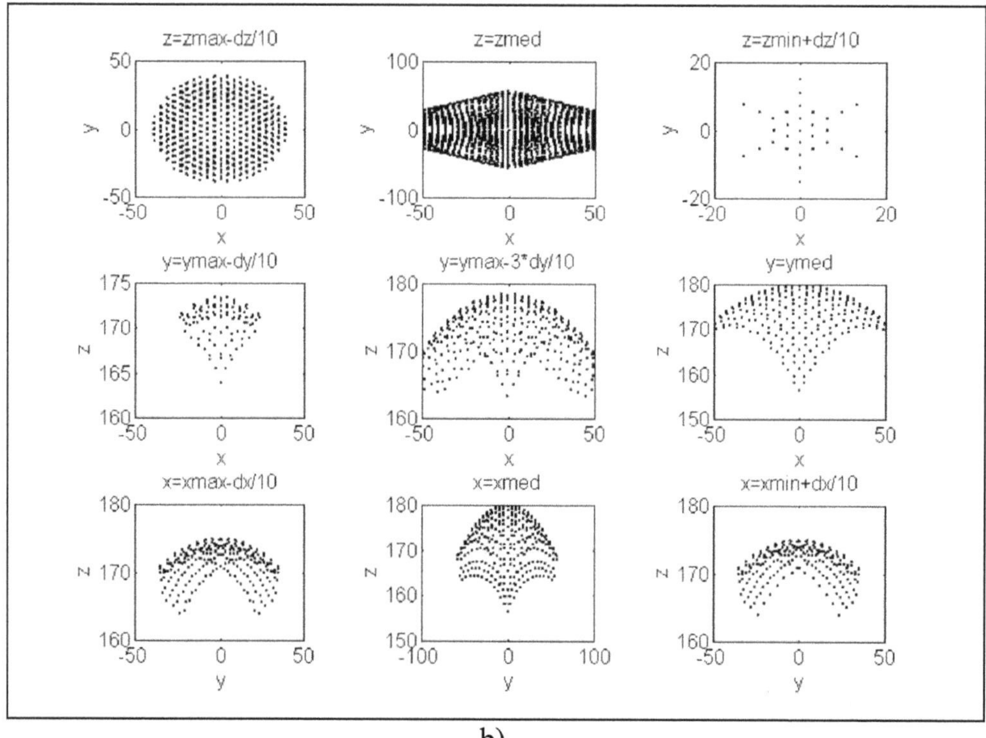

b)

Fig. 3.60: Position workspace of CaPaMan prototype of Fig. 3.58 and Table 3.1: a) a 3D view; b) cross-sections.

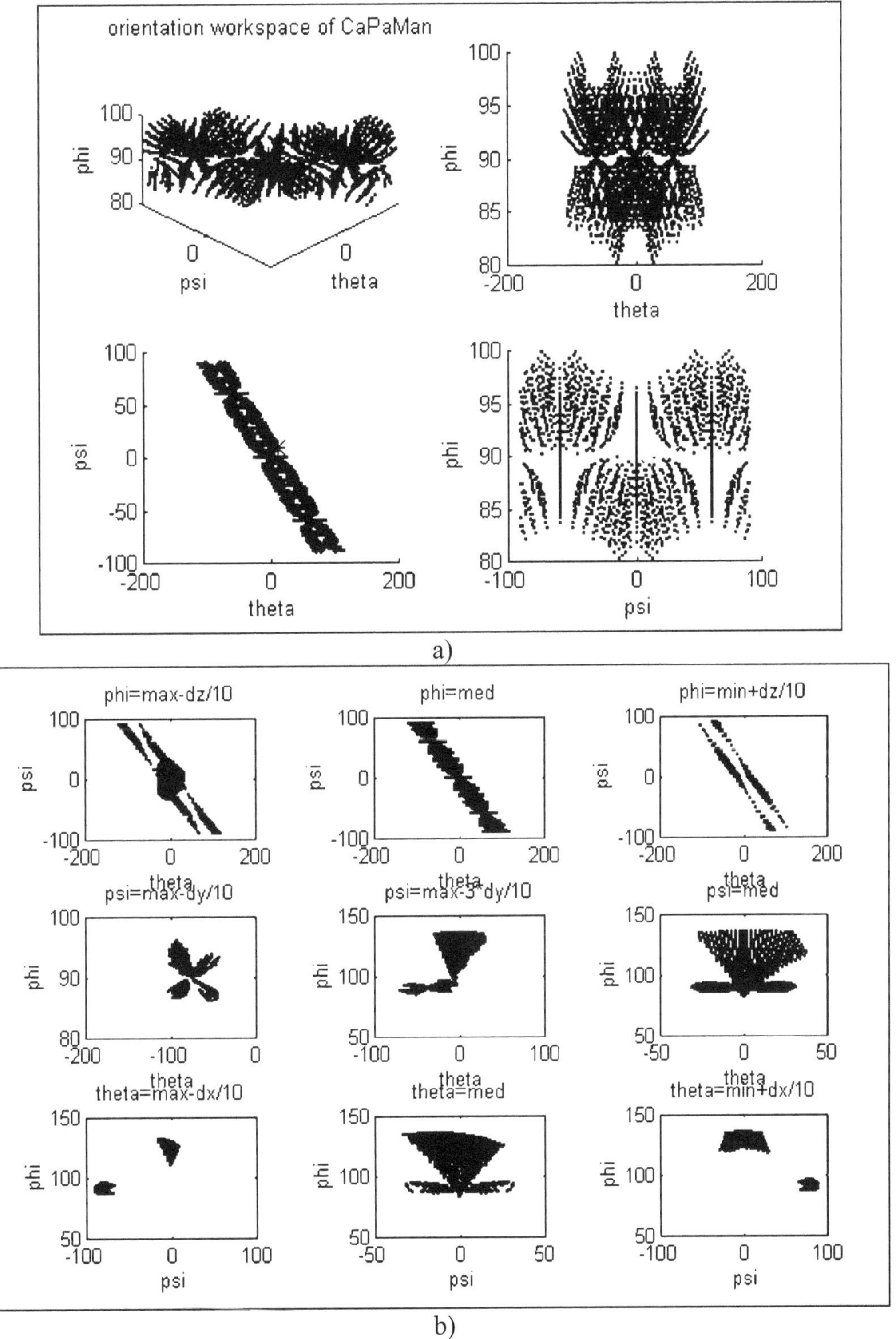

a)

b)

Fig. 3.61: Orientation workspace of CaPaMan prototype of Fig. 3.58 and Table 3.1: a) a 3D view; b) cross-sections.

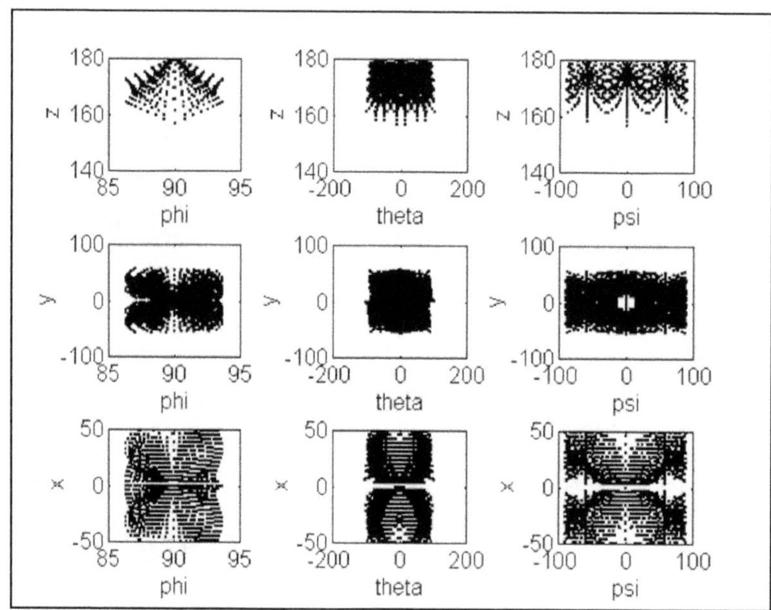

Fig. 3.62: An alternative representation of workspace capability of CaPaMan by combining data from Figs 3.60 and 3.61.

Indeed by properly combining the input motions of the parallelograms, very complex trajectories and orientation changes can be described by the mobile plate MP.
In Fig. 3.64 one can observe numerical results that are related to formulation singularity for orientation path planning. This problem can be easily solved by rearranging the numerical evaluation of the 90 deg. oriented value of the angles in agreement with a continuous path.

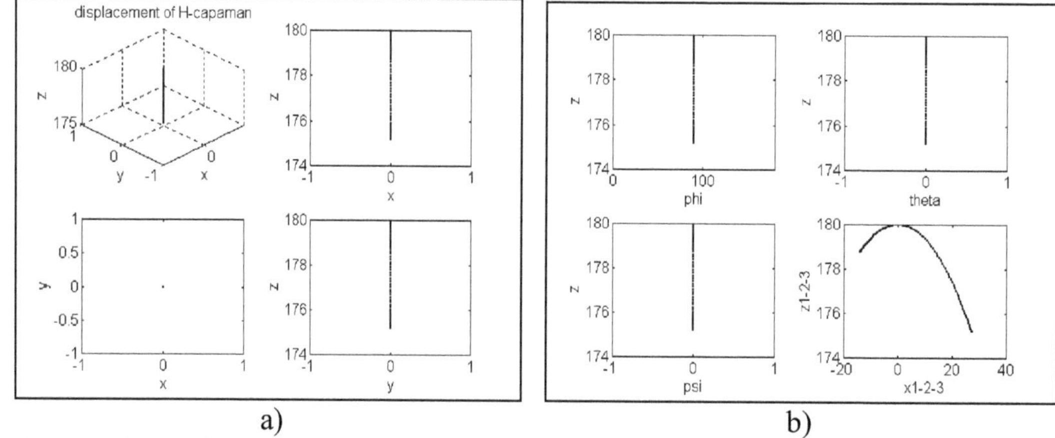

Fig. 3.63: An illustrative example of motion planning of CaPaMan for a straight-line movement of H by using direct kinematics as function of input variables of legmotion with α_{k0} = 70 deg., $\Delta\alpha_k$ = 30 deg., and t_{kf} = 10 sec (k = 1, 2, 3): a) translation path of H; b) orientation movement and input motion path of legs.

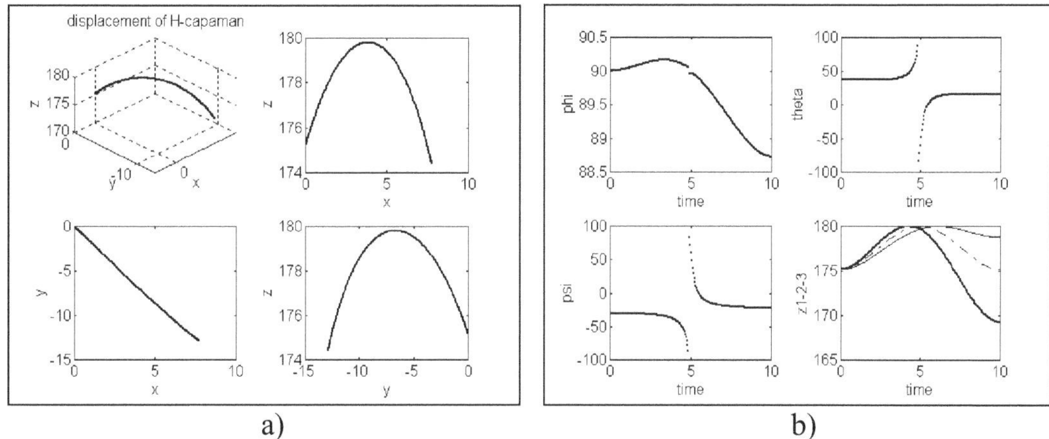

Fig. 3.64: An illustrative example of motion planning of CaPaMan for a straight-line movement of H by using direct kinematics as function of input variables of legmotion with α_{k0} = 70 deg., $\Delta\alpha_1$ = 50 deg., $\Delta\alpha_2$ = 40 deg., $\Delta\alpha_{3f}$ = 30 deg., and t_{kf} = 10 sec (k = 1, 2, 3): a) translation path of H; b) orientation movement and input motion path of legs (k = 1 dotted line; k = 2 dashdot line; k =3 solid line).

Kinematic feasibility can also be proved by obtaining a suitable inverse kinematic formulation, which may help the practical use and control of CaPaMan. The chain peculiarities have been very useful to deduce a closed-form formulation for the reverse position analysis by inverting the expressions for the forward displacement analysis.
Thus, after some algebraic manipulations, the following expression can be obtained for reverse displacement formulation

$$y_1 = k_3$$

$$y_2 = \frac{k_1 - k_2 - k_3}{2} \tag{3.9.31}$$

$$y_3 = \frac{k_1 + k_2 - k_3}{2}$$

where

$$k_1 = \frac{3}{2}\left[r_P \sin(\theta + \psi)(1 + \sin\varphi)\right]$$

$$k_2 = \sqrt{3}\left[2x + r\left(1 - \sin\varphi\right)\cos\left(\psi - \theta\right)\right] \tag{3.9.32}$$

$$k_3 = y + r_P\left(\sin\psi\cos\theta + \cos\psi\sin\theta\sin\varphi\right)$$

and

$$z_1 = k_6 - z_2 - z_3$$

$$z_2 = \frac{2\,k_4\,k_6 - 3(1 + k_4)z_3}{3\,(k_4 - 1)} \tag{3.9.33}$$

$$z_3 = \frac{-\,k_b + \sqrt{k_b^2 - 4k_a k_c}}{2k_a}$$

where

$$k_4 = \sqrt{3}\,\tan\psi$$

$$k_5 = \frac{3}{2}\,r_P\,\cos\varphi \tag{3.9.34}$$

$$k_6 = 3z$$

with

$$k_a = 1 - \frac{(k_4 + 1)}{(k_4 - 1)} + \frac{(k_4 + 1)^2}{(k_4 - 1)^2}$$

$$k_b = \frac{2\,k_4\,k_6}{3\,(k_4 - 1)} + \frac{2\,k_6}{(k_4 - 1)} - \frac{4\,k_4\,k_6\,(k_4 + 1)}{3\,(k_4 - 1)^2} \tag{3.9.35}$$

$$k_c = \frac{k_6^2 - k_5^2}{3} - \frac{2\,k_4\,k_6^2}{3\,(k_4 - 1)} + \frac{4\,k_4^2\,k_6^2}{9\,(k_4 - 1)^2}$$

Finally, the inverse problem for a four-bar linkage can be easily solved to give

$$\alpha_k = \tan^{-1}\left[\frac{(z_k - h_k)}{y_k}\right] \tag{3.9.36}$$

or in the case of $y_k = 0$ as

$$\alpha_k = \sin^{-1}\left[\frac{(z_k - h_k)}{b_k}\right] \tag{3.9.37}$$

Thus, once x, y, z, θ, φ, and ψ are given, Eqs (3.9.31) to (3.9.35) can be solved to give y_1, z_1, y_2, z_2, and y_3, z_3, which then can be used in Eqs (3.936) or (3.9.37) to give α_1, α_2, and α_3 for the input motion.
Practical problems can be numerically solved to calculate trajectory by using the above-mentioned formulation for inverse kinematics in order to move the movable plate through given trajectory points with given orientations.
In particular, a given straight-line motion of the movable plate can be expressed in the form

$$\begin{aligned} x &= x_0 + v_x \,\Delta t \\ y &= y_0 + v_y \,\Delta t \\ z &= z_0 + v_z \,\Delta t \end{aligned} \tag{3.9.38}$$

as well as the movable plate orientation can be expressed as

$$\begin{aligned} \psi &= \psi_0 + v_\psi \,\Delta t \\ \varphi &= \varphi_0 + v_\varphi \,\Delta t \\ \theta &= \theta_0 + v_\theta \,\Delta t\,, \end{aligned} \tag{3.9.39}$$

where x_0 , y_0 , z_0 ,ψ_0 , φ_0 , θ_0 are given values of the initial posture of the movable plate, and the corresponding v_x, v_y, v_z, v_ψ, v_φ, v_θ are desired constant velocity along the fixed frame axes and about the Euler angle rotation axes for a time interval Δt.
The reverse displacement problem for a given motion expressed by Eqs (3.9.38) and (3.9.39) has been solved by using Eqs (3.9.31) to (3.9.37) for the given locations with a scanning interval Δt to obtain input angles.
Figures 3.65 and 3.66 illustrate results of numerical examples. Figure 3.65 shows the input angles as a function of time t for a specific case when a straight-line motion along X axis is required with given orientation of the movable plate. Figure 3.66 has regard to a more complex motion when a constant velocity motion is required both in the H path and the movable plate orientation.
It is worthy of note that the formulation of the inverse kinematics is useful to give feasible solutions, and smooth input motions can be obtained by taking into account the cubic for the input motions.

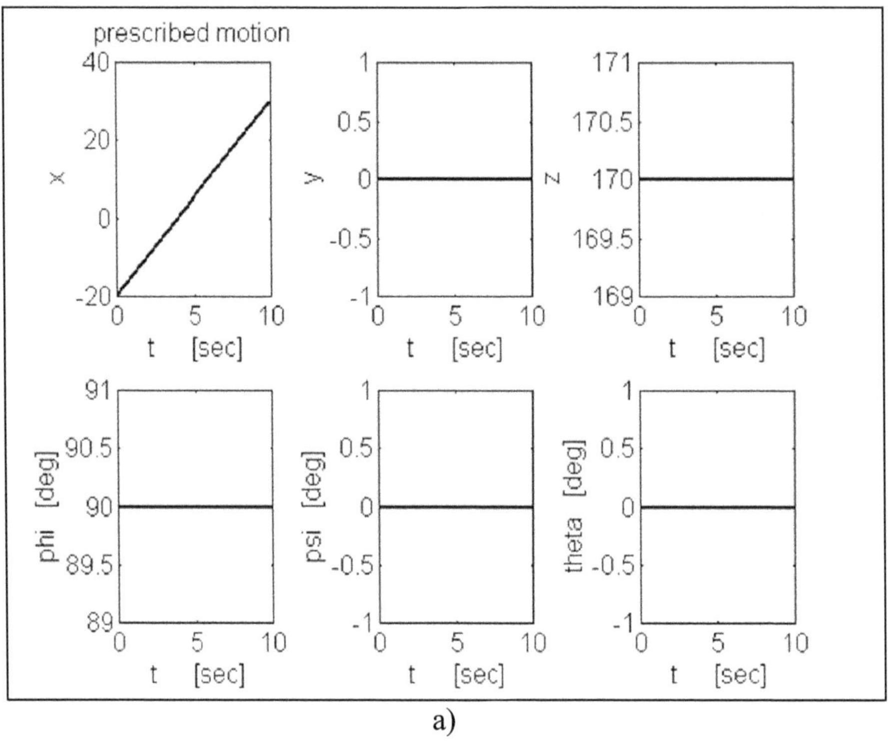

a)

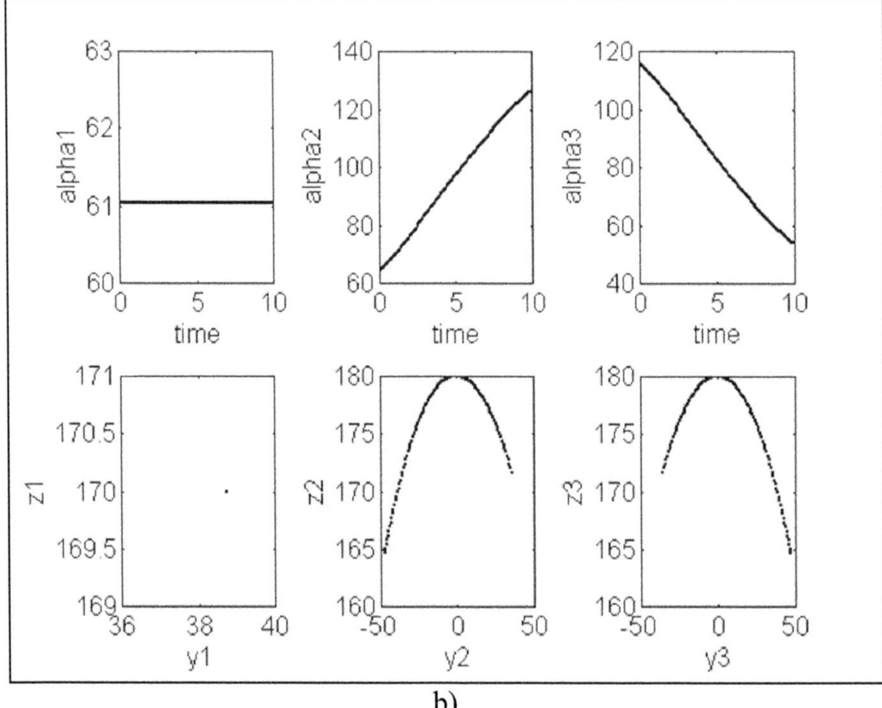

b)

Fig. 3.65: A solution for a problem of inverse kinematics for CaPaMan prototype: a) prescribed motion with v_x = 5 u/sec. (u is a unit length); b) required input motions and displacements for the actuating parallelogram.

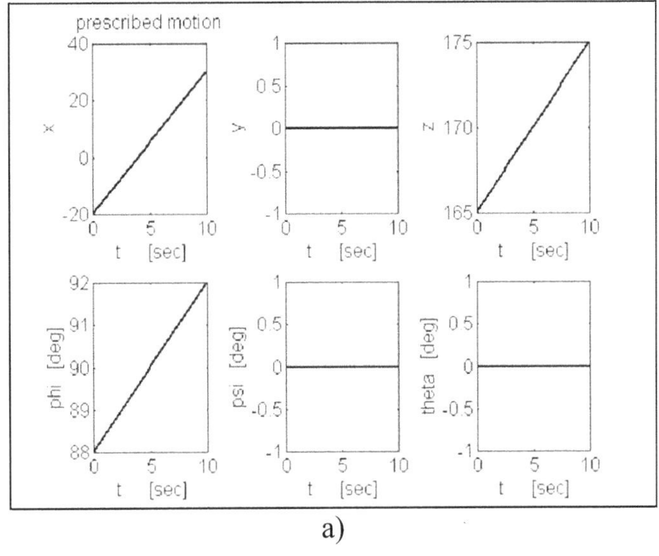

a)

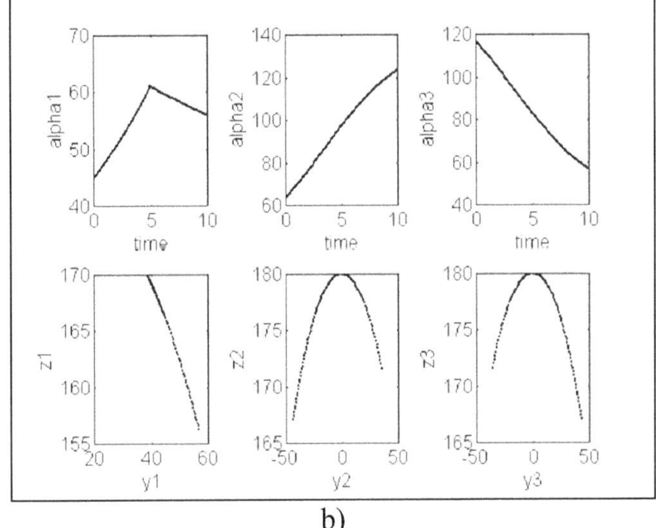

b)

Fig. 3.66: A solution for a problem of inverse kinematics for CaPaMan prototype: a) prescribed motion with $v_x = 5$ u/sec., $v_z = 1$ u/sec., $v_\varphi = 0.4$ rad/sec. (u is a unit length); b) required input motions and displacements for the actuating parallelograms.

Using algebraic formulation or a vector analysis one can perform velocity analysis.
The algebraic formulation for velocity equations consists of differentiating the equations that are obtained from the direct kinematics analysis of the manipulator, as shown in the following for dynamic analysis.
Thus, the differential kinematic relation of Eqs (3.9.8) can be reformulated in the form

$$\mathrm{A}\dot{\boldsymbol{\theta}} = \mathrm{K}\,\mathrm{B}\,\mathbf{t}_{\mathrm{E}} \tag{3.9.40}$$

where $\mathbf{t}_E = [\dot{x} \quad \dot{y} \quad \dot{z} \quad \dot{\varphi} \quad \dot{\psi} \quad \dot{\theta}]^t$ and B is a $[6 \times 6]$ matrix given as

$$B = \begin{bmatrix} I & 0 \\ 0 & R' \end{bmatrix} \tag{3.9.41}$$

in which I is the identity matrix and 0 stands for a $[3 \times 3]$ zero matrix; and matrix R' expresses the relationship between the angular velocity of the movable platform ω and the time derivatives of the Euler angles in the form

$$R' = \begin{bmatrix} s^2\varphi\, s\theta & (c\theta + c\psi\, s\theta)\, c\varphi & 0 \\ -s^2\varphi\, c\theta & (s\theta - c\psi\, c\theta)\, c\varphi & 0 \\ 0 & s\varphi & 1 \end{bmatrix} \tag{3.9.42}$$

Using the algebraic formulation introduces singularities, which can be referred to as formulation singularities. They arise when the matrix R' becomes singular, in other words if φ is equal to 0, 90, or 180 deg. This type of singularity is due to the analysis formulation and a proper consideration of the numerical interpretation or a re-formulation of the expression at the singularity can solve it.
Since the mechanism has 3 d.o.f.s, only three of the six variables can be specified as function of the input crank angles α_k, (k=1,2,3), for describing the configuration of CaPaMan. These independent coordinates can be chosen as the two rotations φ and ψ about two perpendicular axes intersecting at the mobile platform center, and the vertical translation z. The other coordinates x, y, and θ, can be specified by using a proper formulation of the kinematic analysis or determined with constraint equations associated to the mechanism architecture. Thus, Eq. (3.9.40) can be rewritten as

$$A\dot{\boldsymbol{\theta}} = K\, B\, B_R\, t_r \tag{3.9.43}$$

where B_R expresses the relationship between the independent coordinate vector $t_r = [\dot{\varphi} \quad \dot{\psi} \quad \dot{z}]^t$, and dependent coordinates. Because of Eqs (3.9.8) it is possible to specify independent coordinates as function of the input crank angles only as

$$\begin{aligned} z &= \frac{z_1 + z_2 + z_3}{3} \\ \varphi &= \cos^{-1}\left[\pm\frac{2}{3r_P}\sqrt{E}\right] \qquad (z \geq z_1 \Rightarrow +\,;\, z < z_1 \Rightarrow -) \\ \psi &= \tan^{-1}\left[\sqrt{3}\frac{D-F}{D+F}\right] \end{aligned} \tag{3.9.44}$$

with

$$\begin{aligned} E &= z_1^2 + z_2^2 + z_3^2 - z_1 z_2 - z_2 z_3 - z_1 z_3 \\ D &= 2z_2 - z_1 - z_3 \\ F &= 2z_3 - z_1 - z_2 \end{aligned} \tag{3.9.45}$$

when for k = 1,2,3, one considers

$$z_k = b_k \sin\alpha_k \tag{3.9.46}$$

Differentiating Eqs (3.9.43) with respect to time yields

$$\begin{aligned} &\dot{\varphi}\sqrt{E\left(9r_p^2 - 4E\right)} = \dot{z}_1(D-F) + \dot{z}_2(D+2F) - \dot{z}_3(D+2F) \\ &3\dot{z} = \dot{z}_1 + \dot{z}_2 + \dot{z}_3 \\ &\dot{\psi}\frac{6E}{\sqrt{3}} = \dot{z}_1(D+F) + \dot{z}_2 D - \dot{z}_3 F \end{aligned} \tag{3.9.47}$$

with

$$\dot{z}_k = b_k \cos\alpha_k \, \dot{\alpha}_k \tag{3.9.48}$$

From Eq. (3.9.43) the Jacobian matrices associated with the CaPaMan manipulator can be written as

$$K_r = \begin{bmatrix} \frac{6E}{\sqrt{3}} & 0 & 0 \\ 0 & \sqrt{E\left(9r_p^2 - 4E\right)} & 0 \\ 0 & 0 & 3 \end{bmatrix} = K\,B\,B_R \tag{3.9.49}$$

$$A = \begin{bmatrix} (D-F)\,b_1 c\alpha_1 & (D+2F)\,b_2 c\alpha_2 & -(2D+F)\,b_3 c\alpha_3 \\ (D+F)\,b_1 c\alpha_1 & -Db_2 c\alpha_2 & -F\,b_3 c\alpha_3 \\ b_1 c\alpha_1 & b_2 c\alpha_2 & b_3 c\alpha_3 \end{bmatrix} \tag{3.9.50}$$

Equations (3.9.49) and (3.9.50) represent the Jacobian matrices associated to the CaPaMan manipulator, which have been derived by using an algebraic formulation.

They can be used to compute the velocity according to Eq. (3.9.40).
A second approach for deriving the velocity equations consists of writing closed-loop velocity equations as function of linear and angular velocities of the links of the mechanism. This analysis leads to an invariant form of the Jacobian matrices. The velocity vector of an articulation point H_k is formulated from two different loop-closure circuits. Each one consists of the FP, MP, and links of the leg, as shown in Fig. 3.67. The rates of unactuated joints in each leg are passive variables and performing a dot product of the velocity vector-loop equation with an appropriate vector, which is orthogonal to all vectors of unactuated joint rates, and can eliminate them. Thus, the resulting equations can be assembled in the Jacobian matrices. It is assumed that, unless otherwise indicated, all vectors and matrices are represented in the world frame.
By considering Fig. 3.67, for each k-th leg a closed-loop equation can be written as

$$\mathbf{p}_{Hk} = \mathbf{p}_H + R\,\mathbf{r}'_k \tag{3.9.51}$$

$\mathbf{r}'_k$ is the vector connecting H to H_k and it is defined in the moving frame H-$X_P Y_P Z_P$ and R is the rotation matrix which describes the orientation between the moving and fixed frame. By considering Fig. 3.67, Eq. (3.9.56) can also be written as

$$\mathbf{p}_{Hk} = \mathbf{u}_k + \mathbf{l}_k + \mathbf{s}_k + \mathbf{h}_k \tag{3.9.52}$$

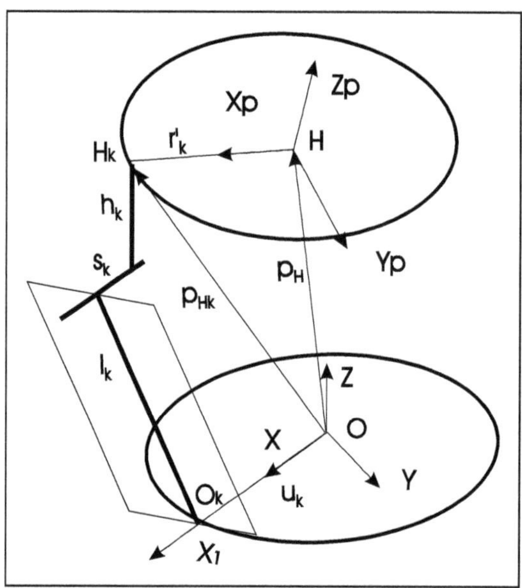

Fig. 3.67: Closed-loop vector equation for the k-th leg.

Differentiating Eqs (3.9.51) and (3.9.52) with respect to time yields

$$\mathbf{v}_H + \Omega\,\mathbf{r}_k = \boldsymbol{\omega}_k \times \mathbf{l}_k + \dot{\mathbf{s}}_k \tag{3.9.53}$$

where $\Omega = \dot{R}R^{t}$; $\boldsymbol{\omega}_k$ is the angular velocity of the k-th leg with respect to the AP plane reference frame; $\mathbf{s}_k$ is the linear velocity of the prismatic joint.
In order to eliminate the passive joint rates in Eq. (3.9.52), we dot-multiply both sides by $\mathbf{h}_k$, which is a constant vector for all the three legs, to obtain

$$\mathbf{h}_k \cdot \mathbf{v}_H + \mathbf{h}_k \cdot \Omega\, \mathbf{r}_k = \mathbf{h}_k \cdot \boldsymbol{\omega}_k \times \mathbf{l}_k \tag{3.9.54}$$

In addition, by considering

$$\mathbf{h}_k \cdot \boldsymbol{\omega}_k \times \mathbf{l}_k = (\mathbf{l}_k \times \mathbf{h}_k)^T \boldsymbol{\omega}_k \tag{3.9.55}$$

and

$$\mathbf{h}_k \cdot \Omega\, \mathbf{r}_k = (\mathbf{r}_k \times \mathbf{h}_k)^T \boldsymbol{\omega} \tag{3.9.56}$$

in Eq. (3.9.54) the closed-loop velocity equations associated to the leg mechanisms can be written as

$$\mathbf{l}_k \times \mathbf{h}_k \cdot \boldsymbol{\omega}_k = \mathbf{h}_k \cdot \mathbf{v}_H + \mathbf{r}_k \times \mathbf{h}_k \cdot \boldsymbol{\omega} \tag{3.9.57}$$

Expressing Eq. (3.9.57) for each leg, the velocity equations associated with the vector analysis can be expressed in a matrix form as

$$A_v \left[\boldsymbol{\omega}_1^t \quad \boldsymbol{\omega}_2^t \quad \boldsymbol{\omega}_3^t\right]^t = K\mathbf{t} \tag{3.9.58}$$

Thus, the Jacobian matrices associated to the CaPaMan manipulator can be written as

$$A_v = \begin{bmatrix} (\mathbf{l}_1 \times \mathbf{h}_1)^t & 0 & 0 \\ 0 & (\mathbf{l}_2 \times \mathbf{h}_2)^t & 0 \\ 0 & 0 & (\mathbf{l}_3 \times \mathbf{h}_3)^t \end{bmatrix} \tag{3.9.59}$$

and

$$K = \begin{bmatrix} (\mathbf{r}_1 \times \mathbf{h}_1)^t & \mathbf{h}_1{}^t \\ (\mathbf{r}_2 \times \mathbf{h}_2)^t & \mathbf{h}_2{}^t \\ (\mathbf{r}_3 \times \mathbf{h}_3)^t & \mathbf{h}_3{}^t \end{bmatrix} \tag{3.9.60}$$

where 0 denotes [1 × 3] zero vector. The obtained Jacobian matrices A_v, which is a [3 ×

9], and K, which is a [3 × 6], can be used for velocity analysis according to Eq. (3.9.40). The above-mentioned expression for Jacobian matrices can also be used for singularity analysis of CaPaMan.
A first type of singularity occurs when the determinant of A vanishes. By considering Eq. (3.9.59) yields the condition, for i = 1 or 2 or 3, in the form

$$(\mathbf{l}_k \times \mathbf{h}_k) = 0 \tag{3.9.61}$$

This type of singularity arises whenever any input crank of the articulated parallelogram is aligned with the connecting bar h_k. In other words whenever any leg is in a fully extended configuration, the manipulator loses 1, 2, or 3 degrees of freedom, depending on the number of legs which are in that configuration.
By considering the algebraic formulation, from Eq. (3.9.50) the expression for the determinant of B can be obtained in the form

$$\det(B) = E\, c\alpha_1 c\alpha_2 c\alpha_3 \tag{3.9.62}$$

From Eq. (3.9.62) it is possible to note that whenever any input crank angle is equal to 90 deg. the manipulator is on the first type of singularity. This result is in agreement with the one obtained with vector analysis by using Eq. (3.9.59).
In order to investigate the condition E = 0, if b_i=b_j, for i≠j, i,j=1,2,3, and assuming u = $c\alpha_1$, v = $c\alpha_2$, w = $c\alpha_3$, E coefficient in Eq. (3.9.62) can be written as

$$E = \frac{1}{2}\left[(u-v)^2 + (u-w)^2 + (v-w)^2\right] \tag{3.9.63}$$

Equation (3.9.63) is equal to zero if u = v = w, whenever the three input crank angles have the same value. According to Eq. (3.9.44) the condition E = 0 can be satisfied if φ is 90 deg. for the specific adopted formulation. This is a formulation singularity due to the fact that for this case, effects on the first and third rotations φ and θ cannot be distinguished. This type of singularity does not have a physical meaning and it can be avoided if the vector analysis is considered.
The second type of singularity arises when matrix K in Eq. (3.9.60) is rank deficient. The rank of a matrix equals the maximal number of independent rows or columns, so the condition is verified by imposing the linear dependence of the columns or rows of K. From Eq. (3.9.60), K becomes singular if its rank is equal to 2 or 1. In particular ($\mathbf{r}_k$ × $\mathbf{h}_k$) represents a vector which lies in the MP plane, so all three vectors, for k = 1,2,3 are linearly dependent because all lie in that plane, but K is not rank deficient. Even if only one of the three vectors has zero components the matrix K has still equal rank to three. If two or all the three vectors have zero components K becomes singular. This condition implies that the platform is aligned with the connecting bars.
In order to analyze the configuration of CaPaMan in its second type of singularity one

can study the condition E = 0 with E expressed as in Eq. (3.9.63). This condition represents a cylinder with elliptical cross-section. The cross-section plane is perpendicular to the unit vector $n=[1,1,1]^t$. An intersection of the elliptic cylinder with a plane x + y + z = 0 is its elliptic directrix with center in point at the origin of the reference frame and semi-major axis p = m/3; semi-minor axis $q = m^2/ 3^{1/2}$ with m = 3 r_P / 2. The elliptic cylinder divides the space into two regions free from singularities: the region inside and outside the cylinder.
Figure 3.68 shows the singular configurations of CaPaMan, as they have been determined with the above-mentioned Jacobian analysis.

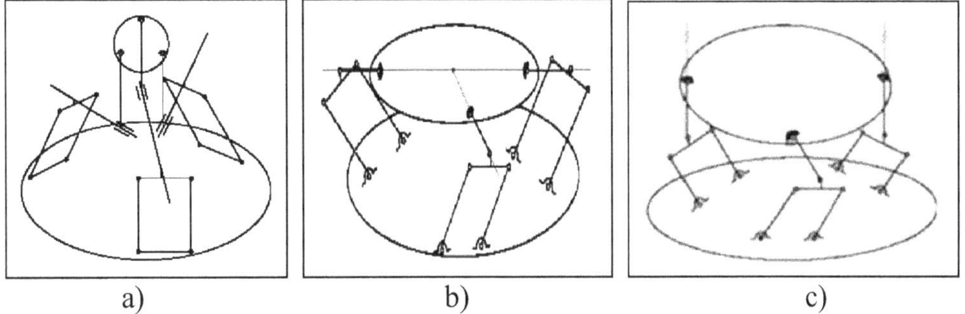

Fig. 3.68: Singular configurations of CaPaMan: a) with φ =90 deg.; b) with the three connecting bars in a coplanar configuration; c) with two aligned connecting bars.

The static behavior of CaPaMan can be investigated in term of force transmission capability through a static analysis. Force transmission is characterized by the parallelogram linkages in the sense that, because of the prismatic joint and assuming friction as negligible, the only force applied to a leg by the movable plate is the one of $\mathbf{R}_k$ which is contained in the plane of the mechanism, as shown in Fig. 3.69. Force $\mathbf{R}_k$ is given by the components R_{ky} and R_{kz} with respect to the k-frame fixed with the linkage plane since $R_{kx} \neq 0$ will determine the sliding of the prismatic joint to a static equilibrium. Thus, the components R_{ky} and R_{kz} will balance the force **F** and the torque **N** acting on the movable plate. The force $\mathbf{F}_P$ and the torque $\mathbf{N}_P$ with respect to the frame fixed to the movable plate can be obtained by using the rotation matrix R as

$$\mathbf{F} = R\,\mathbf{F}_P$$
$$\mathbf{N} = R\,\mathbf{N}_P \tag{3.9.64}$$

A static equilibrium of the movable plate can be expressed for the triangular assembly with the spherical joints in the form

$$\mathbf{F}_P = \mathbf{R}_1 + \mathbf{R}_2 + \mathbf{R}_3$$
$$\mathbf{N}_P = \mathbf{R}_1 \times \mathbf{r}_{PH1} + \mathbf{R}_2 \times \mathbf{r}_{PH2} + \mathbf{R}_3 \times \mathbf{r}_{PH3} \tag{3.9.65}$$

where $\mathbf{R}_k$ (k=1,2,3) are the reaction forces at the articulation points H_k and $\mathbf{r}_{PHk}$ is the vector $\mathbf{HH}_k$.
The geometry of the triangular assembly for the spherical joints can be used to solve Eqs (3.9.65) as a function of the active components R_{ky} and R_{kz} (k = 1, 2, 3) as

$$F_x = R_{2y}\, s_2 + R_{3y}\, s_3$$

$$F_y = R_{1y} + R_{2y}\, c_2 + R_{3y}\, c_3$$

$$F_z = R_{1z} + R_{2z} + R_{3z}$$

$$N_x = R_{2z}\, r_2\, s_2 + R_{3z}\, r_3\, s_3 \qquad (3.9.66)$$

$$N_y = R_{1z}\, r_1 + R_{2z}\, r_2\, c_2 + R_{3z}\, r_3\, c_3$$

$$N_z = R_{1y}\, r_1 + R_{2y}\, r_2 + R_{3y}\, r_3$$

where F_x ,F_y ,F_z , and N_x ,N_y ,N_z , are the components of $\mathbf{F}_P$ and $\mathbf{N}_P$, respectively, in the moving frame; c_k is cos γ_k and s_k is sin γ_k .

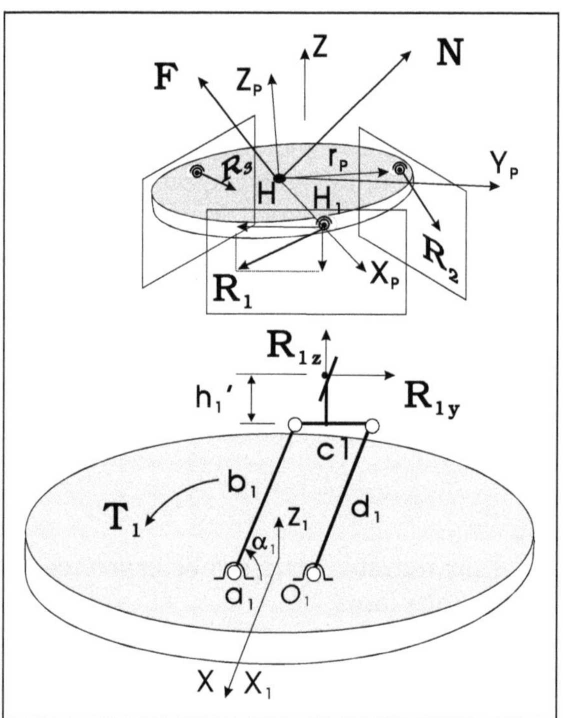

Fig. 3.69: A model of CaPaMan architecture for static and dynamic analysis.

Moreover, Eqs (3.9.66) can be solved in a closed-form to give, after some algebraic manipulations, the following expressions

$$R_{1y} = \frac{1}{3}\left(\frac{N_z}{r_P} + 2\,F_y + 2\sqrt{3}\,F_x \right)$$

$$R_{1z} = \frac{1}{3}\left(F_z + 2\frac{N_y}{r_P} + 2\sqrt{3}\,\frac{N_x}{r_P} \right)$$

$$R_{2y} = \frac{1}{3}\left(\frac{N_z}{r_P} - F_y - \sqrt{3}\,F_x \right) \tag{3.9.67}$$

$$R_{2z} = \frac{1}{3}\left(F_z - \frac{N_y}{r_P} - \sqrt{3}\,\frac{N_x}{r_P} \right)$$

$$R_{3y} = \frac{1}{3}\left(\frac{N_z}{r_P} - F_y - \sqrt{3}\,F_x \right)$$

$$R_{3z} = \frac{1}{3}\left(F_z - \frac{N_y}{r_P} - \sqrt{3}\,\frac{N_x}{r_P} \right)$$

Then, by applying the Principle of Virtual Work for a parallelogram linkage, and taking into account that the force components R_{ky} and R_{kz} act on the projection point of H_k onto the parallelogram plane, the actuator torque on the input crank can be expressed as

$$T_k = R_{ky}\left(b_k \sin\alpha_k + h'_k\right) - R_{kz}\left(b_k \cos\alpha_k + \frac{c_k}{2}\right) \tag{3.9.68}$$

Finally, the actuator torque can be computed by

$$\tau_k = tr_k\;T_k \tag{3.9.69}$$

when the tr_k overall gear ratio is taken into account.
Equations (3.9.67) to (3.9.69) can be useful to evaluate the static performance of CaPaMan as well as to properly select the actuators for given applications.
Figures 3.70 and 3.71 show results for illustrative numerical examples.
An analysis of dynamic behavior of CaPaMan requires the evaluation of inertia actions and computation of the force F and torque N acting on the moving plate, according to

the model of Fig. 3.69 and Eq. (3.9.65), but considering

$$\mathbf{F}_P = \mathbf{F}_{in} + \mathbf{F}_{ex}$$
$$\mathbf{N}_P = \mathbf{N}_{in} + \mathbf{N}_{ex} \tag{3.9.70}$$

in which $\mathbf{F}_{in}$ and $\mathbf{N}_{in}$ are the inertia force and torque; $\mathbf{F}_{ex}$ and $\mathbf{N}_{ex}$ are the external actions due to the payload, the application, or the interaction with the environment.

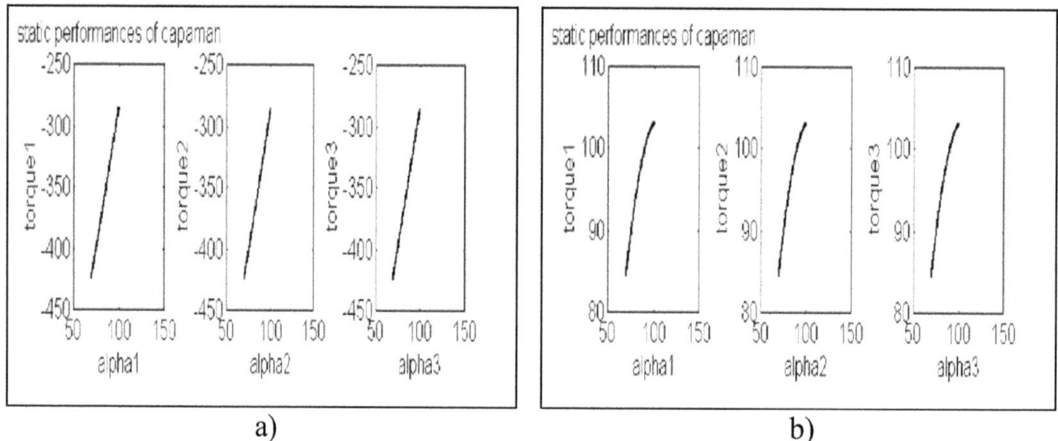

a) b)

Fig. 3.70: Static performances of CaPaMan architecture in terms of actuator torque in N u (u is a unit length) for positioning along the trajectory of Fig. 3.63: a) with a payload of F_z = 10 N; b) for a screwing application with F_z = 1 N and N_z = 1000 N u.

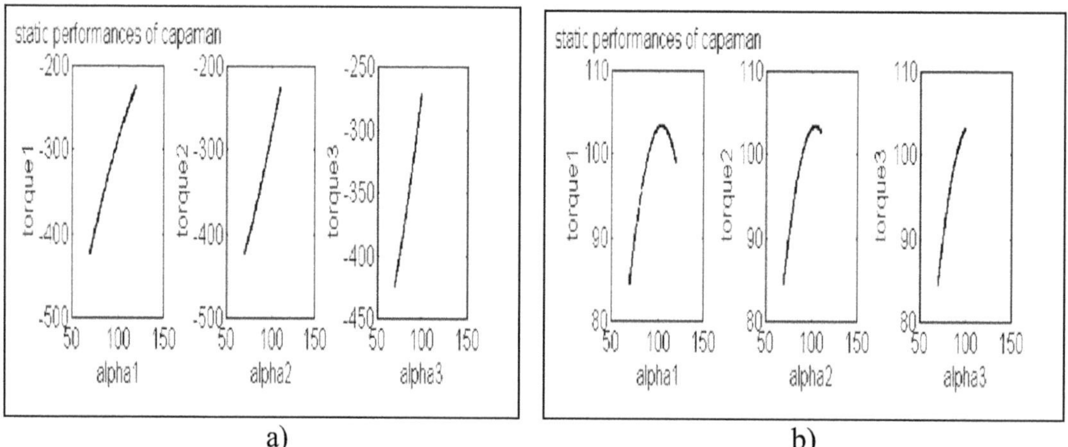

a) b)

Fig. 3.71: Static performances of CaPaMan architecture in terms of actuator torque in N u (u is a unit length) for positioning along the trajectory of Fig. 3.64: a) with a payload of F_z = 10 N; b) for a screwing application with F_z = 1 N and N_z= 1000 N u.

The inertia actions $\mathbf{F}_{in}$ and $\mathbf{N}_{in}$ can be formulated through the mass M and the inertia matrix I^H of the movable plate and payload as

$$\mathbf{F}_{in} = M\,\mathbf{a}_H$$
$$\mathbf{N}_{in} = I^H\,\dot{\boldsymbol{\omega}} + \boldsymbol{\omega} \times I^H\,\boldsymbol{\omega} \qquad (3.9.71)$$

in which $\mathbf{a}_H$ is the acceleration of H center point of MP; $\boldsymbol{\omega}$ is the angular velocity and $\boldsymbol{\omega}$ dot is the angular acceleration of the MP. Then, the actuator torque τ_k is given by

$$\tau_k = tr_k\,T_k + \frac{I_{rk}}{tr_k}\ddot{\alpha}_k \qquad (3.9.72)$$

when the rotor inertia I_{rk} of the actuator and the tr_k overall gear ration of the actuation transmission are taken into account.
In particular, Eqs (3.9.70) and (3.9.71) can be formulated in closed-form expressions with regard to the inverse dynamics because of the algebraic formulation of the kinematics of CaPaMan.
The components of the velocity and acceleration of H point can be obtained by the first and second derivatives, respectively, of the first three expressions in Eqs (3.9.8) as

$$\begin{vmatrix}\dot{x}\\ \dot{y}\\ \dot{z}\end{vmatrix} = \begin{bmatrix}0 & -\frac{1}{\sqrt{3}} & \frac{1}{\sqrt{3}}\\ 1 & 0 & 0\\ 0 & 0 & 0\end{bmatrix}\begin{vmatrix}\dot{y}_1\\ \dot{y}_2\\ \dot{y}_3\end{vmatrix} + \begin{bmatrix}0 & 0 & 0\\ 0 & 0 & 0\\ \frac{1}{3} & \frac{1}{3} & \frac{1}{3}\end{bmatrix}\begin{vmatrix}\dot{z}_1\\ \dot{z}_2\\ \dot{z}_3\end{vmatrix} +$$
$$+\, r_p\begin{bmatrix}-\frac{(1-s\varphi)\,s(\psi-\theta)}{2} & \frac{c\varphi\,c(\psi-\theta)}{2} & \frac{(1-s\varphi)\,s(\psi-\theta)}{2}\\ -(s\varphi\,c\theta\,c\psi - s\theta\,s\psi) & -c\varphi\,s\theta\,c\psi & s\varphi\,s\theta\,s\psi - c\theta\,c\psi\\ 0 & 0 & 0\end{bmatrix}\begin{vmatrix}\dot{\theta}\\ \dot{\varphi}\\ \dot{\psi}\end{vmatrix} \qquad (3.9.73)$$

$$\begin{vmatrix}\ddot{x}\\ \ddot{y}\\ \ddot{z}\end{vmatrix} = \begin{bmatrix}0 & -\frac{1}{\sqrt{3}} & \frac{1}{\sqrt{3}}\\ 1 & 0 & 0\\ 0 & 0 & 0\end{bmatrix}\begin{vmatrix}\ddot{y}_1\\ \ddot{y}_2\\ \ddot{y}_3\end{vmatrix} + \begin{bmatrix}0 & 0 & 0\\ 0 & 0 & 0\\ \frac{1}{3} & \frac{1}{3} & \frac{1}{3}\end{bmatrix}\begin{vmatrix}\ddot{z}_1\\ \ddot{z}_2\\ \ddot{z}_3\end{vmatrix} +$$
$$+\, r_p\begin{bmatrix}-\frac{(1-s\varphi)\,s(\psi-\theta)}{2} & \frac{c\varphi\,c(\psi-\theta)}{2} & \frac{(1-s\varphi)\,s(\psi-\theta)}{2}\\ -(s\varphi\,c\theta\,c\psi - s\theta\,s\psi) & -c\varphi\,s\theta\,c\psi & s\varphi\,s\theta\,s\psi - c\theta\,c\psi\\ 0 & 0 & 0\end{bmatrix}\begin{vmatrix}\ddot{\theta}\\ \ddot{\varphi}\\ \ddot{\psi}\end{vmatrix} +$$

$$+\ r_p \begin{bmatrix} \frac{(1-s\varphi)\,c(\psi-\theta)}{2} & -\frac{s\varphi\,c(\psi-\theta)}{2} & \frac{(1-s\varphi)\,c(\psi-\theta)}{2} \\ s\varphi\, s\theta\, c\psi - c\theta\, s\psi & s\varphi\, s\theta\, c\psi & s\varphi\, s\theta\, c\psi - c\theta\, s\psi \\ 0 & 0 & 0 \end{bmatrix} \begin{vmatrix} \dot\theta^2 \\ \dot\varphi^2 \\ \dot\psi^2 \end{vmatrix} +$$

$$+\,r_p \begin{bmatrix} c\varphi\ s(\psi-\theta) & -(1-s\varphi)\,c(\psi-\theta) & -c\varphi\ s(\psi-\theta) \\ -2\,c\varphi\, c\theta\, c\psi & 2\,(s\varphi\, c\theta\, s\psi + s\theta\, c\psi) & 2\,c\varphi\, s\theta\, s\psi \\ 0 & 0 & 0 \end{bmatrix} \begin{vmatrix} \dot\theta\dot\varphi \\ \dot\theta\dot\psi \\ \dot\varphi\dot\psi \end{vmatrix}$$

where the expressions for first and second derivatives of coordinates y_k and z_k are given as derivatives of Eqs (3.9.27) taking into account the cubic of Eq. (3.9.30).
Similarly, angular velocity and acceleration can be calculated by derivating the expression for Euler angles θ, φ, and ψ to give

$$\dot\psi = \frac{\sqrt{3}}{2\,V_1}\left[(z_2-z_3)\dot z_1-(z_1-z_3)\dot z_2+(z_1-z_2)\dot z_3\right]$$

$$\dot\varphi = \mp\frac{1}{\sqrt{(9r_p^2-4V_1)V_1}}\left[(2z_1-z_2-z_3)\dot z_1+(-z_1+2z_2-z_3)\dot z_2+(-z_1-z_2+2z_3)\dot z_3\right]$$

$$\dot\theta = \frac{2}{(1+\sin\varphi)\sqrt{V_2}}\left[(1+\sin\varphi)(\dot y_1+\dot y_2+\dot y_3)-(y_1+y_2+y_3)\cos\varphi\,\dot\varphi\right]-\dot\psi$$

(3.9.74)

$$\ddot\psi = \frac{\sqrt{3}}{2V_1}\left[(z_2-z_3)\ddot z_1-(z_1-z_3)\ddot z_2+(z_1-z_2)\ddot z_3\right]+$$
$$+\frac{\sqrt{3}}{2V_1^2}\left[-(z_2-z_3)(2z_1-z_2-z_3)\dot z_1^2+(z_1-z_3)(-z_1+2z_2-z_3)\dot z_2^2-(z_1-z_2)(-z_1-z_2+2z_3)\dot z_3^2\right]+$$
$$+\frac{\sqrt{3}}{V_1^2}\left[-(z_1-z_2)(-z_1-z_2+2z_3)\dot z_1\dot z_2+(z_1-z_3)(-z_1+2z_2-z_3)\dot z_1\dot z_3-(z_2-z_3)(2z_1-z_2-z_3)\dot z_2\dot z_3\right]$$

$$\ddot\varphi = \mp\frac{1}{\sqrt{(9r_p^2-4V_1)V_1}}\left[(2z_1-z_2-z_3)\ddot z_1+(-z_1+2z_2-z_3)\ddot z_2+(-z_1-z_2+2z_3)\ddot z_3\right]+$$
$$\mp\frac{1}{2\sqrt{\left[(9r_p^2-4V_1)V_1\right]^3}}\left(V_3\dot z_1^2+V_4\dot z_2^2+V_5\dot z_3^2\right)+ \qquad (z\ge z_1 \Rightarrow "-";\ z<z_1 \Rightarrow "+")$$
$$\mp\left[(V_5-27r_p^2V_1)\dot z_1\dot z_2+(V_4-27r_p^2V_1)\dot z_1\dot z_3+(V_3-27r_p^2V_1)\dot z_2\dot z_3\right]\Big/\sqrt{\left[(9r_p^2-4V_1)V_1\right]^3}$$

$$\ddot{\theta} = \frac{2}{\sqrt{V_2}}(\ddot{y}_1 + \ddot{y}_2 + \ddot{y}_3) - \frac{2(y_1 + y_2 + y_3)\cos\varphi}{(1+\sin\varphi)\sqrt{V_2}}\ddot{\varphi} + \frac{8(y_1 + y_2 + y_3)}{\sqrt{(V_2)^3}}\left(\dot{y}_1^2 + \dot{y}_2^2 + \dot{y}_3^2\right) +$$
$$+\frac{(y_1+y_2+y_3)\left[18r_p^2(1+\sin\varphi)\cos^2\varphi+2V_2\right]}{(1+\sin\varphi)\sqrt{(V_2)^3}}\dot{\varphi}^2 + \frac{16(y_1+y_2+y_3)}{\sqrt{(V_2)^3}}(\dot{y}_1\dot{y}_2+\dot{y}_1\dot{y}_3+\dot{y}_2\dot{y}_3) -$$
$$-\frac{36r_p^2(1+\sin\varphi)^2\cos\varphi}{(1+\sin\varphi)\sqrt{(V_2)^3}}(\dot{y}_1+\dot{y}_2+\dot{y}_3)\dot{\varphi} - \ddot{\psi}$$

The three components of the MP angular velocity $\boldsymbol{\omega}$ can be expressed in terms of Euler angles and their time derivatives as

$$\begin{vmatrix}\omega_x\\ \omega_y\\ \omega_z\end{vmatrix} = \begin{bmatrix}-\cos\varphi\cos\psi & \sin\psi & 0\\ \cos\varphi\sin\psi & \cos\psi & 0\\ \sin\varphi & 0 & 1\end{bmatrix}\begin{vmatrix}\dot{\theta}\\ \dot{\varphi}\\ \dot{\psi}\end{vmatrix} \tag{3.9.75}$$

The components of the angular acceleration can be given by the derivatives of Eqs (3.9.75) as

$$\begin{vmatrix}\dot{\omega}_x\\ \dot{\omega}_y\\ \dot{\omega}_z\end{vmatrix} = \begin{bmatrix}-\cos\varphi\cos\psi & \sin\psi & 0\\ \cos\varphi\sin\psi & \cos\psi & 0\\ \sin\varphi & 0 & 1\end{bmatrix}\begin{vmatrix}\ddot{\theta}\\ \ddot{\varphi}\\ \ddot{\psi}\end{vmatrix} + \begin{bmatrix}\sin\varphi\cos\psi & \cos\varphi\sin\psi & \cos\psi\\ -\sin\varphi\sin\psi & \cos\varphi\cos\psi & -\sin\psi\\ \cos\varphi & 0 & 0\end{bmatrix}\begin{vmatrix}\dot{\theta}\dot{\varphi}\\ \dot{\theta}\dot{\psi}\\ \dot{\varphi}\dot{\psi}\end{vmatrix} \tag{3.9.76}$$

The Newton--Euler equations for CaPaMan can be formulated by considering all the links as rigid bodies and the position, velocity, and acceleration of the movable platform is given by the direct kinematics of the above-mentioned expressions. In order to simplify the mathematical model the effects of link elasticity, viscous damping in the joints, the inertia, and mass of the mechanism legs have been neglected. The simplification in the dynamic analysis consists of neglecting inertia of leg mechanisms in comparison with the inertia of the movable plate. This neglecting can be justified when one considers that the leg motion is quite smoother than that correspondingly obtained for the movable plate. In fact, the motion and mass of the movable plate are more significant with respect to the corresponding leg properties in most of the cases. Indeed, the massive mobile plate plays a major role in inertia effects in the built prototype of Fig. 3.58. The last assumption has been verified on the built prototype during several experiments at the Laboratory of Robotics and Mechatronics in Cassino.
In addition, the simplification is needed to derive the proposed closed-form formulation for the inverse dynamics. Of course, a complete model is required for fast applications of CaPaMan with very fast input motion of the legs. Nevertheless, the proposed

formulation is also helpful to characterize the dynamic performance of CaPaMan by analytical formulation.

Thus, the dynamic equilibrium for the mobile platform MP can be expressed by the Newton–Euler equations in the form of Eqs (3.9.71) when all the terms of force transmission and inertia actions are formulated by using the above.-mentioned expressions. Thus, by considering the geometry of the triangle assembly for the movable plate through

$$\begin{vmatrix} u_{kx} \\ u_{ky} \\ u_{kz} \end{vmatrix} = r_p\, R \begin{vmatrix} \cos\delta_k \\ \sin\delta_k \\ 0 \end{vmatrix} \qquad (k=1,2,3) \tag{3.9.77}$$

Equations (3.9.71) can be solved in a closed-form for CaPaMan architecture to give

$$\begin{vmatrix} R_{1y} \\ R_{2y} \\ R_{3y} \\ R_{1z} \\ R_{2z} \\ R_{3z} \end{vmatrix} = \begin{bmatrix} \frac{\sqrt{3}\left(-u_{2x}+\sqrt{3}\,u_{2y}+u_{3x}+\sqrt{3}\,u_{3y}\right)}{3\left(2u_{1x}-u_{2x}+\sqrt{3}\,u_{2y}+u_{3x}+\sqrt{3}\,u_{3y}\right)} & \frac{\sqrt{3}\left(-u_{2x}+\sqrt{3}\,u_{2y}-u_{3x}-\sqrt{3}\,u_{3y}\right)}{2u_{1x}-u_{2x}+\sqrt{3}\,u_{2y}+u_{3x}+\sqrt{3}\,u_{3y}} & 0 \\ \frac{2\sqrt{3}\left(-u_{1x}+u_{3x}+\sqrt{3}\,u_{3y}\right)}{3\left(2u_{1x}-u_{2x}+\sqrt{3}\,u_{2y}+u_{3x}+\sqrt{3}\,u_{3y}\right)} & \frac{-2u_{1x}}{2u_{1x}-u_{2x}+\sqrt{3}\,u_{2y}+u_{3x}+\sqrt{3}\,u_{3y}} & 0 \\ \frac{2\sqrt{3}\left(u_{1x}-u_{2x}+\sqrt{3}\,u_{2y}\right)}{3\left(2u_{1x}-u_{2x}+\sqrt{3}\,u_{2y}+u_{3x}+\sqrt{3}\,u_{3y}\right)} & \frac{-2u_{1x}}{2u_{1x}-u_{2x}+\sqrt{3}\,u_{2y}+u_{3x}+\sqrt{3}\,u_{3y}} & 0 \\ \frac{V_7}{V_6} & \frac{3V_8}{V_6} & \frac{3V_9}{V_6} \\ \frac{-V_{13}}{V_6} & \frac{-3V_{14}}{V_6} & \frac{-3V_{15}}{V_6} \\ \frac{V_{19}}{V_6} & \frac{3V_{20}}{V_6} & \frac{3V_{21}}{V_6} \end{bmatrix} \begin{vmatrix} F_x \\ F_y \\ F_z \end{vmatrix} +$$

$$+ \begin{bmatrix} 0 & 0 & \frac{2}{2u_{1x}-u_{2x}+\sqrt{3}\,u_{2y}+u_{3x}+\sqrt{3}\,u_{3y}} \\ 0 & 0 & \frac{2}{2u_{1x}-u_{2x}+\sqrt{3}\,u_{2y}+u_{3x}+\sqrt{3}\,u_{3y}} \\ 0 & 0 & \frac{2}{2u_{1x}-u_{2x}+\sqrt{3}\,u_{2y}+u_{3x}+\sqrt{3}\,u_{3y}} \\ \frac{3V_{10}}{V_6} & \frac{3V_{11}}{V_6} & \frac{3V_{12}}{V_6} \\ \frac{-3V_{16}}{V_6} & \frac{-3V_{17}}{V_6} & \frac{-3V_{18}}{V_6} \\ \frac{3V_{22}}{V_6} & \frac{3V_{23}}{V_6} & \frac{3V_{24}}{V_6} \end{bmatrix} \begin{vmatrix} N_x \\ N_y \\ N_z \end{vmatrix} \tag{3.9.78}$$

Terms have been conveniently grouped and the coefficients V_j can be expressed as

$$V_1 = z_1^2 + z_2^2 + z_3^2 - z_1 z_2 - z_1 z_3 - z_2 z_3$$

$$V_2 = 9 r_p^2 (1+\sin\varphi)^2 - 4(y_1 + y_2 + y_3)^2$$

$$V_3 = 16 V_1^2 + 3(z_2 - z_3)^2 \left(9 r_p^2 - 8 V_1\right)$$

$$V_4 = 16 V_1^2 + 3(z_1 - z_3)^2 \left(9 r_p^2 - 8 V_1\right)$$

$$V_5 = 16 V_1^2 + 3(z_1 - z_2)^2 \left(9 r_p^2 - 8 V_1\right)$$

$$V_6 = 3\left[u_{1x}\left(u_{2y} - u_{3y}\right) + u_{2x}\left(-u_{1y} + u_{3y}\right) + u_{3x}\left(u_{1y} - u_{2y}\right)\right]\left(2u_{1x} - u_{2x} + \sqrt{3}u_{2y} - u_{3x} - \sqrt{3}u_{3y}\right)$$

$$V_7 = \sqrt{3}(u_{2x} - u_{3x})$$
$$\left[u_{1x}(-u_{2z} + u_{3z}) - u_{1z}\left(-u_{2x} + \sqrt{3}u_{2y} + u_{3x} + \sqrt{3}u_{3y}\right) + u_{2z}\left(u_{3x} + \sqrt{3}u_{3y}\right) + u_{3z}\left(-u_{2x} + \sqrt{3}u_{2y}\right)\right]$$
$$+3\left(u_{2y} - u_{3y}\right)\left[u_{1x}\left(u_{2z} + u_{3z}\right) + u_{3z}\left(-u_{2x} + \sqrt{3}u_{2y}\right) - u_{2z}\left(u_{3x} + \sqrt{3}u_{3y}\right)\right]$$

$$V_8 = \sqrt{3}\, u_{1x}\left(u_{2y} - u_{3y}\right)\left(u_{2z} - u_{3z}\right) + \left(u_{2x} - u_{3x}\right)\left[-u_{1x}\left(u_{2z} + u_{3z}\right) + u_{1z}\left(u_{2x} - \sqrt{3}u_{2y} + u_{3x} + \sqrt{3}u_{3y}\right)\right]$$

$$V_9 = \left(u_{2x}u_{3y} - u_{2y}u_{3x}\right)\left(2u_{1x} - u_{2x} + \sqrt{3}\, u_{2y} - u_{3x} - \sqrt{3}\, u_{3y}\right)$$

$$V_{10} = \left(-u_{2x} + u_{3x}\right)\left(2u_{1x} - u_{2x} + \sqrt{3}\, u_{2y} - u_{3x} - \sqrt{3}\, u_{3y}\right) \tag{3.9.79}$$

$$V_{11} = \left(-u_{2y} + u_{3y}\right)\left(2u_{1x} - u_{2x} + \sqrt{3}\, u_{2y} - u_{3x} - \sqrt{3}\, u_{3y}\right)$$

$$V_{12} = \sqrt{3}\left(-u_{2y} + u_{3y}\right)\left(u_{2z} - u_{3z}\right) + \left(u_{2x} - u_{3x}\right)\left(-2u_{1z} + u_{2x} + u_{3z}\right)$$

$$V_{13} = \sqrt{3}(u_{1x} - u_{3x})$$
$$\left[u_{1x}(u_{3z} - u_{2z}) - u_{1z}\left(-u_{2x} + \sqrt{3}u_{2y} + u_{3x} + \sqrt{3}u_{3y}\right) + u_{3z}\left(-u_{2x} + \sqrt{3}u_{2y}\right) + u_{2z}\left(u_{3x} + \sqrt{3}u_{3y}\right)\right]$$
$$+3\left[u_{1x}\left(u_{2z} + u_{3z}\right)\left(u_{1y} - u_{3y}\right) + u_{3z}\left(u_{1x} - u_{3y}\right)\left(-u_{2x} + \sqrt{3}u_{2y}\right) - u_{1y}u_{2z}\left(u_{3x} + \sqrt{3}u_{3y}\right) + V_7\right]$$

$$V_{14} = \sqrt{3}\, u_{1x}\left(u_{1y} - u_{3y}\right)\left(u_{2z} - u_{3z}\right) + \left(u_{3x} - u_{1x}\right)\left[u_{1x}\left(u_{2z} + u_{3z}\right) - u_{1z}\left(u_{2x} - \sqrt{3}u_{2y} + u_{3x} + \sqrt{3}u_{3y}\right)\right]$$

$$V_{15} = \left(u_{1x}u_{3y} - u_{1y}u_{3x}\right)\left(2u_{1x} - u_{2x} + \sqrt{3}\, u_{2y} - u_{3x} - \sqrt{3}\, u_{3y}\right)$$

$$V_{16} = \left(-u_{1x} + u_{3x}\right)\left(2u_{1x} - u_{2x} + \sqrt{3}\, u_{2y} - u_{3x} - \sqrt{3}\, u_{3y}\right)$$

$$V_{17} = \left(u_{3y} - u_{1y}\right)\left(2u_{1x} - u_{2x} + \sqrt{3}\,u_{2y} - u_{3x} - \sqrt{3}\,u_{3y}\right)$$

$$V_{18} = \sqrt{3}\left(u_{3z} - u_{2z}\right)\left(u_{1y} - u_{3y}\right) + \left(u_{3x} - u_{1x}\right)\left(2u_{1z} - u_{2z} - u_{3z}\right)$$

$$V_{19} = \sqrt{3}\left\{\begin{array}{l}\left(u_{1x} - u_{2x}\right)\left[\left(u_{3z} - u_{1z}\right)\left(-u_{2x} + \sqrt{3}u_{2y}\right) + \left(u_{2z} - u_{1z}\right)\left(u_{3x} + \sqrt{3}u_{3y}\right)\right] + \\ + u_{1x}\left[u_{1x}\left(u_{3z} - u_{2z}\right) + u_{2x}\left(u_{2z} + u_{3z}\right)\right]\end{array}\right\} +$$
$$+ 3\left(u_{1y} - u_{2y}\right)\left[u_{1x}\left(u_{2z} + u_{3z}\right) + u_{3z}\left(-u_{2x} + \sqrt{3}u_{2y}\right) - u_{2z}\left(u_{3x} + \sqrt{3}u_{3y}\right)\right]$$

$$V_{20} = \sqrt{3}\,u_{1x}\left(u_{1y} - u_{2y}\right)\left(u_{2z} - u_{3z}\right) + \left(u_{1x} - u_{2x}\right)\left[-u_{1x}\left(u_{2z} + u_{3z}\right) + u_{1z}\left(u_{2x} - \sqrt{3}u_{2y} + u_{3x} + \sqrt{3}u_{3y}\right)\right]$$

$$V_{21} = \left(u_{1x}u_{2y} - u_{1y}u_{2x}\right)\left(2u_{1x} - u_{2x} + \sqrt{3}\,u_{2y} - u_{3x} - \sqrt{3}\,u_{3y}\right)$$

$$V_{22} = \left(-u_{1x} + u_{2x}\right)\left(2u_{1x} - u_{2x} + \sqrt{3}\,u_{2y} - u_{3x} - \sqrt{3}\,u_{3y}\right)$$

$$V_{23} = \left(-u_{1y} + u_{2y}\right)\left(2u_{1x} - u_{2x} + \sqrt{3}\,u_{2y} - u_{3x} - \sqrt{3}\,u_{3y}\right)$$

$$V_{24} = \sqrt{3}\left(-u_{2y} + u_{1y}\right)\left(-u_{2z} + u_{3z}\right) + \left(u_{1x} - u_{2x}\right)\left(-2u_{1z} + u_{2x} + u_{3z}\right)$$

Torque τ_k (k=1,2,3) on the input crankshaft of each articulated parallelogram can be obtained from the dynamic equilibrium of the leg mechanism again in the form of Eqs (3.9.68) and (3.9.72).

The proposed formulation is a closed-form and can be very useful to simulate the dynamic operation of CaPaMan and to compute straightforwardly the input torques when a motion is given for the platform or the legs.

Results of a numerical simulation are reported in Fig. 3.72 through time history of computed input torques for a given motion of the movable platform that is characterized by inertia parameters I_{cxx}=I_{cyy}= 12,400 kg mm^2, I_{czz}= 24,600 kg mm^2, I_{cxy}=I_{cxz}=I_{cyz} = 0 . No external force and torque are assumed as applied to the platform, i.e. $\mathbf{F}_{ext}$ = 0 and $\mathbf{N}_{ext}$ = 0. This is the case of an application of the CaPaMan prototype as motion simulator. A specific application of CaPaMan has been successfully experienced as an earthquake simulator.

This illustrative example simulates a motion of the mobile plate of CaPaMan so that the center point H traces a vertical straight-line segment going from the starting point to the end point and backward.

Stiffness characteristics of CaPaMan are related mainly to the stiffness of the links, joints and actuators.

Referring to the mechanical design of Figs 3.57 and 3.58 you can model the mechanical system by lumped spring parameters for each CaPaMan leg, as shown in Fig. 3.73.

The stiffness parameters are indicated as k_b for the crank link, k_d for the driven link, k_c for the coupler link, k_h for the connecting coupler rod, and k_T for the actuating system.

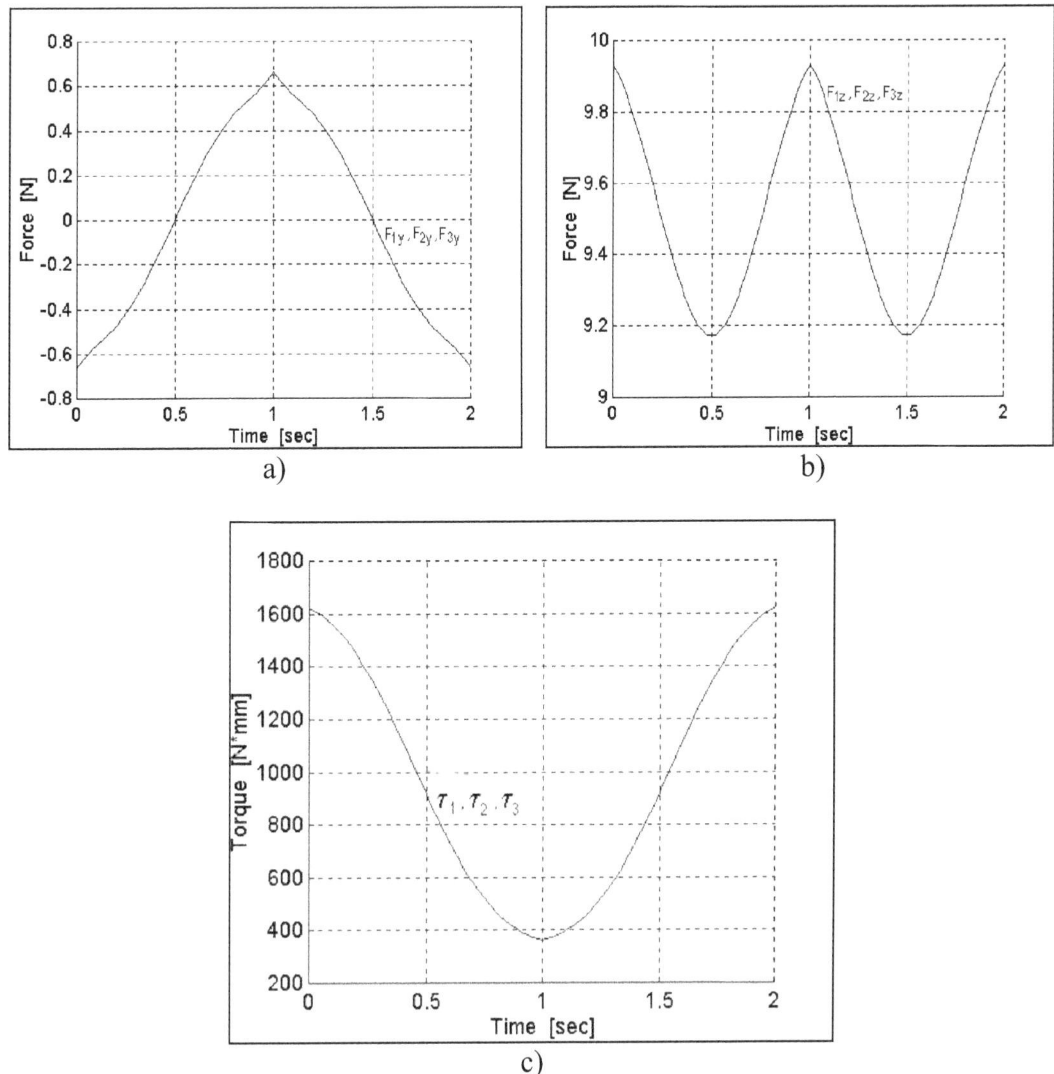

Fig. 3.72: Numerical results of dynamic simulation of CaPaMan for a straight-line motion of H with α_{ki} = 50 deg., $\Delta\alpha_{ki}$ =80 deg., and Δt = 2 sec: a) horizontal joint forces R_{ky}; b) vertical joint forces R_{kz}; c) torques about the input crankshafts of legs.

The crank link b can be modeled as a beam because driving torque and reaction components act upon it so that stiffness can be evaluated through equivalent axial stiffness coefficient k_b. The spring coefficient k_b describes link axial stiffness and support radial stiffness. k_T describes the crank link flexural stiffness and the torsion stiffness of bearing, transmission, and actuator. The driven link d behaves as a rod and its stiffness can be suitably described by the coefficient k_d which takes into account the axial stiffness of the link and the radial stiffness of the joints only. The coupler link c carries the prismatic sliding joint that frees the parallelogram of transversal force components. Thus, the coupler link is loaded and may have flexion deformation. Nevertheless its stiffness coefficient k_c can be considered due only to axial stress since the link is designed for flexion rigidity. In fact generally the coupler link can be considered not deformable since it is massive in order to ensure no flexion deformation for payload capability and high precision positioning of CaPaMan. However k_c can be used to conveniently consider the stiffness of the coupler link and the transversal compliance of the sliding joint.

Because of the ball joint SJ in the mechanical design of the CaPaMan mobile plate, only force acts on each CaPaMan leg and particularly this force lies in the plane of the parallelogram with components R_y and R_z, according to the scheme of Fig. 3.69.

Therefore k_h describes primarily the axial stiffness of the connecting rod and radial stiffness of the joints, although it may take into account the compliance of the prismatic guide when the sliding joint works at its extremity.

A stiffness evaluation of CaPaMan can be obtained by computing in a general way displacements Δx, Δy, Δz, $\Delta\varphi$, $\Delta\psi$, $\Delta\theta$, occurring at the movable plate at a static configuration when a force $F = (F_x , F_y , F_z)$ and/or a torque $T = (T_x , T_y , T_z)$ act upon it, according to the procedure outlined in 3.7.

Referring to Fig. 3.73 and assuming as rigid the coupler c and the connecting rod h, the forces acting for elastic deformation can be expressed as

$$N_b = k_b \,\Delta b$$

$$N_d = k_d \,\Delta d \tag{3.9.80}$$

$$T = \frac{k_T}{b}\,\Delta\alpha_2$$

Thus, the static equilibrium of a CaPaMan leg can be expressed by referring to the equilibrium of the coupler in the form

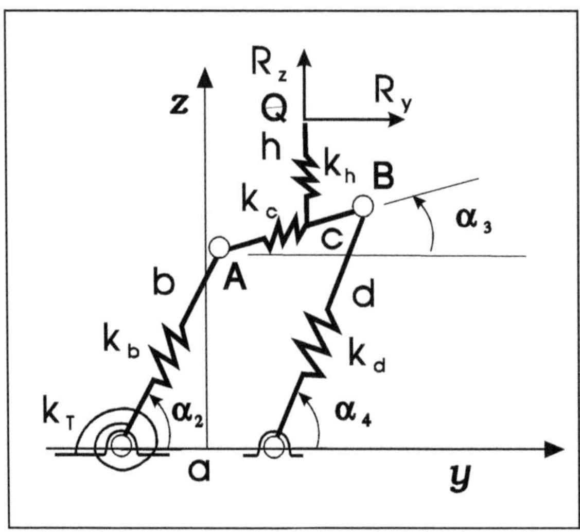

Fig. 3.73: Stiffness parameters in a schematic model for a CaPaMan leg.

$$\begin{vmatrix} R_y \\ R_z \\ 0 \end{vmatrix} = \begin{vmatrix} k_b\, c\alpha_2 & k_d\, c\alpha_4 & -\frac{k_T}{b} s\alpha_2 \\ k_b\, s\alpha_2 & k_d\, s\alpha_4 & \frac{k_T}{b} c\alpha_2 \\ k_b\, r_b & -k_d\, r_d & \frac{k_T}{b} r_T \end{vmatrix} \begin{vmatrix} \Delta b \\ \Delta d \\ \Delta\alpha_2 \end{vmatrix} \tag{3.9.81}$$

with

$$r_b = \frac{c}{2}\sin(\alpha_2+\alpha_3) + h\cos(\alpha_2+\alpha_3)$$

$$r_d = -\frac{c}{2}\sin(\alpha_2+\alpha_4) + h\cos(\alpha_2+\alpha_4) \tag{3.9.82}$$

$$r_T = \frac{c}{2}\cos(\alpha_2-\alpha_3) + h\sin(\alpha_2-\alpha_3)$$

where the moment of the forces has been computed about point Q by considering null the external moment, since the ball joint on the top of the leg mechanism does not transmit torque. The coordinates of Q, coinciding with H_j (j=1,2,3) of the corresponding leg, can be computed by the geometry of Fig. 3.73 in a general form

$$y = -\frac{a}{2} + b\cos\alpha_2 + \frac{c}{2}\cos\alpha_3 - h\sin\alpha_3$$

$$z = b\sin\alpha_2 + \frac{c}{2}\sin\alpha_3 + h\cos\alpha_3 \tag{3.9.83}$$

when the angle α_3 is zero Eqs (3.9.83) become Eqs (3.9.26)
Compliant displacements can be computed by using Eqs (3.9.83) for Δy and Δz and by considering $\Delta\alpha_3$ as orientation angle variation for the coupler so that the following expression can be obtained

$$\begin{vmatrix} \Delta y \\ \Delta z \\ \Delta\alpha_3 \end{vmatrix} = \begin{vmatrix} \cos\alpha_2 - \frac{c-2h}{2c}\sin\alpha_2 & \frac{c-2h}{2c}\sin\alpha_2 & -b\frac{3c-2h}{2c} \\ \frac{3c-2h}{2c}\sin\alpha_2 & -\frac{c-2h}{2c}\sin\alpha_2 & b\frac{3c-2h}{2c} \\ \frac{\sin\alpha_2}{c} & -\frac{\sin\alpha_2}{c} & \frac{b}{c} \end{vmatrix} \begin{vmatrix} \Delta b \\ \Delta d \\ \Delta\alpha_2 \end{vmatrix} \tag{3.9.84}$$

when small deformations are assumed to allow a linearized formulation with $\alpha_2 = \alpha_4$.

Angle $\Delta\alpha_3$ has been computed by considering the position of points A and B in Fig. 3.73 to fulfil the condition

$$\sin \alpha_3 = \frac{b \sin\alpha_2 - d \sin\alpha_4}{c} \tag{3.9.85}$$

By substituting Eq. (3.9.84) into Eq. (3.9.81) one can obtain

$$\mathbf{F}_R = K_L \, \Delta\mathbf{v} \tag{3.9.86}$$

where $\mathbf{F}_R$ is the vector of the actions on the CaPaMan leg mechanism, $\Delta\mathbf{v}$ is displacement vector, and K represents the 3×3 stiffness matrix of a CaPaMan leg which can be computed according to Eq. (3.9.12) as

$$K_L = C_{LF} K_{LP} C_{LK} \tag{3.9.87}$$

with

$$C_{LF} = \begin{vmatrix} \cos\alpha_2 & \cos\alpha_2 & -\sin\alpha_2 \\ \sin\alpha_2 & \sin\alpha_2 & \cos\alpha_2 \\ r_b & -r_d & r_T \end{vmatrix} \tag{3.9.88}$$

$$C_{LK} = \begin{vmatrix} \cos\alpha_2 - \dfrac{c-2h}{2c}\sin\alpha_2 & \dfrac{c-2h}{2c}\sin\alpha_2 & -b\dfrac{3c-2h}{2c} \\ \dfrac{3c-2h}{2c}\sin\alpha_2 & -\dfrac{c-2h}{2c}\sin\alpha_2 & b\dfrac{3c-2h}{2c} \\ \dfrac{\sin\alpha_2}{c} & -\dfrac{\sin\alpha_2}{c} & \dfrac{b}{c} \end{vmatrix} \tag{3.9.89}$$

$$K_{LP} = \begin{vmatrix} k_b & 0 & 0 \\ 0 & k_d & 0 \\ 0 & 0 & \dfrac{k_T}{b} \end{vmatrix} \tag{3.9.90}$$

Then, CaPaMan stiffness matrix can be evaluated by considering the three legs's architecture and the general force transmission formulation.
In particular the static equilibrium of the movable plate can be expressed as

$$\begin{vmatrix} F_x \\ F_y \\ F_z \\ N_x \\ N_y \\ N_z \end{vmatrix} = \begin{vmatrix} \sin\delta_1 & 0 & \sin\delta_2 & 0 & \sin\delta_3 & 0 \\ \cos\delta_1 & 0 & \cos\delta_2 & 0 & \cos\delta_3 & 0 \\ 0 & 1 & 0 & 1 & 0 & 1 \\ 0 & r_1 \sin\delta_1 & 0 & r_2 \sin\delta_2 & 0 & r_3 \sin\delta_3 \\ 0 & r_1 \cos\delta_1 & 0 & r_2 \cos\delta_2 & 0 & r_3 \cos\delta_3 \\ r_1 & 0 & r_2 & 0 & r_3 & 0 \end{vmatrix} \begin{vmatrix} R_{y1} \\ R_{z1} \\ R_{y2} \\ R_{z2} \\ R_{y3} \\ R_{z3} \end{vmatrix} \tag{3.9.91}$$

which can be written in a synthetic matrix form of Eq. (3.9.14) as

$$\mathbf{W} = C_F \, \mathbf{W}_L \tag{3.9.92}$$

where **W** represents the external actions, force and torque, acting on the movable plate; C_F is the so-called 'transmission matrix' describing the equilibrium relations and **R** is the vector of force components R_y and R_z acting on the three leg mechanisms.
R can be obtained from $\mathbf{F}_R$ when the third component is neglected since the rotation equilibrium of the coupler link of the leg mechanisms does not affect the static equilibrium of the movable plate because of the ball joint. Then, **R** can be obtained by F_{R1} , F_{R2} , F_{R3} as

$$\mathbf{R} = \begin{vmatrix} R_{p1} & R_{p2} & R_{p3} \end{vmatrix}^t \tag{3.9.93}$$

where R_{pj} (j=1,2,3) represents the partitioned vector F_{Rj}.
In an analogous way the meaningful compliant displacements are Δy and Δz, and angle $\Delta\alpha_3$ can be neglected so that the displacement vector $\Delta\mathbf{v}$ can be written as

$$\Delta\mathbf{v} = \begin{vmatrix} \Delta y_1 & \Delta z_1 & \Delta y_2 & \Delta z_2 & \Delta y_3 & \Delta z_3 \end{vmatrix}^t \tag{3.9.94}$$

Consequently it yields to

$$R_{pj} = K_{pj} \begin{vmatrix} \Delta y_j & \Delta z_j \end{vmatrix}^t \tag{3.9.95}$$

where K_{pj} is a 2×2 matrix which has been obtained as a suitable portion of K expressed by Eqs (3.9.87) and (3.9.90). Finally, K_p of the legs can be formulated as

$$K_p = \begin{vmatrix} K_{p1} & 0 & 0 \\ 0 & K_{p2} & 0 \\ 0 & 0 & K_{p3} \end{vmatrix} \tag{3.9.96}$$

to give

$$\mathbf{R} = K_p \, \Delta\mathbf{v} \tag{3.9.97}$$

In addition, the compliant displacements $\Delta\mathbf{v}$ produce compliant displacement of the movable plate $\Delta\mathbf{S} = (\Delta x, \ \Delta y, \Delta z, \Delta\varphi, \Delta\theta, \Delta\psi)^t$ that can be computed by using the direct kinematic formulation of Eqs (3.9.8) and (3.9.26) in the form of Eq. (3.9.16) to give

$$C_K = \begin{vmatrix} C_x & 0 & -\frac{1}{\sqrt{3}} & 0 & \frac{1}{\sqrt{3}} & 0 \\ 1 & 0 & C_y & 0 & 0 & 0 \\ 0 & \frac{1}{3} & 0 & \frac{1}{3} & 0 & \frac{1}{3} \\ C_\varphi & -\frac{2}{3r_p} & 0 & \frac{2}{3r_p} & 0 & \frac{2}{3r_p} \\ 0 & 0 & 0 & -\frac{\sqrt{3}}{C_\psi} & 0 & \frac{\sqrt{3}}{C_\psi} \\ \frac{1}{C_\theta} & 0 & \frac{1}{C_\theta} & \frac{\sqrt{3}}{C_\psi} & \frac{1}{C_\theta} & -\frac{\sqrt{3}}{C_\psi} \end{vmatrix} \tag{3.9.98}$$

with

$$C_x = -\frac{r_p}{2\Delta y_1}(1 - \sin\Delta\varphi)\cos(\Delta\psi - \Delta\theta)$$

$$C_y = -\frac{r_p}{\Delta y_2}(\sin\Delta\psi \ \cos\Delta\varphi + \cos\Delta\psi \ \sin\Delta\varphi \sin\Delta\theta)$$

$$C_\varphi = \frac{1}{\Delta y_1} \tag{3.9.99}$$

$$C_\psi = 2\,\Delta z_1 - \Delta z_2 - \Delta z_3$$

$$C_\theta = \frac{3r_p}{2}(1 + \sin\Delta\varphi)$$

Thus, the stiffness matrix of CaPaMan is computed as

$$K_{CaPaMan} = C_F \, K_p \, C_K \tag{3.9.100}$$

A simplified analysis for CaPaMan stiffness can be useful for an easy characterization of CaPaMan.
The basic idea consists of deriving an equivalent mechanical system by using a suitable scheme for stiffness composition and equivalence. A simple way to build an equivalent scheme is to use the basic concept of stiffness composition for serial and parallel spring systems. These well-known rules give the equivalent stiffness for a series of springs as the sum of the stiffness parameters of components and for a parallel system of springs as the sum of the inverse of the stiffness parameters of components.
Thus, from the leg mechanism scheme of Fig. 3.73 with torsion stiffness parameters k_{Tb} and k_{Td} that are considered in both the frame joints, one can easily obtain the equivalent scheme of Fig. 3.74 by using the above-mentioned rules. In particular one can compute a stiffness coefficient of a CaPaMan leg mechanism from Eq. (3.9.86) in the form

$$R_p = K_{Leq} \begin{vmatrix} \Delta y & \Delta z \end{vmatrix}^t \tag{3.9.101}$$

with the so-called 'equivalent spring matrix' that is defined as

$$K_{Leq} = \begin{vmatrix} k_y & 0 \\ 0 & k_z \end{vmatrix} \tag{3.9.102}$$

with

$$k_y = \left(\frac{1}{k_b} + \frac{1}{k_d} \right) \cos \alpha_2 + k_c + \frac{k_{Ty}}{b}$$

(3.9.103)

$$k_z = \left(\frac{1}{k_b} + \frac{1}{k_d} \right) \sin \alpha_2 + k_h + \frac{k_{Tz}}{b}$$

in which

$$k_{Ty} = \left(\frac{1}{k_{Tb}} + \frac{1}{k_{Td}} \right) \sin \alpha_2$$

(3.9.104)

$$k_{Ty} = \left(\frac{1}{k_{Tb}} + \frac{1}{k_{Td}} \right) \cos \alpha_2$$

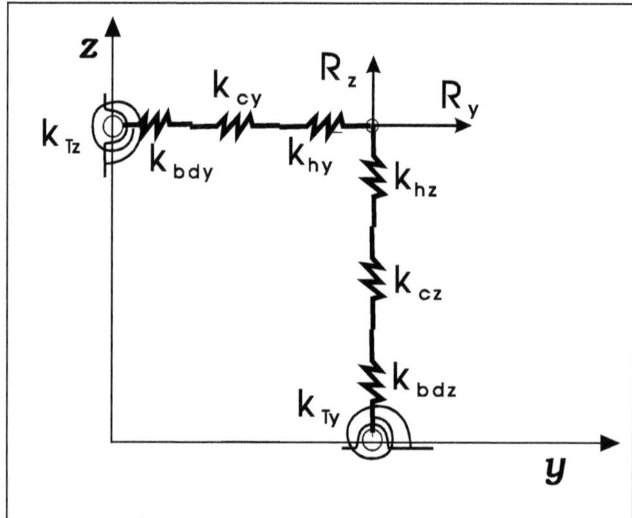

Fig. 3.74: An equivalent model for stiffness evaluation of the CaPaMan leg mechanism.

Furthermore, by again using the composition rules for parallel spring systems' stiffness coefficients for CaPaMan can be computed for a 3D Cartesian scheme as

$$k_{CaPaMan-xs} = \frac{\sin\delta_1}{k_{y1}} + \frac{\sin\delta_2}{k_{y2}} + \frac{\sin\delta_3}{k_{y3}}$$

$$k_{CaPaMan-ys} = \frac{\cos\delta_1}{k_{y1}} + \frac{\cos\delta_2}{k_{y2}} + \frac{\cos\delta_3}{k_{y3}} \tag{3.9.105}$$

$$k_{CaPaMan-zs} = \frac{1}{k_{z1}} + \frac{1}{k_{z2}} + \frac{1}{k_{z3}}$$

Although this proposed procedure is very approximate and inncomplete, the equivalent scheme may help to understand the role of the stiffness of components on the overall system behavior in a simple, straightforward way.

A practical performance index for stiffness evaluation of robots is established in standard codes, such as function of the measurements of linear compliant displacements of end-effector reference point when a force is acting on the robot.

Thus, the basic idea is to have a synthetic scalar index for stiffness for each coordinates direction. In order to extend this idea it can be thought convenient to formulate scalar quality index $C_{CaPaMan}$ which synthetically describes stiffness characteristics of CaPaMan from more detailed formulation as the herein-proposed stiffness matrix.

Indeed the proposed analysis with the equivalent stiffness parameters can be considered for an approach to give a synthetic, easily computable quality index in the form

$$C_{CaPaMan} = k_{CaPaMan-xs} + k_{CaPaMan-ys} + k_{CaPaMan-zs} \tag{3.9.106}$$

However, even by using the results of the proposed analysis for CaPaMan stiffness matrix one can formulate a quality index C_{leg} for the leg mechanism's stiffness from Eq. (3.9.87) in the form

$$C_{leg} = \det K_L \tag{3.9.107}$$

and a quality index $C_{CaPaMan}$ for the overall parallel manipulator from Eq. (3.9.106) as

$$C_{CaPaMan} = \det K_{CaPaMan} \tag{3.9.108}$$

where det is the determinant of a matrix.
The proposed indices of Eqs (3.9.106) to (3.9.108) give a synthetic characterization of CaPaMan by considering the stiffness of a given CaPaMan configuration so that they can be considered as a local numerical evaluation.
Since the manipulation aim can be described by several precision points, it can be of interest to define a so-called 'global stiffness performance index' as a function of CaPaMan configurations within a given mobility range. By using any of the previous proposed indices a global stiffness quality index $GS_{CaPaMan}$ for CaPaMan can be formulated in the form

$$GS_{CaPaMan} = \frac{\Delta C_{CaPaMan}}{\left|C_{CaPaMan}\right|_{max}} \tag{3.9.109}$$

where $\Delta C_{CaPaMan}$ is the variation of the stiffness quality index $C_{CaPaMan}$ within the mobility range and the numerator term is the maximum value of the considered index.
The proposed indices can be useful to characterize CaPaMan and can be used to classify different designs or suggest design changes for optimum stiffness behavior.
The CaPaMan manipulator has also been used to test the engineering feasibility of the formulation for optimum design of parallel manipulators. Indeed, by using the existent prototype, simulations have been carried out to validate the proposed optimum design by considering several guess solutions and imposing workspace and stiffness characteristics of the built prototype. The numerical results of the tests have been obtained according to the dimensions of the prototype.
Singularity analysis for the CaPaMan manipulator can be characterized by a singularity condition that can be expressed as

$$G = \frac{9r_p^2}{2b_k^2} \tag{3.9.110}$$

Equation (3.9.110) represents the radius of a cylinder, which divides the configuration space into two regions free from singularities: the region inside and outside the cylinder. Indeed, by properly choosing design parameters it is possible to have a G value greater than 8 to obtain an architecture for the CaPaMan manipulator which is free from singularities.

Thus, Eq. (3.9.110) can be used conveniently as objective function for prescribing singularity-free conditions.

Stiffness analysis of CaPaMan has been formulated by modeling each leg of CaPaMan as explained previously according to the scheme of Fig. 3.73. The lumped stiffness parameters can be assumed as k_{bk}=k_{dk}=2.625 × 10^6 N/m and k_{Tk}= 58.4 × 10^3 Nm/rad.

Position and orientation workspace volumes have been conveniently evaluated by using the kinematics of the CaPaMan manipulator and they have been characterized by using the schemes of Fig. 3.75 for the approximation in the optimization numerical process.

Thus, the design problem for the CaPaMan manipulator has been formulated as minimizing a function F (X), which is the weighted sum of the objective functions as in Eqs (3.9.26) but grouping those criteria for facilitating the numerical procedure to have

$$
\begin{aligned}
f_1(\mathbf{X}) &= \left| 1 - \frac{V_{pos}{}^*}{V_{pos}{}'} \right| \\
f_2(\mathbf{X}) &= \left| 1 - \frac{V_{or}{}^*}{V_{or}{}'} \right| \\
f_3(\mathbf{X}) &= \left(1 - \frac{|\Delta x_d|}{|\Delta x_g|} \right) + \left(1 - \frac{|\Delta y_d|}{|\Delta y_g|} \right) + \left(1 - \frac{|\Delta z_d|}{|\Delta z_g|} \right) + \\
&\quad + \left(1 - \frac{|\Delta \varphi_d|}{|\Delta \varphi_g|} \right) + \left(1 - \frac{|\Delta \psi_d|}{|\Delta \psi_g|} \right) + \left(1 - \frac{|\Delta \theta_d|}{|\Delta \theta_g|} \right) \\
f_4(\mathbf{X}) &= |1 - G/8|
\end{aligned}
\tag{3.9.111}
$$

with the constraints

$$
\begin{aligned}
& x_{max} \le x'_{max}\,;\, y_{max} \le y'_{max}\,;\, z_{max} \le z'_{max} \\
& \varphi_{max} \le \varphi'_{max}\,;\, \psi_{max} \le \psi'_{max}\,; \\
& (\theta_{max})_\psi \le (\theta'_{max})_\psi
\end{aligned}
\tag{3.9.112}
$$

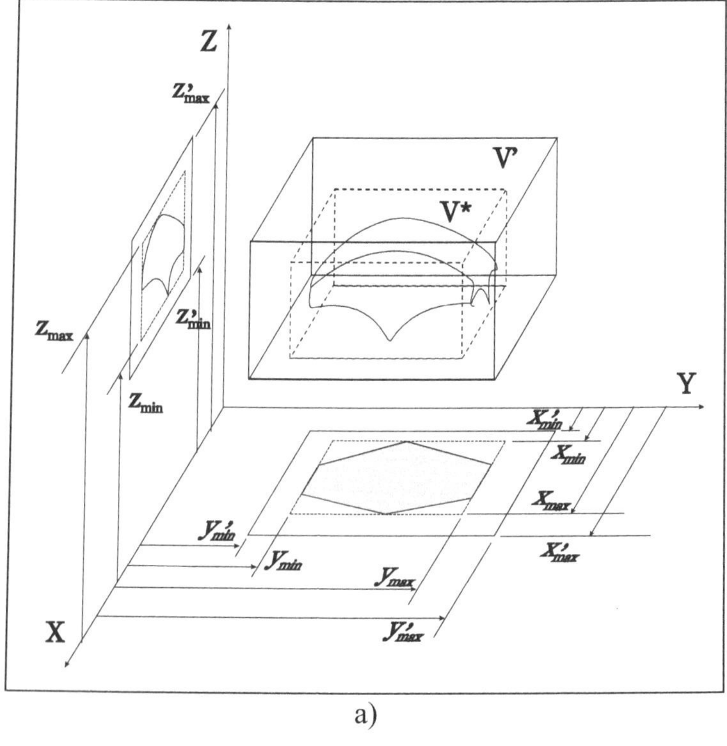

a)

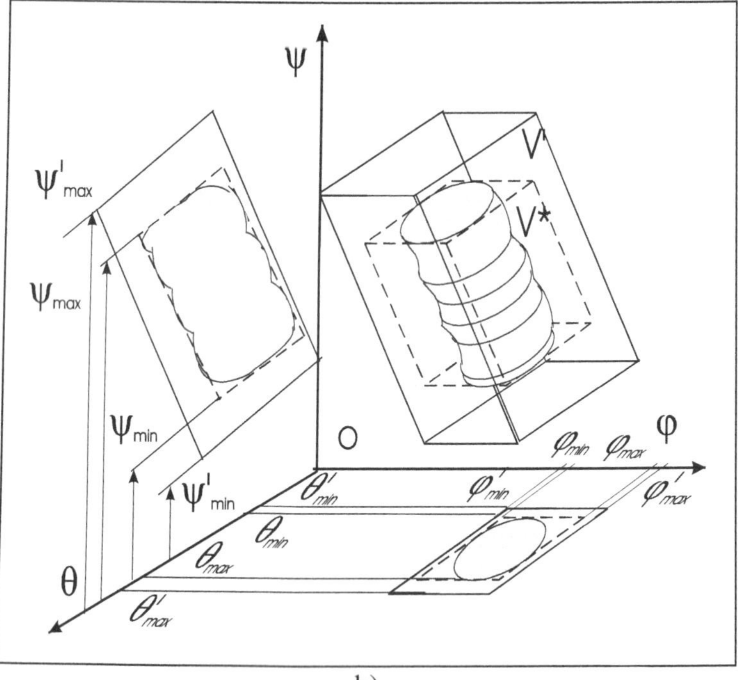

b)

Fig. 3.75: A numerical approximation for workspace evaluation of the CaPaMan manipulator in terms of: a) volume of the position workspace; a) volume of the orientation workspace.

when weighting factors w_i can be assumed as $w_1 = w_2 = 3$; $w_3 = 2$; $w_4 = 1$. Design parameters of the optimum solution for CaPaMan architecture have been obtained as $a_k = 199.0$ mm; $b_k = 90.0$ mm; $h_k = 100.0$ mm; $r_P = 109.3$ mm; $s_k = 0–50.0$ mm; and $\alpha_k = 80–135$ deg.

Figure 3.76 a) shows the objective function F (X) and each component f_i of the objective function during the numerical process, which takes 20 iterations. Figure 3.76 b) shows the evolution of some of the constraints in Eqs (3.9.11).

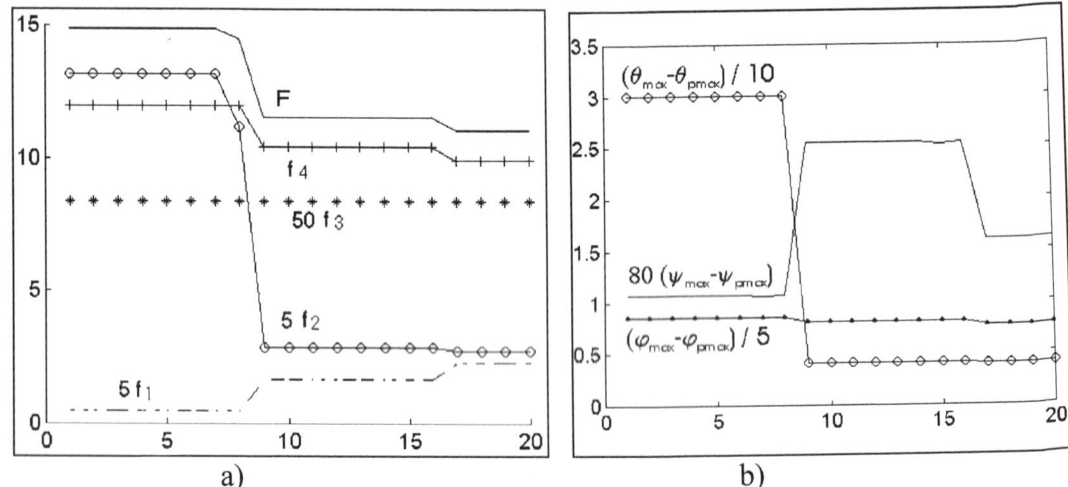

Fig. 3.76: Results of a numerical example of the proposed multi-objective design formulation for CaPaMan: a) plots of the evolution of the objective functions; b) plots of the evolution of the design constraints.

Chapter 4:
Fundamentals of the mechanics of grasp

4.1 Gripping devices and their characteristics

The tool extremity of a manipulator, which is generally denominated as the end-effector, has the fundamental role of interaction with the environment and objects that are manipulated. Indeed, the success of a manipulator strongly depends on the end-effector efficiency that is based on its design characteristics and operation performances.

The variety of grasping tasks requires a variety of mechanical designs of end-effectors. They are usually designed for very specific applications, when the scope is well identified in the size, shape, and material of objects, as well as in a given manipulator operation. But they are also given for universal aims that are achieved through anthropomorphic devices that mimic the human hand with high flexibility and versatility. Intermediate solutions are also available and designed in a great variety.

In addition, a distinction is usually considered between proper grasping devices, whose purpose consists of grasping an object, and so-called 'general end-effectors', whose action differs from grasping, as for example a screwing operation.

Indeed, end-effectors are specific tools that are devoted, and/or applied to robots for specific operations and they are often adapted to robots through suitable attachments on the wrist flange only. Those end-effector tools can also be designed and built for specific applications in specific robots with the aim of specializing and/or optimizing the use of a robotic system.

Grasping devices are considered of primary importance since they can complete the manipulation capability of a robot by giving the practical success of a manipulation.

The main characteristics of a grasping device can be recognized in the following aspects:

a) grasping capability;
b) maintaining and release of grasp;
c) mechanical design and actuation;
d) compatibility with robots.

The aspects a) and b) are concerned with the operation of grasping devices in their function of interacting with the environment through the grasped objects and they can be analyzed by means of the mechanics of grasp, as in the following chapter.

In particular, the grasping capability is related to the possibilities of the grasping devices in terms of shape, dimension, and weight of the objects that can be grasped. The hold and release phases of grasping are concerned with the change of the mechanics of grasp during the operation and manipulation of a robot with grasped objects.

The design aspects of item c) are related to the mechanical designs and operation performances of grasping devices that are available in a great variety.

The compatibility with robots at item d) can be considered from interface viewpoints both for electronic/electric hardware and mechanical design. But it is mainly concerned with the operation of the grasping device that should not limit or complicate the operation of the robot in its practical application. With this aim, the industrial grasping devices are usually made with one d.o.f. (degree of freedom) only and they are controlled directly by the robot control unit.

The different mechanical designs of grasping devices can be classified as a function of grasp types as:

- anthropomorphic grasp, which is similar to human actions when the grasp of objects is obtained by means of bilateral contacts or more general multilateral contacts, as shown in the examples of Fig. 4.1a);
- non-anthropomorphic grasp, when the grasp is performed through configurations which are not possible for human hands, such as in unilateral contacts and/or by using specific systems and phenomena for specific material and geometry of an object, as shown in the examples of Fig. 4.1b) and c).

In addition, the grasping devices can be classified as compared to human hands which they try to substitute or mimic. Indeed, even the terminology for the gripper parts refers to human anatomy: fingers indicate the elements that execute the grasp on objects, and fingertips are the region and material of the fingers that are in contact with the objects. Therefore, a grasping device can be also named as:

- anthropomorphic, when the grasp is performed by operating as human fingers;
- non-anthropomorphic, when the grasp is performed in other ways.

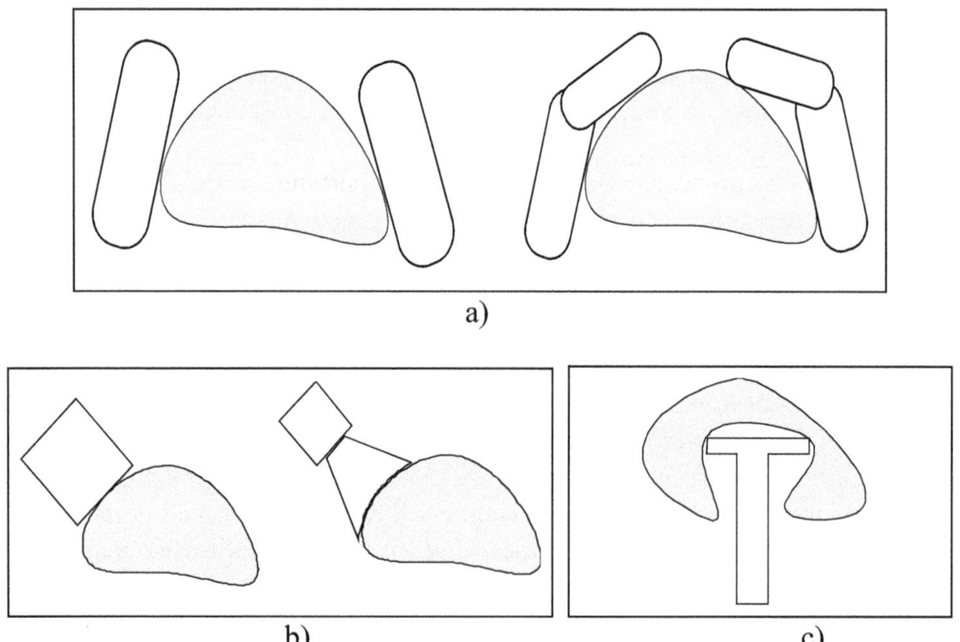

Fig. 4.1: Examples of grasp types: a) anthropomorphic grasp with bilateral contacts; b) non-anthropomorphic grasp with unilateral contacts; c) non-anthropomorphic grasp with internal contacts.

In addition, the anthropomorphic grasping devices can be differentiated in:
- grippers, when the fingers do have not an anthropomorphic design but they are rigid elements of a mechanism;
- hands, when the fingers are made with anthropomorphic mechanical designs.

Generally, industrial grippers are designed as two-finger systems, which are powered and controlled for the grasping action by one actuator only.

The hands are usually made of complex architectures with a more complicated operation that is powered by several actuators. The anthropomorphic grasping hand devices can be designed with two or more articulate fingers, which show two or more phalanges.

Besides the grasping devices, specific systems have been developed with specific design to grasp specific objects by using alternative grasping methodologies. They operate by using specific phenomena with peculiar design architectures, such as pneumatic systems, vacuum devices, and electrostatic or magnetic devices. In theses cases they can be properly named 'end-effectors' in order to distinguish them from the above-mentioned grippers and hands.

In particular, pneumatic end-effectors make use of pneumatics in order to operate fingertips that can be expansive tubes for internal or external grasp, mainly for delicate objects. Figure 4.2 shows examples of these end-effectors. In particular, in Fig. 4.2a) a finger is shown when it is filled and ready to contact delicate objects; in Fig. 4.2b) an application is reported in which a multi-fingered end-effector grasps an electronic card with several expanded fingers.

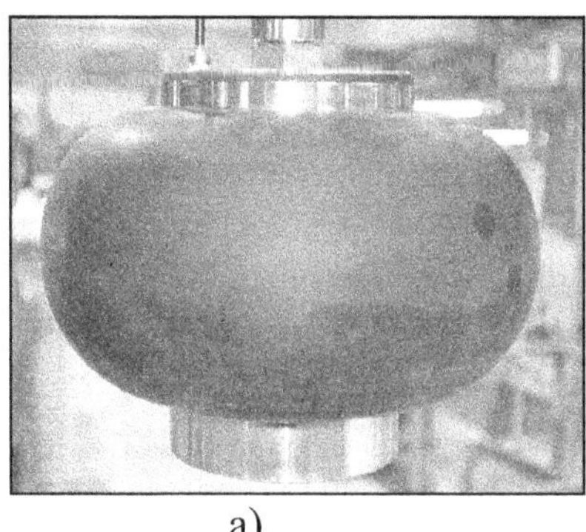

a)

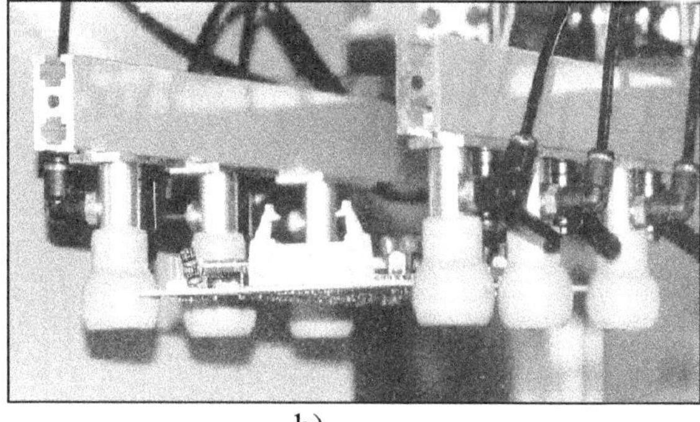

b)

Fig. 4.2: Examples of pneumatic expanding finger end-effectors; a) one finger for delicate objects; b) an application with several expanded fingers grasping an electronic card.

The vacuum systems are widely used with cups that can be properly shaped and sized as a function of the grasped objects that must have a suitable smooth surface.

In Fig. 4.3 examples of suction-cups are shown as those that are used in robot pneumatic end-effectors. Examples of pneumatic vacuum end-effectors are shown for the case of a multi suction-cup end-effector in Fig. 4.4a) and for a palm end-effector with several suction-cups in Fig. 4.4b).

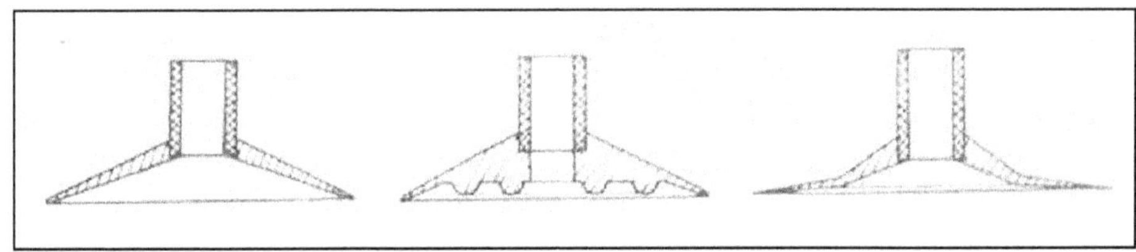

Fig. 4.3: Examples of suction-cups for robotic end-effectors.

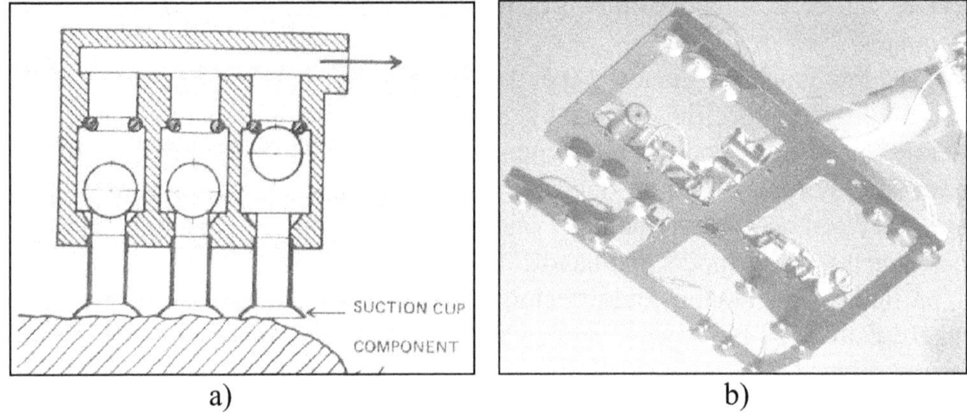

Fig. 4.4: Examples of end-effectors with suction-cups: a) a scheme for a multi suction-cup end-effector; b) a palm end-effector with several suction-cups.

The electrostatic and magnetic systems can be used for objects with ferromagnetic properties so that they can be grasped through the action of electromagnetic forces.
In Fig. 4.5 a scheme for a typical electromagnetic end-effector is shown together with a slider-crank mechanism, which is useful for smooth releasing.

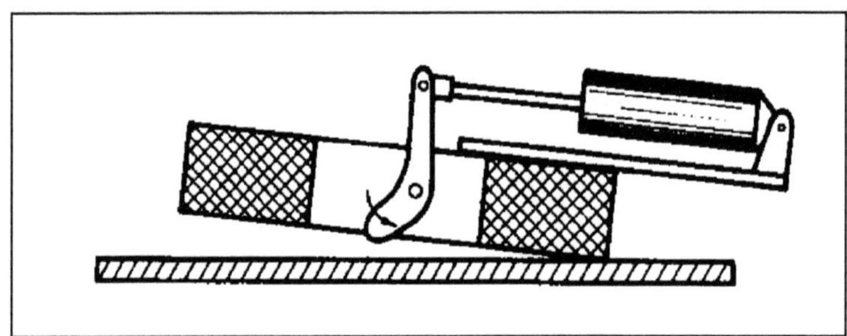

Fig. 4.5: A scheme for an electromagnetic end-effector with a releasing mechanism.

However, all grasping devices should fulfil the following requirements, mainly for installation on robots: low-cost, robust design and operation, simplicity in the mechanical design and operation, and easy integration.
In particular, robotic systems require low-cost grasping devices in order to confirm the economical advantage of a robotization. The gripper design can be considered of

fundamental importance since a grasping device is the robot extremity that interacts with the environment and performs tasks. Therefore it is responsible for the success of a robotized manipulation as characterized by movement and action on objects. A robust operation can refer mainly to high reliability of the operation of a grasping device, even in unusual situations with critical aspects.
The simplicity of the mechanical design can be considered as a fundamental requirement in order not to complicate further a robotic system with the goal of a specific grasping task. Such design simplicity can help the maintenance of a robotic system, and also its operation both in terms of programming and integrated performance with robotic arm. Indeed, such integration can be considered essential to achieve a suitable versatility of the robotic system when it is completed with a grasping device.
Usually, grasping devices and particularly grippers are designed with a certain modularity so they can be easily adjusted to different grasping situations, mostly by interchanging fingers or fingertips as shown in the illustrative cases of Fig. 4.6.

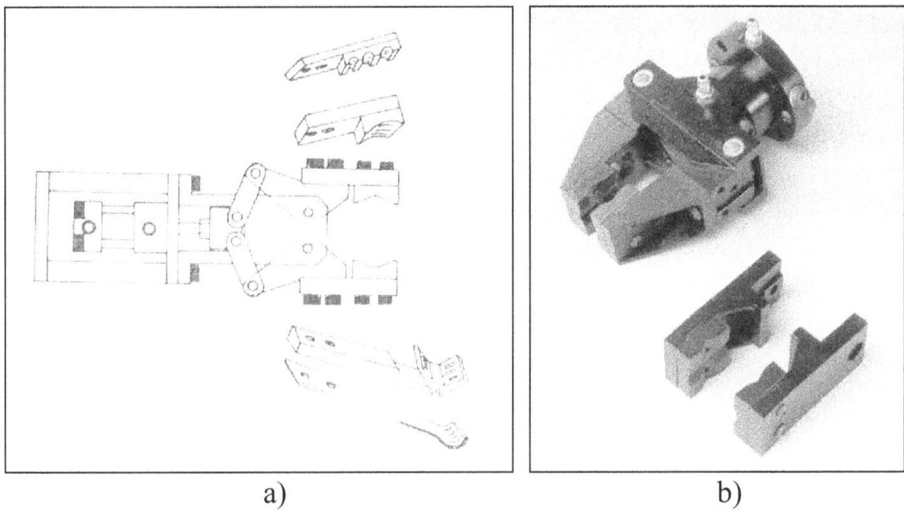

a) b)

Fig. 4.6: A gripper with interchangeable fingers with different fingertips: a) a scheme for a mechanical design; b) an industrial gripper for robots.

The above-mentioned requirements are usually obtained in such a way that industrial grasping devices are not equipped with sensors for controlling the grasp that can be ensured by a previous analysis of the grasp mechanics and grasping operation/strategy.
Increasingly, an industrial finger can be sensored through a force sensor or rarely by means of tactile sensors in order to control grasping. However, most of the industrial grippers are made to operate with two fingers, since most of the grasps can be performed with two fingers only and two-finger grippers are the smallest suitable mechanical architectures for grasping hand devices.
In fact, the widest applied architecture in industrial applications is the two-finger gripper with a planar mechanism as gripping mechanism, since it operates successfully with the simplest and robust mechatronic design. The wide diffusion of two-finger

grippers has also been determined by the fact that they can be easy to manufacture, install, and run. Moreover, in some cases they have the easy possibility of being interchanged during the operations, and to be modified and enhanced in their components with no great additional efforts.

Although the human hand is provided with five articulated fingers, The daily experience shows that two fingers perform most of the grasps, mainly with regular objects. In addition, even the five fingers are used in combination by acting as two fingers only.

Regular objects are those objects having a shape of simple geometry that can be represented by analytical expressions.

Figure 4.7 gives an illustrative example of frequency of different grasps by human hand but specifically for the case of a cylindrical object of a size comparable with the hand dimension. Of course the human hand has many other possibilities because of the 19 d.o.f.s for the five fingers and other d.o.f.s for the palm. In fact, the human hand can perform a great variety of different grasp configurations by using the fingers in several ways to adjust a grasp to an object mainly in termsof its shape.

Nevertheless, the two-finger grasp is the most used even for complex tasks, as for example in the case of Fig. 4.8 referring to food manipulation by Chinese chopsticks. This two-finger grasp is well-established in Asiatic populations (with billions of users!) for grasping not only solid objects, and it can be dated to several centuries B.C.

Therefore this chapter is focused mainly on two-finger grippers that can be designed and operated in industrial and non-industrial applications with the aforementioned characteristics for high reliability and robustness.

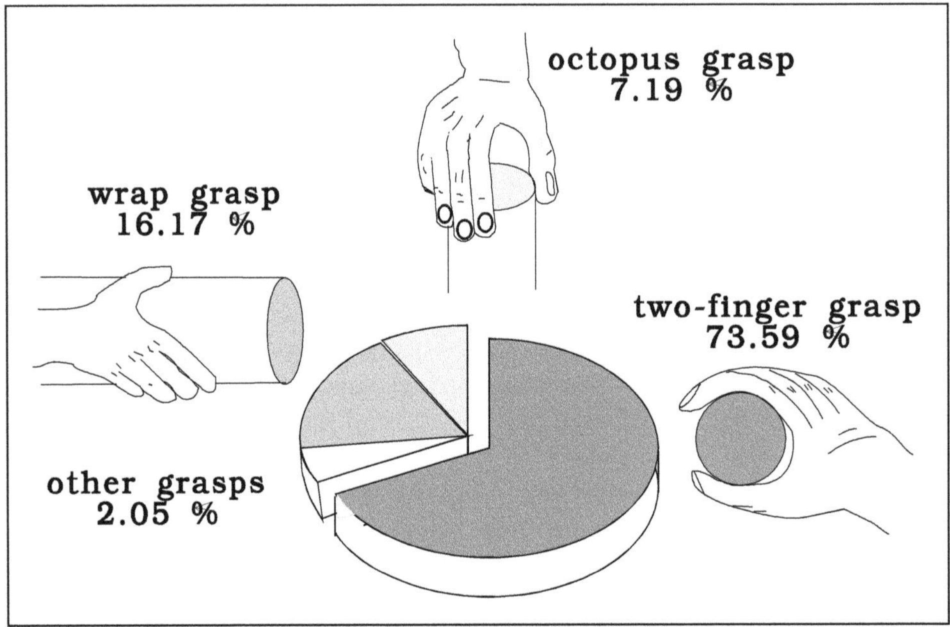

Fig. 4.7: Statistical data on human grasping modes for a cylinder object.

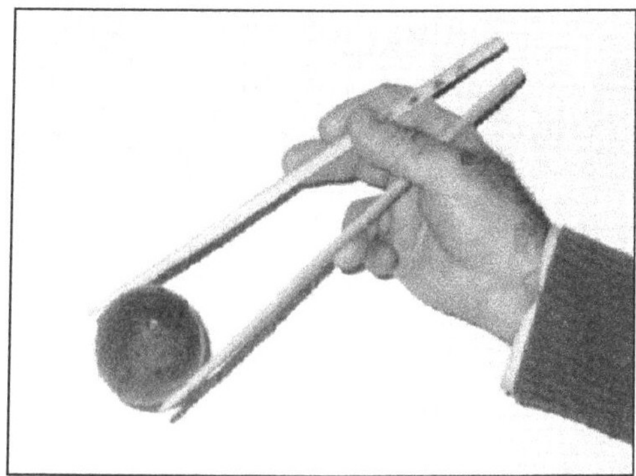

Fig. 4.8: A typical two-finger grasp: the Chinese chopsticks.

In addition, two-finger grippers are also useful for an easy application for more complicated tasks. Figures 4.9 and 4.10 show examples of such typical applications: Fig. 4.9 shows how two fingers can be properly designed and located to grasp long objects by their extremities only; Fig. 4.10 shows alternative grasps by using two or more two-finger grippers to grasp long objects and even large 3D shaped objects.

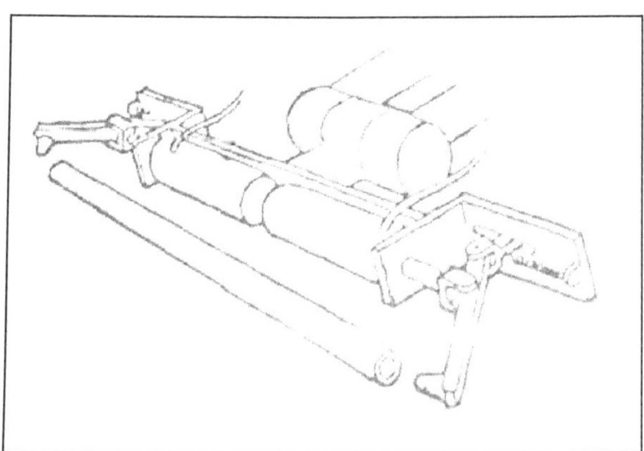

Fig. 4.9: An example of a specific two-finger gripper for long objects.

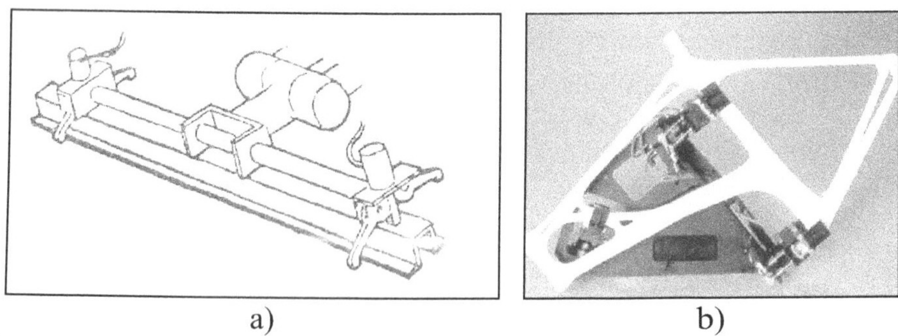

a) b)

Fig. 4.10: Examples of multi two-finger end-effectors with: a) two two-finger grippers for long objects; b) three two-finger grippers for large objects.

In Fig. 4.11 a general scheme is shown for an industrial gripper with two fingers composed of:
- an actuation system, that usually is of a pneumatic type with suitable pipelines and electrovalve;
- a transmission system, connecting the actuator to a driving mechanism;
- a driving mechanism that transmits movement and force to the fingers;
- fingers and fingertips, that can be specifically designed and shaped for contacting specific objects.

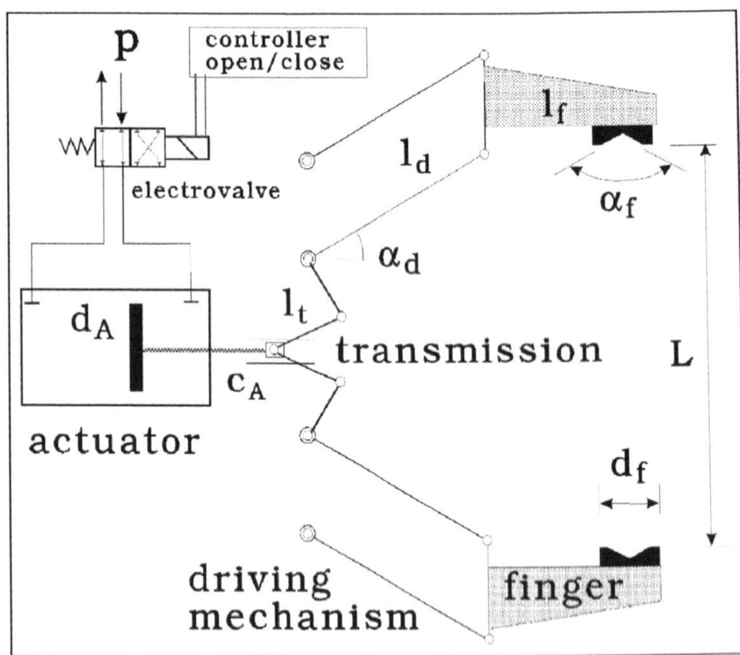

Fig. 4.11: A scheme for a general design and components of a two-finger gripper.

In Fig. 4.11 design parameters are also indicated but the controller is only mentioned as the robot control unit usually carries out the control action yet.
Indeed, the interactions among the gripper components are so strong that a gripper can be properly considered as a mechatronic system since its design and operation depend both on the mechanics of grasping and operation of the electropneumatic sub-system. Therefore, the modeling and formulation of mechanics of grasp and actuation operation of electropneumatic components can be considered of fundamental importance and a mechatronic approach is needed both for operation and design purposes.

4.2 A mechatronic analysis for two-finger grippers

Figure 4.12 shows a general diagram for mechatronic layout of grippers by stressing main components and operations.

The goal of a gripper is the grasp of an object but the successful grasping action includes keeping the grasp performances during the manipulation and release of the grasped object. Grasping objects can be considered a complex operation, which basically requires an in-depth study of mechanics since it is mainly related to mechanical interaction between object and fingers, but the grasping operation and efficiency strongly depend on the operation of all gripper components. The mechanical aim is performed by fingers, which are driven by suitable gripping mechanisms (that include both transmission and driving mechanisms). The purpose of a gripping mechanism is to transmit from the actuator to the fingers, suitable motion during the approaching movement of the fingers to the object and then suitable force during the grasping.

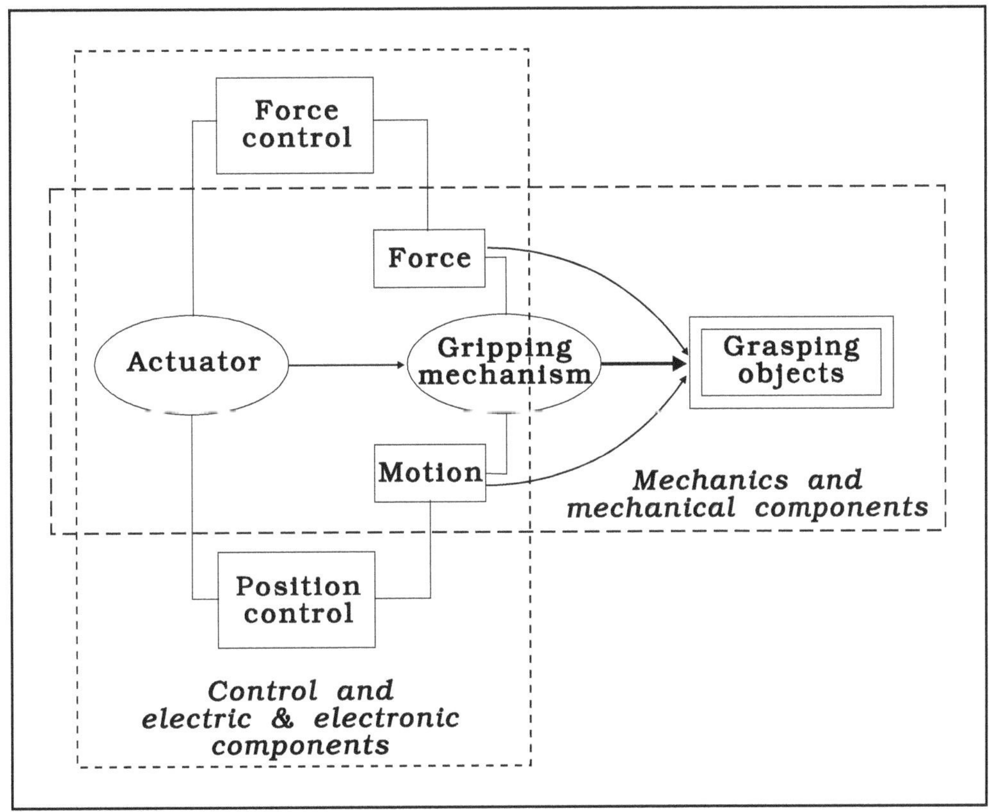

Fig. 4.12: Mechatronic layout for gripper design and operation.

Thus, the performance of a gripping mechanism is basically as a mechanical operator, but it can also be provided with suitable additional components and sensors to enhance its capability. Motion and force actions for grasping can be considered complex tasks involving a mechatronic concept of multidiscipline integration, since a control of these operations can be obtained by application of suitable technologies from the electric and electronic fields.

In Fig. 4.12 each block can individuate hardware of components as well as their basic purposes and operations. In particular, the blocks motion and force refer to these basic characteristics of a gripper when it is properly equipped with suitable sensors. Each block corresponds to a control hardware and software. But it is possible to have only position control in many cases, when the grasping force has been suitably sized by means of previous analysis or it has been calibrated by *ad hoc* calibrations.

The dotted large box in Fig. 4.12 indicates two main sub-systems and corresponding behaviors: mechanics and control with electronics/electric hardware. The overlapping of the two boxes, which includes the blocks motion, force, and control, and partially gripping mechanism, will indicate the mechatronics integration that is needed for a successful but efficient operation of a gripper.

The actuator is the source for the energy and movement of the fingers and therefore ultimately it is responsible for the grasping action. Consequently, its operation is a function of the force and transmission level, which are usually under control or pre-programmed. While a gripping mechanism is basically a mechanical component, an actuator can be electric, hydraulic, or pneumatic depending on the energy that it uses to obtain the gripper operation. The mechanical transmission again makes important mechanical considerations even for the operation of the actuator itself.

Usually industrial sensored grippers have a mechatronic design, such as the one shown in Fig. 4.13 and they are composed of:

- two fingers, which can be interchangeable as a function of the objects to be grasped;
- a gripping mechanism, which transmits the motion and force from an actuator to the fingers;
- a hydraulic or pneumatic actuator, generating a required actuating force;
- a hydraulic or pneumatic circuit, providing a control of the gripper performance, mostly in terms of the force.

The pneumatic/hydraulic circuit regulates the flow and pressure of the fluid, which is used in the actuator. Usually, the circuit is made as a common industrial circuit for automatic devices, whose main components are valves and electrovalves, which can be properly selected from the market production in order to obtain a desired operation of the actuator and then fingers.

In Fig. 4.13 the shown circuit is drawn by using ISO standards for representation of the components, which are a filter, a compressor, a tank with its safety valve, and a four-wayelectrovalve for three positions. The electrovalve is operated through electric signals that are generated by the control block. The control block represents both the hardware equipment and software algorithms for controlled operation of the actuator. Usually the hardware equipment consists of a PLC (Programmable Logic Controller), which can be of industrial type in order to give signals to the electrovalve, to receive signals from sensors on fingers and actuator, and to elaborate a control program that is stored in its memory. The sensors for grippers can be force sensors that are usually installed on the fingertips to regulate the grasping force. But usually industrial grippers are regulated on the basis of position control only, by controlling the finger configuration through the position of the actuator. The software algorithms for gripper control are usually designed specifically for specific sensor equipment and actuator capability. In the case of grasping force control, characteristic parameters, such as the gains G_c, k_T, and K_F, for the gripping system are needed to achieve suitable efficiency of the control operation. The control algorithms are stored in the memory of a PLC, which runs the program to operate the gripper. Instead of a PLC, the control of the gripper can be run directly by the robot controller by using some specific input and output channels of its additional control capability.

In Fig. 4.13 each sub-system has been indicated by a dotted-line box with the aim of emphasizing the variety of systems in a gripper. But an efficient operation requires a suitable design or choice of components so that a proper integration and correct interface will run the gripper successfully.

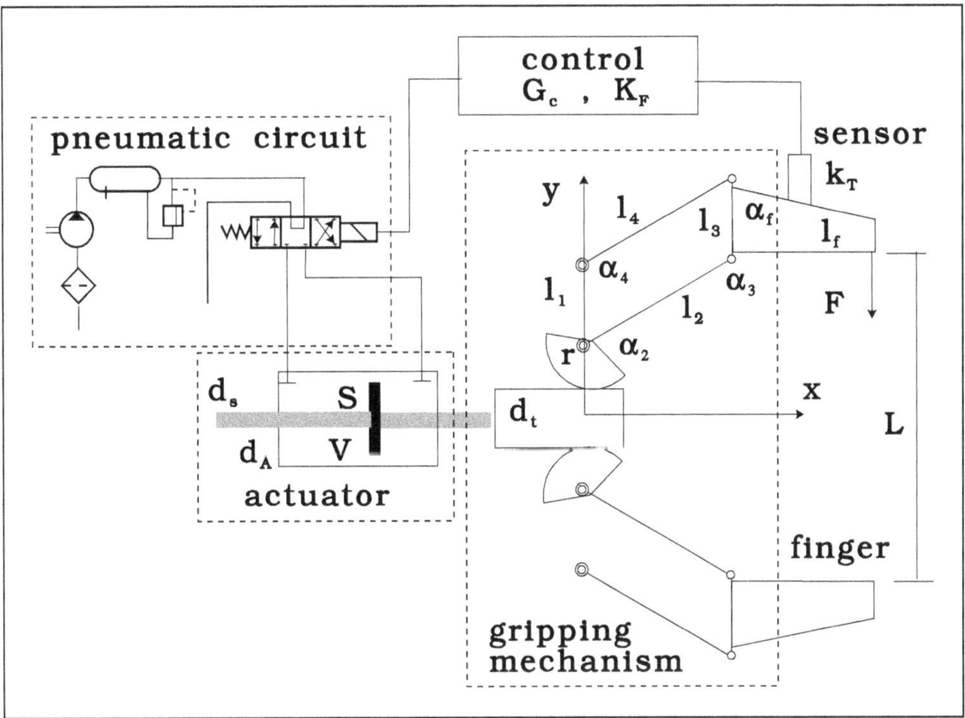

Fig. 4.13: A mechatronic design for two-finger grippers with characteristic parameters

4.3 Design parameters and operation requirements for grippers

The ultimate operation requirement of a gripper consists of grasping an object with a suitable force to achieve a stable prehension. From a mechanical viewpoint, the grasping capability can be described through a grasping force F, exerted by the fingers, and by the grasping size L, which is the distance between fingers when grasping the largest object, as shown in Fig. 4.13.

A two-finger gripper for industrial applications can be designed according to the schemes of Figs 4.11 and 4.13, where basic components are indicated as: shaped fingers; gripping mechanisms; connecting transmissions; actuator; electrovalve and related circuit, sensors, and control devices. Indeed industrial applications in a structured environment may not require sensorial means so that a gripper structure can be simplified.

The design and operation parameters can be grouped in sets containing geometrical parameters, fluid and circuit characteristics, and control variables.

Referring to the case of study of Fig. 4.13 geometrical parameters are those shown in the figure, where S is the net cross-section area of the actuating piston and V is its volume. In addition, k_1, k_2, x, and s_n are the characteristics of the hydraulic/pneumatic circuit; and G_c, k_T, K_F, are the parameters of the control scheme. The design parameters can be even indicated straightforwardly as the hydraulic or pneumatic pressure p, the electrovalve type and size, the actuator dimensions through stroke c_A and diameter d_A, the transmission characteristics as the input–output ratio τ_t and size l_t, the gripping mechanism characteristics as input–output ratio τ_g, size l_g and grasping configuration α_d, the finger shape with α_f and size l_f, and the grasping size L.

Similarly an electric gripping system can be sketched, analyzed, and formulated by using the characteristic parameters of electric components.

Strong constraints for designing gripping devices can be considered as lightness, small dimensions, rigidity, multi-task capability, simplicity, and lack of maintenance.

A basic requirement for an industrial gripper design can be recognized in a low-cost and reliable design. These design requirements can be achieved by considering specific end-effectors or two-finger grippers. In the last case the minimum number of fingers corresponds to a minimum complexity of a grasping device.

An increasingly demanding requirement for an industrial application consists also of light mechanical gripper devices as payload usually includes the gripper weight. This requires using light and new materials for gripper components.

Thus, a mechatronic design can be a suitable solution fulfilling the above-mentioned requirements for grippers. Mechatronic design for gripping devices is achieved when relations are considered among the components and also with the application and environmental features.

Nevertheless, basic questions may be deduced for an optimal design of a gripper by taking into account both the general scheme of Fig. 4.12 and specific architectures of Figs 4.11 and 4.13. They can be formulated as:

- How can the fingers be sized and shaped?
- Which gripping mechanisms can the designer properly select?
- What formulation can be deduced to properly design gripping mechanisms for a given application or a multi-task purpose?
- How can the actuator and hydraulic/pneumatic/electric circuit be selected from the commercially available means?
- How can a multi-task capability be achieved?
- What control scheme and sensors can be used and/or designed for a given grasping action?

In order to answer the above-mentioned questions, the following observations can be and usually are deduced for designing each component, since a first gripper design solution can be obtained by means of a step-by-step procedure considering one

component at each step. Nevertheless, a mechatronic design can be achieved when all the design requirements and constraints are considered and formulated at the same time by using a concurrent design approach in agreement with the scheme of Fig. 4.12.
A systematic approach for the design problem of grippers may also need an index formulation of the fundamental properties, both of the system and the application. This may give rational design formulations in the form of specific gripping mechanism design or grippers operation. This indexing activity can be obtained from a mechanical point of view by looking at the mechanics of the components, but it should include more of the mechatronic layout by also formulating the behavior of the control system.
The performance index formulation can also be useful for a comparable evaluation of grippers so that a synthetic indexing of the mechatronic layout is helpful for designers and mainly for users in practical selection among different design solutions and operations.
The above-mentioned requirements and considerations can also be satisfied through specific solutions when constraints and data are expressed for a specific application. Nevertheless, when a more general mechatronic viewpoint is considered, it can give awareness of the gripper capabilities as well as an insight into possible enhancements or adjustments when the application may be changed.
Industrial robots are usually arranged so that any gripper may be installed since they are supplied with pipelines and a given pneumatic or hydraulic pressure flow. Therefore, a design problem for robot application can be stated as: given a grasp force and an input displacement a designer must select an actuator with a suitable size for *a priori*-determined pressure flow. Further problems may arise in conceiving a hydraulic/pneumatic circuit and at most in selecting a suitable electrovalve to give a desirable response both from kinematic and force viewpoints. Thus a hydraulic/pneumatic system for a gripper can be supplied through a component selection among available commercial means which are designed for automation purposes.
A multi-task capability may also be requested of a two-finger gripper and usually it is obtained by simply giving the possibility to fix the fingers by means of screws or similar simple attachments in different positions on the driving mechanism. Therefore, another design problem is related to the question: is it possible to design and properly maintain the grasping mechanical characteristics within finger interchangeable solutions?
At the most a multi-task capability for a two-finger gripper consists of gripping several objects with different shape and size, but within limited ranges. This can be achieved by using properly shaped fingers and from a general viewpoint by adopting interchangeable fingers each of which is designed for a specific grasping task, as the case shown in Fig. 4.6. Indeed, fingertips are the only finger part that can be interchangeable as function of the size and shape of the objects.
In addition, when a multi-task purpose is concerned with different mechanical operations a solution may be adopted by using a two-finger gripper as an intermediate device to temporally install a required specific end-effector.
Another fundamental aspect for gripper design can be identified in the type of finger motion. Two types of finger motion may occur in gripping mechanisms, as shown in Fig.

4.14: a swinging motion during which a finger rotates, as in the example of Fig. 14 a); a parallel motion consisting of a translation movement so that a finger maintains its orientation with respect to the gripper frame, as in the example of Fig. 14 b). Although parallel motion is desirable since it ensures a constant grasp configuration avoiding squeezing forces, swinging motion can be more convenient since it may give larger grasping capability. A challenging question is how to design a gripping mechanism according to a suitable finger motion. The choice can be obtained by looking at the specific characteristics of a specific application in terms of grasping accuracy, which makes preferable a parallel motion, or in terms of a grasping capability, makes preferable a swinging motion.
However, most multi-task capabilities can be devoted to a two-finger design solution whose successful operation strongly depends on a suitable programming of robot and gripper, once the grasping action has been analyzed in all its phases, as illustrated in the next section.

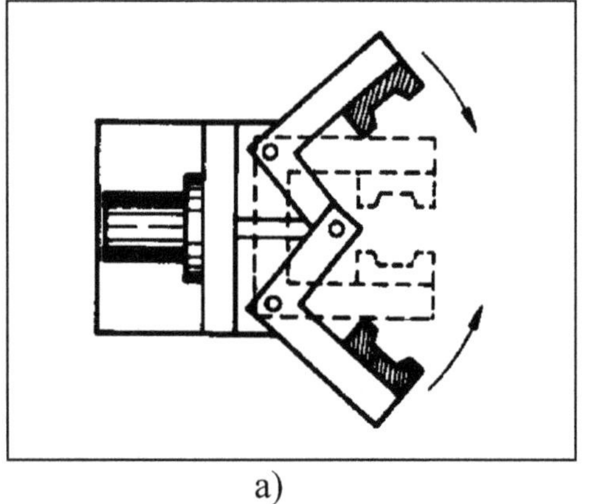

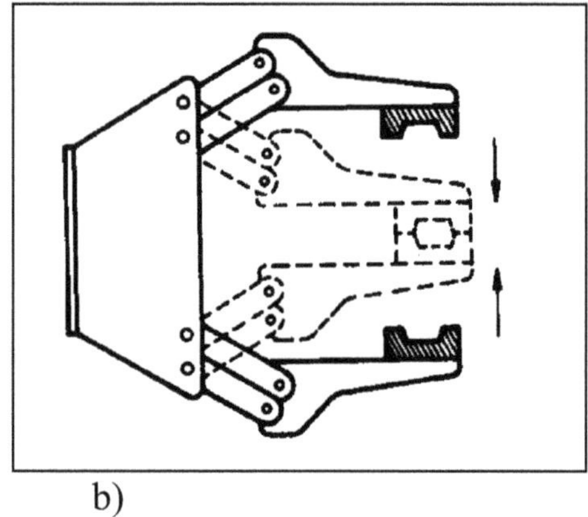

a) b)

Fig. 4.14: Basic movements for gripper fingers: a) swinging motion; b) translation parallel motion.

4.4 Configurations and phases of two-finger grasp

Two-finger grippers are widely used because two-finger grasp can be used in most cases. Most of the objects that are used in human or industrial manipulations have a size that is comparable with the size of the human hand or gripping devices and they have a so-called 'regular geometry', which describes the shape of basic easily graspable figures.
Figure 4.15 summarizes feasible mechanical designs for two-finger grippers. In general, fingers can be designed as articulated fingers to mimic the human hand, as shown in Fig. 4.15. b) and c), or as coupler links to use planar linkages as gripping mechanisms, as shown in Fig. 4.15. a).
A further peculiarity can be recognized in flexible tips for the contacts, which are installed on fingertips and finger phalanges with a suitable curvature in order to be easily adapted to the shape of specific objects, as schematically shown in Fig. 4.15 by small grey tips. These tips can be of flexible material to help the shape adaptation and mainly to enlarge the contact surface so that the contact pressure can be limited to avoid surface

damage of the grasped object. Indeed, the need for a suitable low value of contact pressure is a requirement to adopt two, three, four, and even more contacts, although static equilibrium for grasping can be ensured with two contacts only in planar cases.

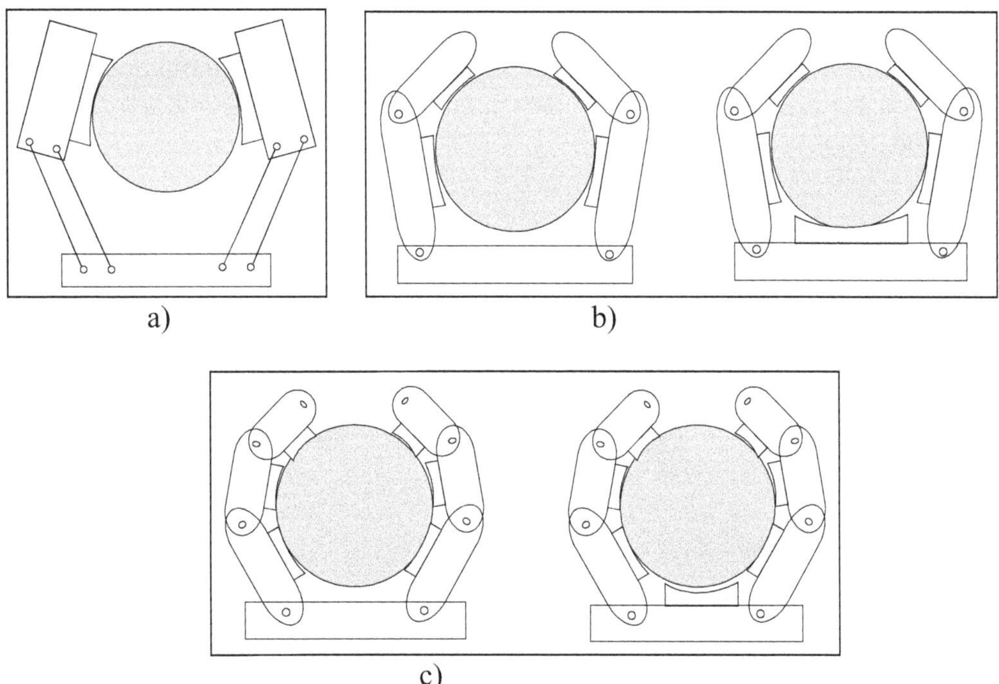

Fig. 4.15: Grasping configurations and mechanical designs for two-finger grippers: a) rigid fingers for two-grasp contacts; b) articulated fingers without and with palm for four-grasp contacts; c) articulated fingers without and with palm for six-grasp contacts.

Since the main purpose of a two-finger gripper can be recognized as performing a planar grasp, some fundamental design considerations may be deduced by approaching specifically a two-finger grasp and its mechanics.

The fundamental characteristics of a two-finger grasp can be modeled as in Fig. 4.16, which has been obtained to describe all the phases of a grasping action as:

in Fig. 4.16a) one finger impacts the object and starts the grasping while the fingers are moving to close with a suitable approaching motion;

in Fig. 4.16b) the finger pushes the object against the other finger while the motion of the fingers continues the closing;

in Fig. 4.16c) the second final impact of the fingers to the object concludes the approaching and pushing motion;

in Fig. 4.16d) the object is grasped and a static equilibrium of the object between the fingers is ensured;

in Fig. 4.16e) dynamic force may change the equilibrium but still the object is statically grasped by the fingers;

in Fig. 4.16a, f) an external disturbance, such as an impact with another object in the

environment, may change the equilibrium and the object may move to another static equilibrium configuration between the fingers.
The above-mentioned analysis can be outlined similarly for any other case of grasping devices, even with multiple contacts, such as those in Fig. 4.15.
The analysis has been focused on the interactions among object and fingers through a purely mechanical viewpoint, but the feasibility of finger configurations strongly depends on the mechatronic gripper design and operation that guide the finger motion and action.

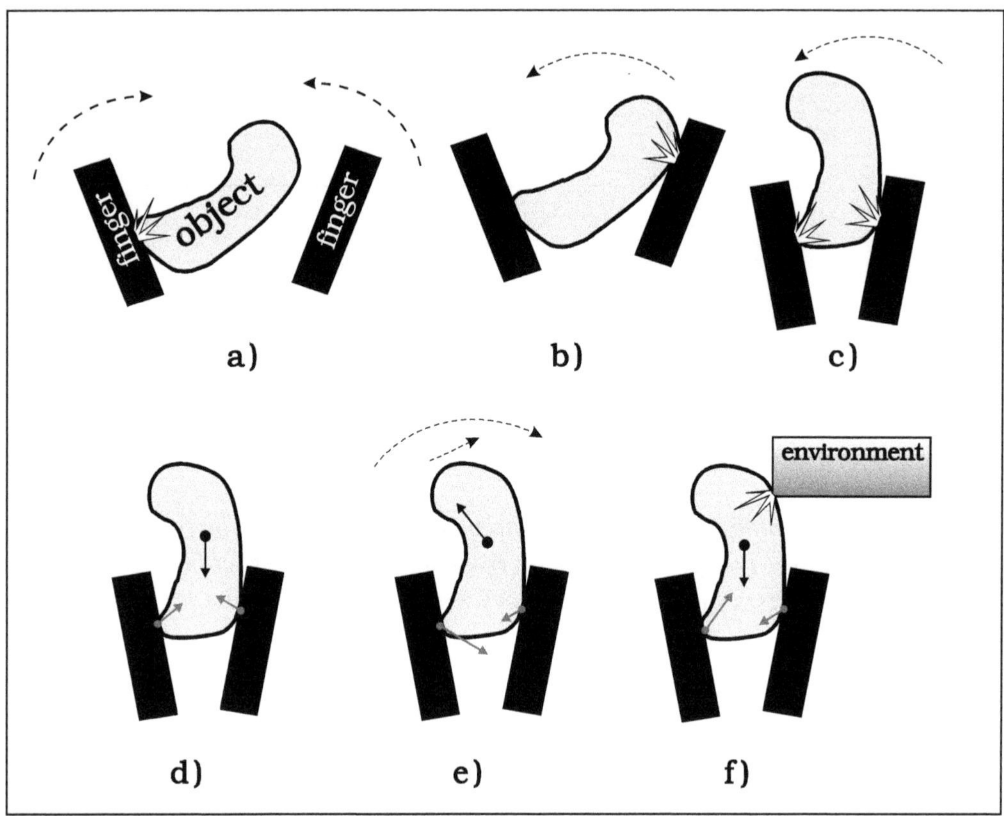

Fig. 4.16: A representation of the phases for a two-finger grasp: a) first grasp impact; b) contact with both fingers; c) final grasp impact; d) static grasp; e) dynamic grasp; f) external disturbance of the grasp equilibrium.

The analysis of Fig. 4.16 refers to planar grasp but it can also be extended to three-dimensional cases by looking similarly at the phases in planes of a Cartesian frame reference for a three-dimensional grasp.

4.5 Model and analysis of two-finger grasp

Figure 4.17 has been modeled to summarize all the phases of a grasping action and considers the mechanics of grasp through all the aspects shown in Fig. 4.16.

A reference frame can be identified for the relative motion of the object with respect to the fingers by using those characteristic lines that can be determined, as shown in the scheme of Fig. 4.17. In particular, the so-called 'contact line' can be identified as the line joining points A and B, which are the application points of the grasping forces F_1 and F_2 exerted by the fingers. The so-called 'squeezing line' can be identified as the orthogonal line to the contact line that passes across the gravity center G of the object. The so-called 'slipping line' is the line that is orthogonal to the plane, which is determined by the planar grasp configuration that is identified by the contact and squeezing lines.
Referring to Fig. 4.17, a configuration of a two-finger grasp can be characterized by possible elementary motions of a grasped object among the fingers as:
- slipping movement along the slipping-line direction, which is orthogonal to a plane of the gripping forces exerted by the fingers;

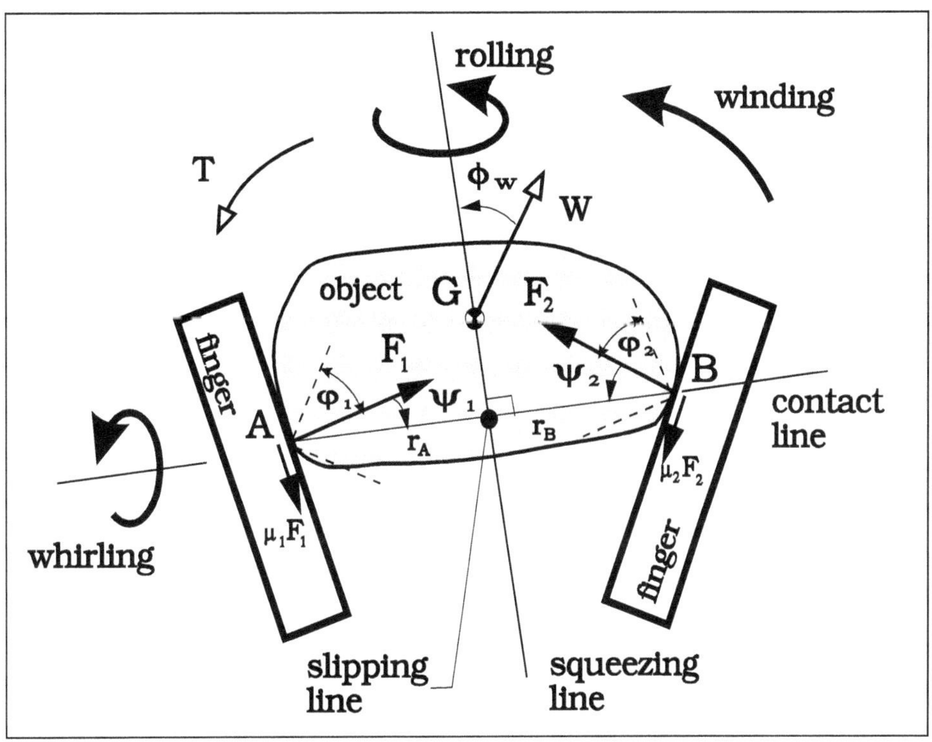

Fig. 4.17: Configuration for a planar grasp with two fingers and its parameters.

- squeezing movement, which may occur along the squeezing-line in sliding the object outward or inward to the gripper itself due to an effect of force components pushing the object (in this case an increase or a decrease of the gripping forces by fingers may produce a squeezing motion or even a release of the object);
- rolling rotation about the squeezing line consisting of a motion of revolution of the object among the fingers due to an external torque or a force couple;

- winding rotation about the slipping line consisting of a motion of revolution of the object among the fingers due to an external torque or a force couple;
- whirling rotation about the contact line, which can also be due to a torque by the object's weight when its mass center does not lay on the contact line.

No relative translation motion is assumed along the contact line since the object and fingers are usually assumed as rigid bodies along this direction. Otherwise a compression of the object may occur along the contact line direction.

Although two contact points cannot be considered enough from a kinematic viewpoint even for a planar grasp, a two-finger gripper can perform a suitable grasp when force constraints are taken into account so that four conditions can be achieved for a firm grasp that avoids the 3 d.o.f.s motion in a planar grasp.

In fact, all the above-mentioned elementary motions of the object may be avoided when suitable grasping force F_1 and F_2 are exerted by the fingers so that friction forces $\mu_1 F_1$ and $\mu_2 F_2$ may arise to completely balance the external action on the object as being two additional constraints.

In Fig. 4.17 F_1 and F_2 are the grasping forces, which are exerted by the fingers and usually they are not equal, since the contact points A and B (which are assumed as application points for the forces) are not generally located in the same relative position on the two fingers. However, the difference in position can be very small because of symmetry of the mechanical design and even of the grasping configuration that is usually desired and achieved by also properly shaping the fingertips. Consequently, the grasping forces can be thought of with a common value F.

Similar observations can be developed for the friction evaluation at the points A and B through the friction coefficients μ_1 and μ_2, which can be assumed with a common value μ, since they have the same situation and materials at the contacts. The angles φ_1 and φ_2 are the angles of the friction cones.

The grasping configuration of the object with respect to the fingers gives the angles ψ_1 and ψ_2 as the angles measured between the grasp force directions and contact line. These angles strongly depend on the orientation of the fingers, the position of the contact points, and the shape of the object and fingers. Although the contact points can be located in the same position on the fingers, these angles can differ from each other. However, because of the symmetry of a two-finger gripper, in a design procedure it is convenient to assume them as equal to the most unfavorable value, which refers to the case that gives a squeezing component from the grasping force itself. The grasping forces can exert a pushing action through the component F sin ψ.

In addition, in Fig. 4.17 r_A and r_B represent the distances of A and B, respectively, from the squeezing line; W is the weight of the object and it is oriented with an angle Φw with respect to the squeezing line; and T is an external torque acting on the object and it includes the inertial actions due to the manipulator movement. W may include the inertial and external forces so that the model of Fig. 4.17 can describe all the situations, which may occur to an object grasped by a two-finger gripper as shown in Fig. 4.16.

The static equilibrium of a grasped object between the fingers can be expressed by using the model of Fig. 4.17 along the directions of the contact and squeezing lines in terms of forces as

$$F_1 \cos\psi_1 - F_2 \cos\psi_2 + \mu_1 F_1 \sin\psi_1 - \mu_2 F_2 \sin\psi_2 + W \cos\Phi_w = 0$$
$$F_1 \sin\psi_1 + F_2 \sin\psi_2 - \mu_1 F_1 \cos\psi_1 - \mu_2 F_2 \cos\psi_2 + W \sin\Phi_w = 0 \quad (4.5.1)$$

and in terms of torque about the squeezing line as

$$T - r_G W \sin\Phi_w - r_A F_1(\sin\psi_1 - \mu_1 \cos\psi_1) + r_B F_2(\sin\psi_2 - \mu_2 \cos\psi_2) = 0 \quad (4.5.2)$$

Grasping forces F_1 and F_2 can also be evaluated by considering a force closure to formulate conditions for a stable grasp neglecting the friction forces in the form

$$F_G \geq S \quad (4.5.3)$$

and

$$T_G \geq N \quad (4.5.4)$$

where F_G and T_G represent the active components of force and torque, respectively, in the grasp; S and N are disturbing actions. They can be expressed by using a scheme as in Fig. 4.17 and for example, in the case of a planar grasp analysis, in order to avoid squeezing and winding they assume the form

$$F_G = F_1 \tan\varphi_1 \cos\psi_1 + F_2 \tan\varphi_2 \cos\psi_2$$
$$S = W \cos\phi_W + F_1 \sin\psi_1 + F_2 \sin\psi_2 \quad (4.5.5)$$
$$T_G = F_1 r_A \tan\varphi_1 \cos\psi_1 - F_2 r_B \tan\varphi_2 \cos\psi_2$$
$$N = T + F_2 r_B \sin\psi_2 - F_1 r_A \sin\psi_1$$

Indeed, the direction of the friction forces as depending on signs in Eqs (4.5.1) – (4.5.5) will be determined to be contrary to the relative motion object-finger, when the static equilibrium is numerically solved.
Dynamic variations can be modeled by assuming T and W as variables as functions of the motion of the gripper and object.
Neglecting friction forces in the computations ensures more stable behavior, since the friction forces will act to increase the constraints for the object motion with respect to the fingers.
The model of Fig. 4.17 refers to the case of a planar grasp but it can even be used for a general three-dimensional grasp. This can be characterized as having the actions F_1, F_2,

W, and T with components out of the plane that is defined by the contact line and squeezing line. In three-dimensional cases the static equilibrium and conditions for stable grasp can be deduced through a similar formulation by considering the equilibrium equation components along and about all the axes of the reference frame, including the slipping line.

Problems for a correct analysis of a grasp may arise by looking at the relative posture of fingers and object. The relative posture can be analyzed through the location of the contact line by means of the position of points A and B. The orientation of the contact line can be determined through the angles ψ_1 and ψ_2. The position of A and B, and the angles ψ_1 and ψ_2, can be considered the unknowns for a grasp problem together with the grasping forces F_1 and F_2, so that Eqs (4.5.1) and (4.5.2) cannot be enough to solve the complete general problem of a grasp.

Moreover, instability may arise for force or motion perturbation and trajectory evolution of the gripper due to the dynamics of the arm on which the gripper is installed. In this case some grasping adjustments can be successfully obtained by using the friction force capability or a greater actuating force. In fact, the adhesion friction increases the stability of the prehension because the friction cones reduce the sensitivity of constraint reactions to the external force perturbations. Therefore, the grasping and friction forces can be considered conveniently as variables but depending on the actuation and configuration of the grippers.

Thus, the mechanics of the grasp still presents research interest to obtain proper formulation of kinematic constraints and force closure for a stable grasp with a minimum complexity in the finger shape and action.

Kinematic aspects of the grasp can be taken into account by assuming as variable the parameters ψ_1 and ψ_2, r_A, r_B, and r_G. Indeed, these parameters cannot be determined *a priori* in a real grasp since they describe the grasp configuration in its firm and stable status. Therefore, for a gripping mechanism design they can be assumed as given, and Eqs (4.5.1) to (4.5.5) can be used to obtain the grasping forces F_1 and F_2 to be exerted by the gripping mechanism through the fingers. Thus, the values of F_1 and F_2 can be used to size the gripping mechanism and actuator.

Alternatively, given the grasping forces by also shaping the fingertip *a priori*, Eqs (4.5.1) to (4.5.5) can be used to design the grasping configuration in order to determine the kinematic parameters only.

Therefore, one can advise two design problems for grippers, namely:

- a so-called ‘problem for grasp action’, whose aim is the design of a gripping mechanism in terms of type and dimensional syntheses of mechanisms;
- a so-called ‘problem for grasp configuration’, whose aim is to ensure a firm grasp of the object between the fingers by designing the actuator and fingertips in terms of shape and size.

A grasp action of a gripper is related mainly to the grasping forces that can be exerted by the fingers by considering the operation of the actuator, gripping mechanism, and sensors.

A grasp configuration of a gripper is related mainly to the posture (location and orientation) of the object between the fingers, but also of the fingers. The posture of the object and fingers depends on the kinematic operation of the gripper system.

4.6 Mechanisms for grippers

Usually transmission and driving mechanisms can be considered together as gripping mechanisms. Their main purpose consists of operating suitable motion and force transmission from the actuator to the fingers.

Most of the gripper efficiency is due to grasping force efficiency. Grasping force efficiency can be evaluated both in terms of mechanism operation and grasping action.

The mechanism operation can be evaluated by looking at its mechanical efficiency and motion capability.

The mechanical efficiency of a mechanism is defined as the ratio between the input force and output force for the aim of the mechanism that in this case is the grasping force exerted by the coupler as finger link.

The motion capability can be evaluated by looking at the type of the coupler motion, which refers to the controlled finger motion/configuration to/at grasp, and its extension.

The above-mentioned evaluations require suitable analysis and design procedures, as explained in the next sections, also they are used for catalogue and classification purposes of the gripping mechanisms that are used in industrial grippers.

In general special attention is focused on the grasping efficiency in order to obtain suitable light mechanical design of grippers. This requires that a gripping mechanism usually actuates as a force multiplier and a good solution may be obtained with planar linkage with several links. However, need of limited encumbrance, stiff design, and easy operation may limit the number of links and indeed most industrial gripping mechanisms show kinematic chains based on four-bar or slider-crank linkages, as for example in Fig. 4.27. The coupler is the finger link.

Figure 4.27 shows examples of the most used gripping mechanism in industrial grippers, which are actuated by a linear actuator, although a rotative actuator can be installed alternatively on a rotative joint. One or two actuators can actuate this kind of mechanism so those fingers can operate simultaneously or independently from each other with a symmetrical or unsymmetrical behavior, respectively

Many other types of mechanism are used as gripping mechanisms in industrial grippers in order to achieve suitable mechanical designs with high grasping efficiency, small-sized robust design, and light and low-cost devices.

In Fig. 4.19 a few illustrative examples are shown for the several types of mechanisms that can be used successfully as gripping mechanisms.

In particular, Fig. 4.19 a) shows mechanisms that are based on revolute and prismatic joints, with one or two actuators. The fingers can move with a swinging or a parallel motion. In Fig. 4.19 b) the gripping mechanisms are based on cam transmissions and spring elements. The cam transmissions permit very high force transmission ratio and compacted design. Springs are also used to have underactuated releasing actions.

In Fig. 4.19 c) the core of the gripping mechanism is made of gears with very stiff operation.

In Fig. 4.19 d) the transmission from the actuator is obtained by using screw elements and the rest of the mechanism chain can have the usual architectures for gripping mechanisms.

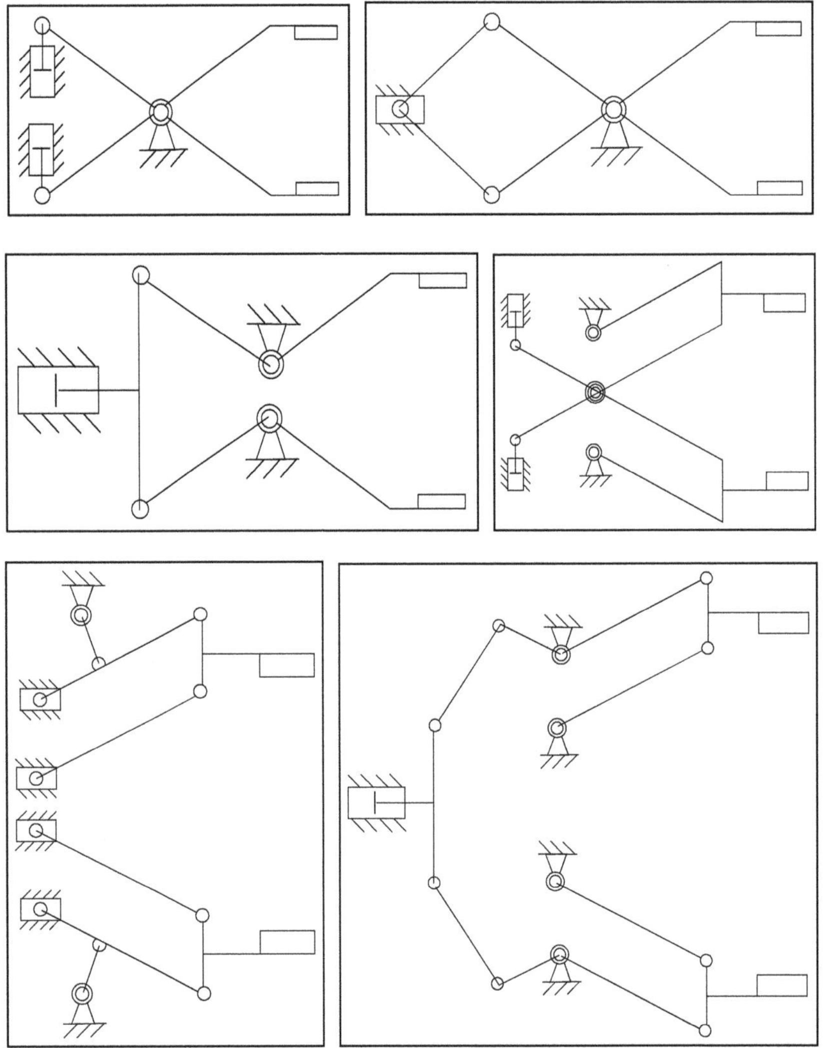

Fig. 4.18: Gripping mechanisms whose chain architecture is based on four-bar and slider-crank linkages.

Of course many other mechanisms can be used as gripping mechanisms, not only of the above-mentioned types. Indeed, further combination of those mechanism architectures can be conceived, but usually the specific purposes of a gripper together with design constraints for light and small devices give a preferable choice of simple mechanism types, such as the above-mentioned mechanisms.

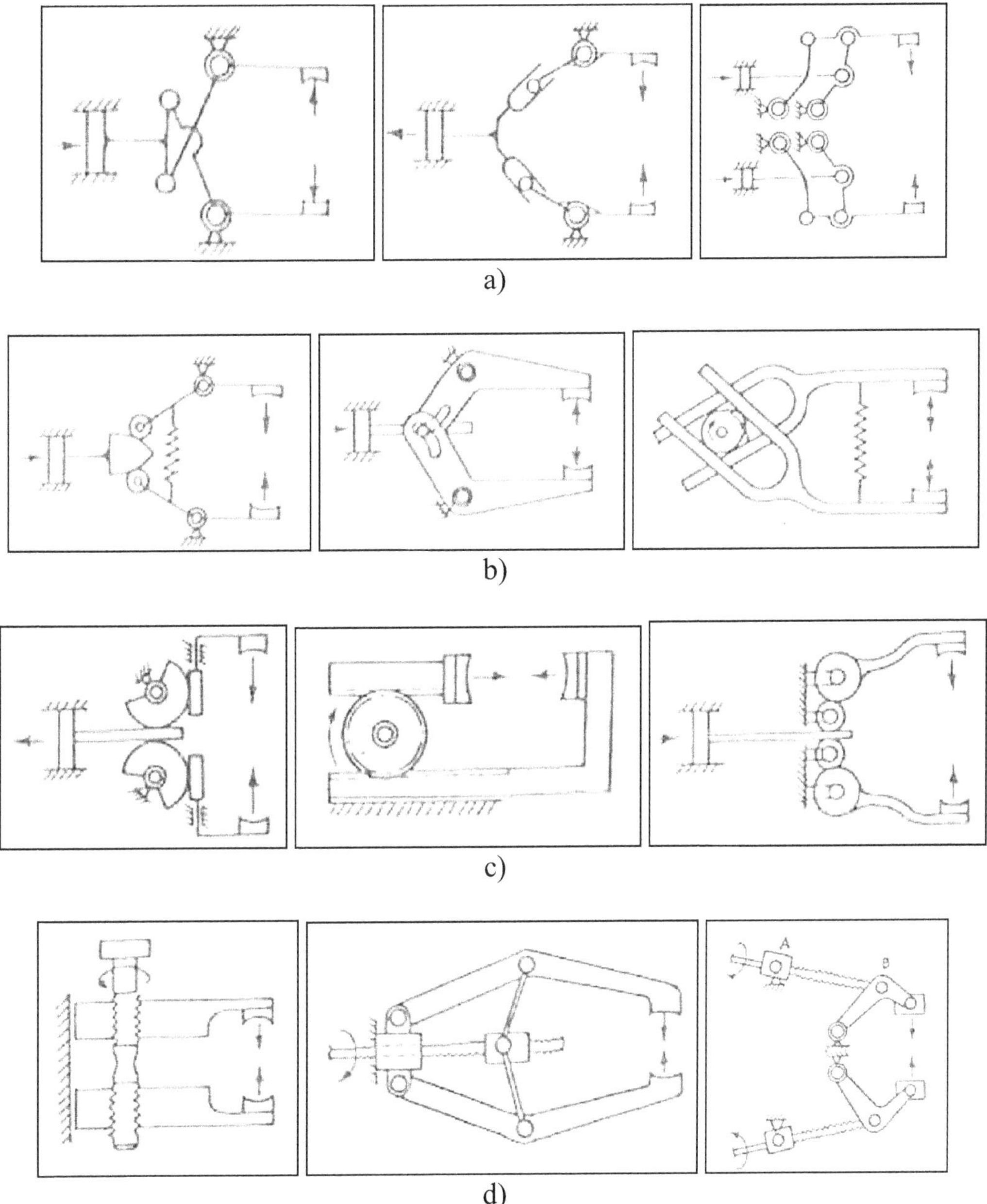

Fig. 4.19: Examples of mechanisms used as gripping mechanisms: a) linkage type; b) cam-and-follower type; c) gear-and-rack type; d) screw-driven type.

4.6.1 Modeling gripper mechanisms

A fundamental problem both for designing and operating a gripping mechanism can be recognized in the modeling of the basic characteristics of a gripper in order to identify the mechanism chain and its kinematic characteristics, but by considering the integration with other components of the mechatronic design.

Nevertheless, the modeling activity for gripping mechanisms can be restricted to

kinematics view of mechanisms and then numerical evaluation of the behavior will take into account the overall system. In the modeling of gripping mechanisms the analysis problem can be very different to the design problem.
In the analysis modeling, an existing gripping mechanism must be recognized in a built mechanical design by determining the mechanism architecture and its kinematic parameters. Generally, the result of the identification activity gives a well-determined mechanism, and problems may arise only in the numerical determination of the kinematic parameters and mainly in the range mobility.
In the design problem the kinematic scheme of a chosen mechanism can be built with different mechanical designs, although the first way consists of constructing the kinematic architecture by using its geometry.
Figures 4.20, 4.21, and 4.22 are illustrative examples for analysis modeling of existing commercial grippers but they can also be used as examples for design modeling with further considerations.
In particular, Fig. 4.20 shows a complex mechanical design in which the identification of the kinematic chain can be difficult if the gripper is not observed during its operation. In fact, the moving prismatic joints and rolling contact can be difficult to identify by means of drawing or vision inspection only. In addition, the analysis of the mechanism links can also be not quite evident in terms of numerical determination.
In the case of design modeling, the identification procedure can be even more doubtful, when one will think to build a mechanical design corresponding to the kinematic sketch in Fig. 4.20 b), since it can give more than one mechanical solution, even very different form that in Fig. 4.20 a).
Indeed, this is an illustrative example that shows that a gripper design can strongly depend on the specific expertise of a designer in mechanisms and grippers.
In Fig. 4.21 the mechanical design in a) shows the same geometry as the kinematic diagram in b) so that the modeling is straightforward both for analysis and design purposes. The use of cam connections can be very useful to have small-sized designs with an easy understanding of the gripper operation.

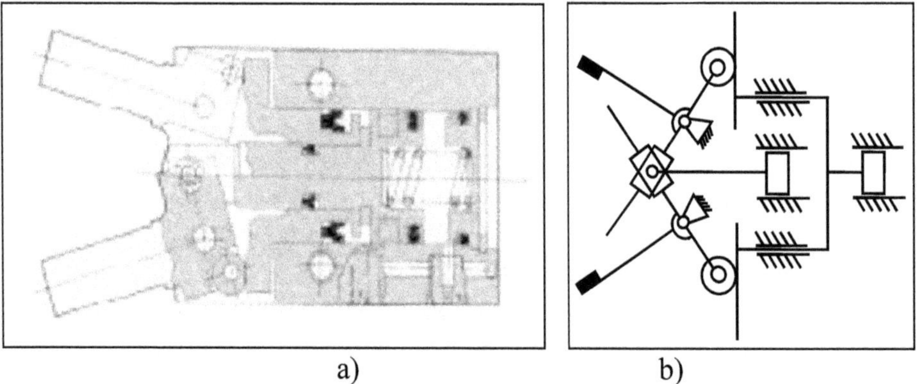

a) b)

Fig. 4.20: An example of a commercial industrial two-finger gripper: a) the mechanical design; b) its kinematic chain.

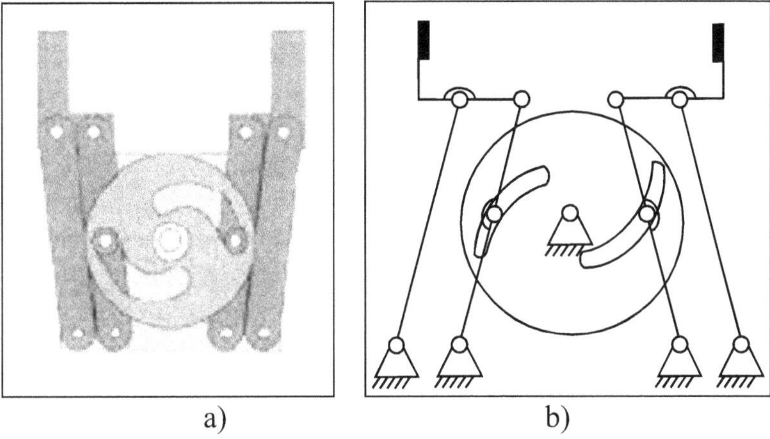

Fig. 4.21: An example of a commercial industrial two-finger gripper: a) the mechanical design; b) its kinematic chain.

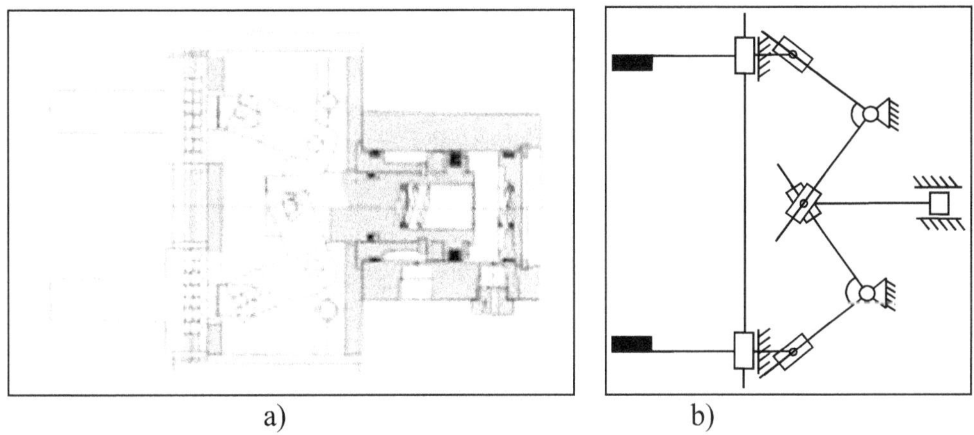

Fig. 4.22: An example of a commercial industrial two-finger gripper: a) the mechanical design; b) its kinematic chain.

In Fig. 4.22 the mechanical design of the gripper shows how to obtain a parallel motion of the fingers from a gripping mechanism with finger swinging motion by adding a sub-chain to the original kinematic chain, when one refers to the case of Fig. 4.20. In fact, the gripping mechanism of Fig. 4.22 shows the same kinematic architecture as the case of Fig. 4.20 with an addition of a few links and joints. The addition of these elements makes the gripping mechanism larger but improves the kinematic behavior of the gripping mechanism.

Once the kinematic scheme of a gripping mechanism is identified together with its kinematic parameters by analysis modeling or design choice, a numerical evaluation can be required to evaluate the performances and efficiency of the mechanism. The classical methods for mechanism analysis in terms of kinematics, statics, and dynamics can be used successfully. Nevertheless, the peculiar application of the gripping mechanism can be considered to deduce an evaluation that can be useful not only to classify the solutions but also to help for design purposes.

4.6.2 An evaluation of gripping mechanisms

In order to evaluate performances and efficiency of a gripping mechanism it is convenient to analyze its behavior by simulating its operation.
An additional evaluation of a gripping mechanism can be formulated by using the peculiarity of the mechanism for its grasping purposes.
A suitable simulation of a gripping mechanism can be obtained by means of kinematic and static analysis through classical methods for studying mechanisms.
Although a grasping efficiency is based on the statics aspects of the gripper system, great importance can also be addressed to the motion behavior during the phases of approaching a suitable grasp configuration, since a grasp strongly depends on the finger position and orientation through position of points A and B and angles ψ_1 and ψ_2 , as it can be observed from Fig. 4.17.
Among the several methods that are available for kinematic analysis of mechanisms, in the case of gripping mechanisms the technique of closure equations can be considered convenient even for design purposes, since it permits the deduction of closed-form expressions for the kinematics of a mechanism.
Referring to the case of a four-bar linkage with the kinematic parameters shown in Fig. 4.23, the method is outlined as reviewing the fundamentals and deducing useful formulation for the analysis of gripping mechanisms. Figure 4.23 shows a general four-bar linkage in which l_1 denotes the frame length, l_2 and l_4 are respectively the input and the output length links, and l_3 is the length of the coupler on which the finger is installed.

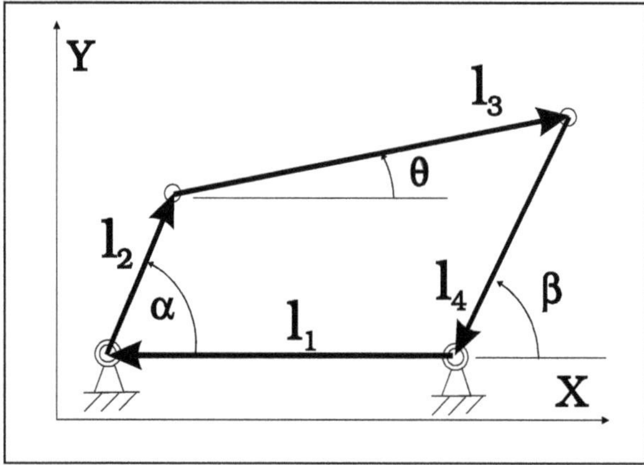

Fig. 4.23: A scheme for loop closure of a four-bar linkage mechanism.

The loop closure can be determined by looking at the vectors that can be identified on the links themselves, as shown in Fig. 4.23. These vectors can identify a closed polygonal circuit that can be formulated as

$$\mathbf{l}_1 + \mathbf{l}_2 + \mathbf{l}_3 + \mathbf{l}_4 = 0 \tag{4.6.1}$$

as the sum of the vectors of the closed circuit. Indeed, one can formulate many closure equations, such as Eq. (4.6.1) as many circuits can be identified in the kinematic chain of a mechanism. But then, independent circuits should also be identified in order to write a suitable number of independent equations that will be equal to the number of unknowns in order to have a solvable system.
The vector equation can be expressed in scalar form by using its components with respect to a reference frame as functions of the kinematic parameters of the mechanism. In the case of Eq. (4.6.1) for Fig. 2.23 the scalar equations can be written as

$$
\begin{aligned}
& l_2 \cos\alpha + l_3 \cos\theta + l_4 \cos\beta = l_1 \\
& l_2 \sin\alpha + l_3 \sin\theta + l_4 \sin\beta = 0
\end{aligned}
\tag{4.6.2}
$$

by referring to OXY frame.
In an analysis procedure, the sizes of a mechanism are assumed as given and the unknowns of the problem are the angles, which determine the mechanism configuration. Since the four-bar linkage is a 1 d.o.f. mechanism, one angle can be assumed as a given input and the other two angles are the unknowns for the two Eqs (4.6.2). The system of equations can be solved as outlined in 3.1.2 to obtain Eqs (3.1.25) to (3.1.27).
The above-mentioned formulation refers to the position problem for a planar mechanism and Eqs (4.6.1) and (4.6.2) are named as position closure equations.
The velocity analysis of mechanisms can be carried out by differentiating Eq. (4.6.1) and (4.6.2) with respect to time in order to obtain the vector expression of the velocity closure equation as

$$
\dot{\mathbf{l}}_1 + \dot{\mathbf{l}}_2 + \dot{\mathbf{l}}_3 + \dot{\mathbf{l}}_4 = 0 \tag{4.6.3}
$$

and its scalar components in the form

$$
\begin{aligned}
& l_2 \dot{\alpha} \sin\alpha + l_3 \dot{\theta} \sin\theta + l_4 \dot{\beta} \sin\beta = 0 \\
& l_2 \dot{\alpha} \cos\alpha + l_3 \dot{\theta} \cos\theta + l_4 \dot{\beta} \cos\beta = 0
\end{aligned}
\tag{4.6.4}
$$

where the dot symbol represents the time derivative.
Equations (4.6.3) express the derivative of the sum of the vectors of the closed circuit. Indeed, one can formulate many velocity closure equations, such as Eq. (4.6.3) as many circuits can be identified in the kinematic chain of a mechanism. But then, independent circuits should also be identified in order to write a suitable number of independent equations that will be equal to the number of unknowns in order to have a solvable system.
Equations (4.6.4) are linear equations as functions of the unknowns that are the angular

velocities of the links and they can be solved straightforwardly, once the position problem has been preliminarily solved.
The acceleration analysis can be carried out by differentiating Eq. (4.6.3) and (4.6.4) with respect to time in order to obtain the vector expression of the acceleration closure equation as

$$\ddot{\mathbf{l}}_1 + \ddot{\mathbf{l}}_2 + \ddot{\mathbf{l}}_3 + \ddot{\mathbf{l}}_4 = 0 \tag{4.6.5}$$

and its scalar components in the form

$$\begin{aligned} &l_2\left(\ddot{\alpha}\sin\alpha + \dot{\alpha}^2\cos\alpha\right) + l_3\left(\ddot{\theta}\sin\theta + \dot{\theta}^2\cos\theta\right) + l_4\left(\ddot{\beta}\sin\beta + \dot{\beta}^2\cos\beta\right) = 0 \\ &l_2\left(\ddot{\alpha}\cos\alpha - \dot{\alpha}^2\sin\alpha\right) + l_3\left(\ddot{\theta}\cos\theta - \dot{\theta}^2\sin\theta\right) + l_4\left(\ddot{\beta}\cos\beta - \dot{\beta}^2\sin\beta\right) = 0 \end{aligned} \tag{4.6.6}$$

where the double dot symbol represents the second time derivative.
Equations (4.6.5) express the derivative of the sum of the acceleration vectors of the closed circuit. Indeed, one can formulate many acceleration closure equations, such as Eq. (4.6.5) as many circuits can be identified in the kinematic chain of a mechanism. But then, independent circuits should be also identified in order to write a suitable number of independent equations that will be equal to the number of unknowns in order to have a solvable system.
Equations (4.6.6) are linear equations as functions of the unknown angular accelerations of the links and can be solved straightforwardly, once the position and velocity problems have been preliminarily solved.
The method of closure equations can be used also to derive numerical procedures for the case of more complicated mechanisms with several loops when the kinematic circuits give a system of equations that are not suitable for hand computation or closed-form formulation. Several techniques are available from the field of mechanism analysis.
Since in general gripping mechanisms are not very complicated, it can be convenient to deduce analysis formulation by using straightforwardly the closure loop equations as in the following example for a four-bar linkage mechanism.
Figure 4.24 shows a general four-bar linkage in which a denotes the length of the frame, b and d are respectively the input and the output length links, and c is the length of the coupler whose fingertip point M is located by the length p and the angle Γ.
Referring to Fig. 4.24, if OXY is a fixed reference and assuming that α is the input angle, positive counter-clockwise, β the output angle, θ the angle between the generic position of c and X-axis, from the loop closure equations it follows

$$\beta = 2\tan^{-1}\frac{\sin\alpha - (\sin^2\alpha + B^2 - C^2)^{1/2}}{B + C} \tag{4.6.7}$$

$$\theta = 2\tan^{-1}\frac{\sin\alpha - (\sin^2\alpha + B^2 - D^2)^{1/2}}{B + D}$$

where

$$B = \cos\alpha - \frac{a}{b}$$

$$C = \frac{a^2 + b^2 - c^2 + d^2}{2bd} - \frac{a}{d}\cos\alpha \qquad (4.6.8)$$

$$D = \frac{a}{c}\cos\alpha - \frac{a^2 + b^2 + c^2 - d^2}{2bc}$$

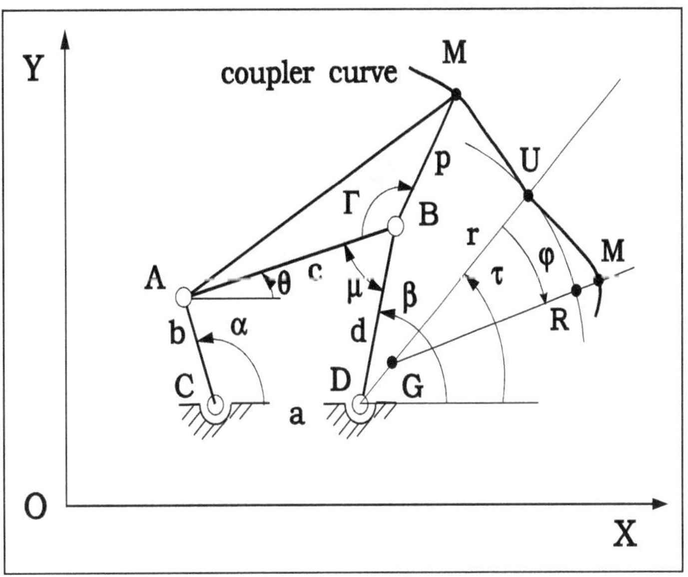

Fig. 4.24: A scheme for kinematic and accuracy analysis of a four-bar linkage gripping mechanism with coupler fingertip point M.

The coordinates of the fingertip point M can be computed as

$$x = b\cos\alpha + c\cos\theta - p\cos(\Gamma - \theta)$$

$$y = b\sin\alpha + c\sin\theta + p\sin(\Gamma - \theta) \qquad (4.6.9)$$

Differentiating and solving the loop closure equations for velocity and acceleration, angular velocities of the follower and coupler are given as

$$\dot{\beta} = \frac{b}{d}\frac{\sin(\alpha - \theta)}{\sin(\beta - \theta)}\dot{\alpha}$$

$$\dot{\theta} = \frac{b}{c}\frac{\sin(\alpha - \beta)}{\sin(\beta - \theta)}\dot{\alpha} \tag{4.6.10}$$

and angular accelerations can be expressed as

$$\ddot{\beta} = \frac{b\ddot{\alpha}\sin(\alpha - \theta) + b\dot{\alpha}^2\cos(\alpha - \theta) + c\dot{\theta}^2 - d\dot{\beta}^2\cos(\beta - \theta)}{d\sin(\beta - \theta)}$$

$$\ddot{\theta} = \frac{b\ddot{\alpha}\sin(\alpha - \beta) + b\dot{\alpha}^2\cos(\alpha - \beta) - d\dot{\beta}^2 + c\dot{\theta}^2\cos(\beta - \theta)}{c\sin(\beta - \theta)} \tag{4.6.11}$$

The velocity V and the acceleration A of fingertip point M can be computed in the form

$$V = (\dot{x}^2 + \dot{y}^2)^{1/2} \tag{4.6.12}$$

$$A = (\ddot{x}^2 + \ddot{y}^2)^{1/2} \tag{4.6.13}$$

where the velocity components can be expressed as

$$\dot{x} = -b\dot{\alpha}\left[\sin\alpha + \frac{c\sin\theta + p\sin(\Gamma - \theta)}{c\sin(\beta - \theta)}\sin(\alpha - \beta)\right]$$

$$\dot{y} = b\dot{\alpha}\left[\cos\alpha + \frac{c\cos\theta - p\cos(\Gamma - \theta)}{c\sin(\beta - \theta)}\sin(\alpha - \beta)\right] \tag{4.6.14}$$

and the acceleration components can be given as

$$\ddot{x} = -b\left(\ddot{\alpha}\sin\alpha + \dot{\alpha}^2\cos\alpha\right) - c\left(\ddot{\theta}\sin\theta + \dot{\theta}^2\cos\theta\right) + p\left[\ddot{\theta}\sin(\Gamma - \theta) - \dot{\theta}^2\cos(\Gamma - \theta)\right]$$

$$\ddot{y} = b\left(\ddot{\alpha}\cos\alpha - \dot{\alpha}^2\sin\alpha\right) + c\left(\ddot{\theta}\cos\theta - \dot{\theta}^2\sin\theta\right) - p\left[\ddot{\theta}\cos(\Gamma - \theta) + \dot{\theta}^2\sin(\Gamma - \theta)\right]. \tag{4.6.15}$$

The Eqs (4.6.7) to (4.6.15) allow a kinematic analysis of a four-bar linkage gripping mechanism whose size is assigned.
In addition, using the computations of the above-mentioned formulation one can carry out accuracy analysis.
Accuracy of a gripping mechanism is usually referred to as the position error of a reference coupler fingertip point in reaching prescribed trajectory points during the motion, and it can be evaluated as the maximum of these deviation errors along a determined coupler segment.
Referring to Fig. 4.24, let M be the fingertip point lying on the coupler curve, and R be the corresponding point on the circle of center G when the GR line intercepts M.
In this case, the accuracy of a coupler segment, which is traced by M, can be investigated by means of the geometric ratio

$$\varepsilon = \frac{MR}{r} \tag{4.6.16}$$

in which r = GR is the radius of the osculating circle. Referring to the scheme of Fig. 4.24, the geometrical deviation MR can be computed as

$$MR = MG - r \tag{4.6.17}$$

where

$$MG = \left[(x - x_G)^2 + (y - y_G)^2 \right]^{1/2} \tag{4.6.18}$$

and

$$r = \frac{Po\,U^2}{U'U} \tag{4.6.19}$$

with Po the instantaneous center of velocity and U' the inflection point relating to U. MR is positive or negative when a coupler curve segment circumscribes or inscribes its osculating circular arc, respectively.
Moreover, assuming τ as the angle between X-axis and GU, positive counter-clockwise, one can compute the positive clockwise angle between GM and GU as

$$\varphi = \tau - \tan^{-1} \frac{y - y_G}{x - x_G} \tag{4.6.20}$$

Thus, the accuracy analysis of the fingertip motion can be carried out by using Eqs

(4.6.16) to (4.6.20), once the kinematic analysis of the gripping mechanism has been solved.

In the accuracy computation, a problem can be advised to identify the main point U of the fingertip motion as a reference point for the accuracy errors, but also to determine the desired coupler curve for the fingertip motion. In the above-mentioned formulation the osculating circle has been used for a general-purpose formulation, but an accuracy analysis can be developed as referring to specific curves with specific expressions with related geometric characteristics.

In addition, the peculiar characteristics of a mechanism as gripping mechanism can be used to characterize but to formulate performance indices of the grasping purposes of the mechanism.

A performance index can be defined as a synthetic measure of the operation efficiency and therefore it should be formulated by taking into account all parameters that can describe the successful operation of a gripper. Indeed, one can formulate one or more criteria for performance evaluation of grippers by taking into account the characteristics and considerations that have been discussed in previous sections.

Nevertheless, a synthetic formula for an index of merit can be useful to give a measure both for analysis and design purposes.

In analysis procedure a one-value performance index can be useful to qualify a gripper and even to compare it with other grippers. In fact, a choice of a gripper among many available solutions can be obtained by looking at analysis results mainly in terms of performance indices. Of course, a performance index should be formulated by considering the main aspects of gripper design and operation through as many parameters as possible. Among the many possibilities, let us discuss a formulation of a few indices as examples of the indexing synthetic evaluation of grippers.

Among the many possible choices, two performance indices are introduced as related to the main characteristics of grippers in terms of approaching motion and configuration of fingers, and statics action of the grasp.

The finger grasp action can be evaluated both from the kinematics and statics viewpoints.

The kinematics of the grasp can be thought of as related to the position and orientation of the finger with respect to the grasped object, and a satisfactory grasp is achieved when the capability is conveniently exploited and the finger motion to the grasp does not require manipulator small motion adjustment. In particular, significant are the angular excursion of the finger and trajectory accuracy of a reference point of the finger. The trajectory accuracy can be evaluated referring to a linear path that can be considered optimal finger trajectory on the basis of human grasp experience.

Moreover, the swinging angle of the input link is important in order to take into account a measure of mechanism excursion and also for actuator characteristic requirements, which are fundamentally power and motion capabilities.

For the motion aspects of the gripping mechanism a so-called 'Capability Index' (C.I.) can be introduced in the form

$$\text{C.I.} = \frac{L\, r}{l_F\, \varepsilon_{max}} \cos \Delta\alpha_F \tag{4.6.21}$$

where:
- L is the capability of the gripper under evaluation, which is related to the gripping mechanism between the fully open and fully closed configurations;
- r is the input angle ratio given as

$$r = \frac{\alpha_{1L} - \alpha_{1C}}{\alpha_{1L} + \alpha_{1C}} \tag{4.6.22}$$

  which is the ratio that is computed from the angle swept by the input link of the gripping mechanism from the maximum capability configuration α_{1L} and minimum configuration α_{1C} , and the average angle of the aforementioned input range angle;
- l_F is the finger link length;
- ε_{max} is the maximum geometric deviation of the trajectory of the fingertip point M with respect to the ideal linear path within the permitted finger motion;
- $\Delta\alpha_F$ is the angular excursion of the finger link during the motion from the minimum capability configuration to the maximum capability of the gripping mechanism.

The C.I. can be considered as a characteristic of two-finger grippers depending on the gripping mechanism proportions and grasped object size.
Assuming this C.I. formulation, a gripping mechanism is evaluated as more effective than another when the C.I. is calculated greater, which may occur when:
- the configuration capability L is greater since the gripping mechanism allows the grasping of larger objects;
- the input angle ratio r is greater since the position control on the actuator can be transmitted with a greater precision on the finger motion;
- the size l_F of the finger link is smaller, since a smaller and lighter gripper is required in order to reduce as much as possible the manipulator payload and versatility;
- the deviation ε_{max} is smaller, since linear finger motion can be considered as desirable also because no small manipulator adjustment motion is needed, mostly in industrial applications with small pieces grasps.

For statics grasp action of gripping mechanisms a so-called ‘Grasping Index’ (G.I.) can be introduced in the form

$$\text{G.I.} = \frac{F_A \cos\psi_A + F_B \cos\psi_B}{P} \tag{4.6.23}$$

in which P is the force exerted by the actuator; the grasping configuration is considered through ψ_A and ψ_B ; and the grasping action by means of F_A and F_B.
The G.I. performance index can be further simplified by assuming $\psi_A = \psi_B = \psi$ and F_A

$= F_B = F$ to give

$$G.I. = \frac{2\,F\cos\psi}{P} \tag{4.6.24}$$

Equations (4.6.23) and (4.6.24) take into account both the aspects of force and motion transmission, which can be recognized as fundamental for the design and working of a gripping mechanism. Indeed the formulation of G.I. takes into account mechatronics aspects when one considers that: actuator characteristics are included through P evaluation; grasp control design is considered through a regulation of grasp force F; mechanics of grasp is described through the grasp action F, but the grasp configuration through the angle ψ; and the efficiency of the gripping mechanism is evaluated through the mechanism efficiency F/P. In addition the G.I. in the form of Eq. (4.6.24) is also very useful because it can be easily evaluated for several gripping mechanisms by using the Principle of Virtual Work. In fact, by assuming as negligible the friction forces in the joints of a mechanism, the expression of the efficiency ratio F/P can be conveniently expressed as

$$\frac{F}{P} = \frac{2\left(\dot{r}\cos\psi + r\,\dot{\psi}\cos\psi\right)}{v_P} \tag{4.6.25}$$

in which v_P is the velocity of the actuator, and r dot and ψ dot are the position and angular velocities of the fingers. The right side of the expression (4.6.25) can be computed by using a velocity analysis of gripping mechanisms so that the G.I. can be evaluated by performing a kinematic analysis only.

With the formulation of G.I. a gripping mechanism is evaluated as more effective than another when its G.I. is computed greater for a determined grasping configuration. Moreover the G.I. is a characteristic, which is independent of the size of the gripper mechanism, due to its dimensionless definition. In fact, from Eq. (4.6.24) the ratio of F to P can be expressed as the ratio of v_P to v_F.

The actuator velocity v_P is related to the angular velocity α dot of the input link of the gripping mechanism through its kinematics, and it can be determined in the form

$$v_P = \tau\, r\, \dot{\alpha} \tag{4.6.26}$$

where τ represents the transmission factor and r is the distance from which the input link receives the motion.

A general summing up formula, as outlined in the analysis of velocity closure Eqs (4.6.3) and (4.6.4) can represent the finger contact point velocity v_F as

$$v_F = \sum_{j=1}^{j=n} c_j l_j \dot{\alpha}_j \tag{4.6.27}$$

where l_j is the length of link j ,which can be referred through the link proportion K_j to the input link length l_2 as

$$l_j = K_j l_2 \tag{4.6.28}$$

n is the number of the links in the gripping mechanism kinematic chain; c_j is the corresponding coefficient of the term l_j and it assumes a transcendental function form; and dot α_j is the angular velocity of the link j whose expression takes the form

$$\dot{\alpha}_j = m_j \dot{\alpha}_2 \tag{4.6.29}$$

when m_j is the corresponding dimensionless coefficient depending on the mechanism configuration.
Therefore, the finger contact point velocity can also be expressed as

$$v_F = C l_2 \dot{\alpha}_2 \tag{4.6.30}$$

where C takes into account all the computational factors which are independent of the gripping mechanism scale and input angular velocity.
Then, the angle ψ, which can be considered as an orientation angle of the finger link body, can be deduced from

$$\cos\psi = H \alpha_2 \tag{4.6.30}$$

where H is a dimensionless factor, which is independent of the mechanism size. Its expression depends on the input angle α_2.
Making use of the F computation and G.I. formulation it is possible to deduce design charts in order to derive the optimal linkage proportions for a given gripping mechanism kinematic chain, and to operate dimensional modifications in an optimization procedure.
In Fig. 4.25 examples of analytical expressions for the above-mentioned indices are shown as referring to illustrative gripping mechanisms. The differences among these grippers can be remarked by and through the formulas of the proposed performance indices.

4.6.2.1 A numerical example of index evaluation

In Fig. 4.26 the kinematic chain of a gripping mechanism for a widely-used two-finger gripper is shown as referring to one finger only, due to the symmetry of the mechanical

design.
The plots in Fig. 4.27 show the numerical results of the kinematic analysis of the gripping mechanism of Fig. 4.26 by using the formulation and indices that are outlined in section 4.6.
It is worth noting how the design and operation charts for C.I. and G.I. indicate optimal design sizes and operation angles for optimum performances.

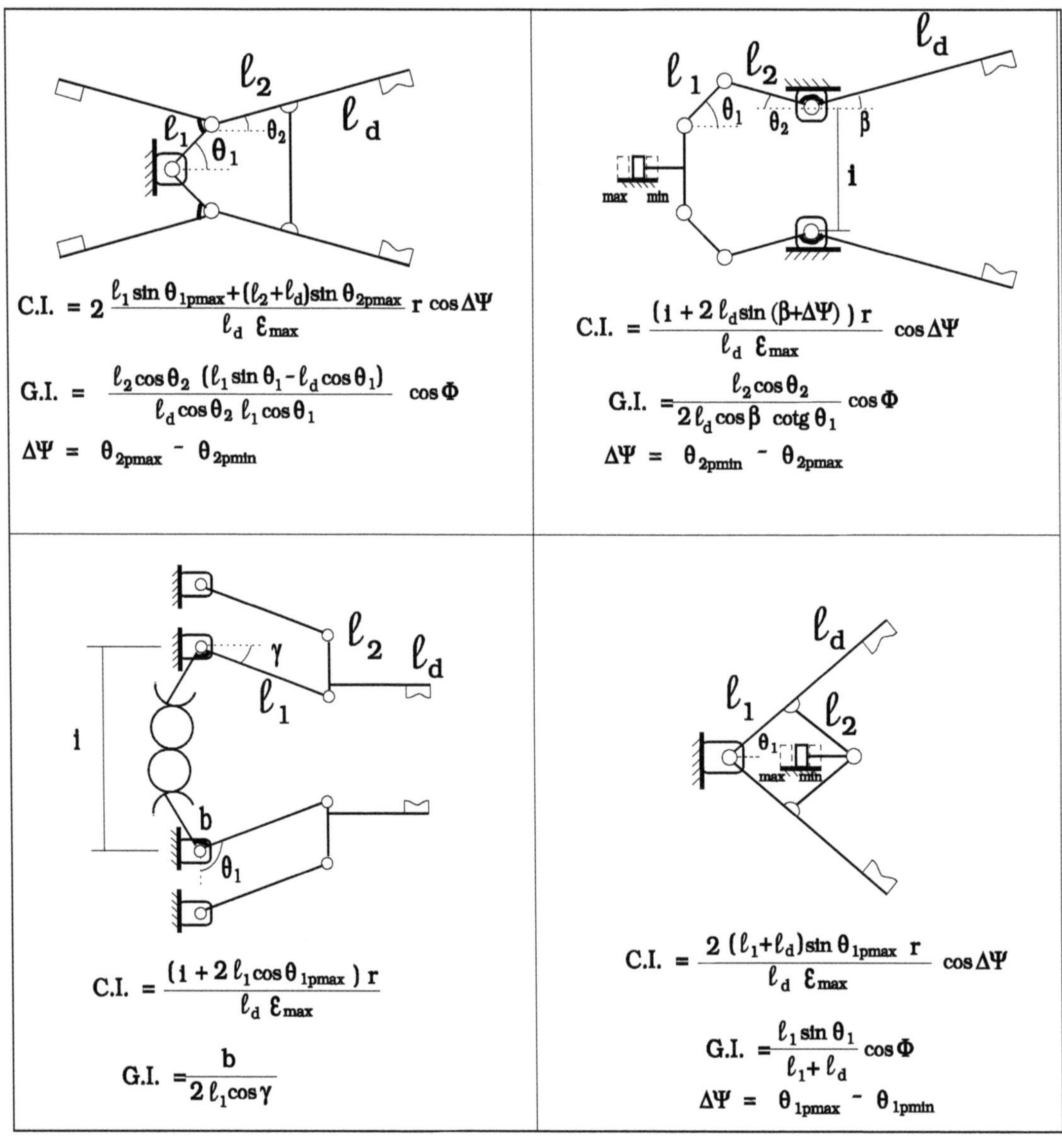

Fig. 4.25: Examples of an analytical evaluation of gripping mechanisms through the Capability Index (C.I.) and the Grasping Index (G.I.).

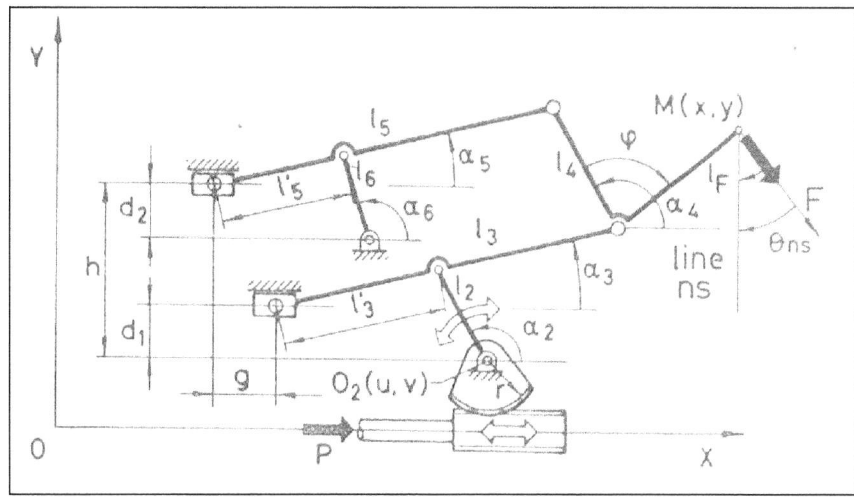

Fig. 4.26: A gripping mechanism and its parameters for a two-finger gripper.

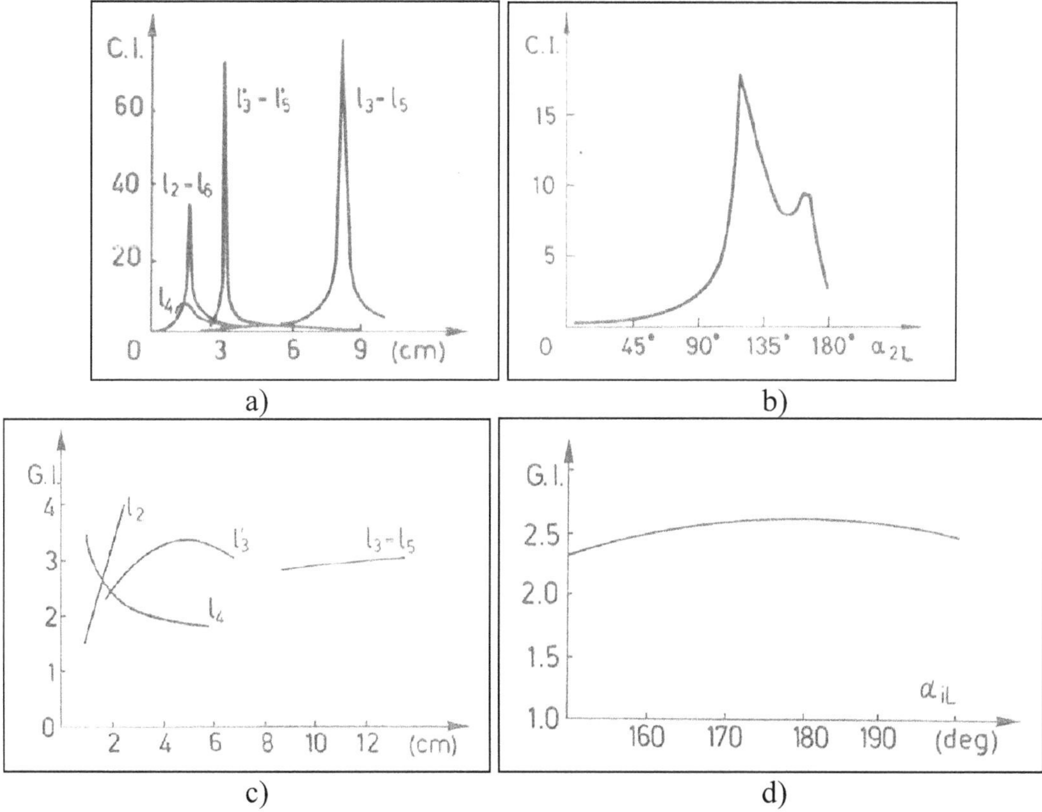

Fig. 4.27: Numerical evaluation of performance indices of the gripping mechanism of Fig. 4.26: a) Capability Index (C.I.) versus mechanism sizes; b) Capability Index (C.I.) versus input crank angle at the open configuration; c) Grasping Index (G.I.) versus mechanism sizes; d) Grasping Index (G.I.) versus input crank angle at the open configuration.

4.7 Designing two-finger grippers

The design problem for grippers consists of sizing all the components of the gripper to ensure suitable grasping performances for a given class of objects with a certain weight and dimension.

Usually, design procedures can provide separate designs for the components and the system is then adjusted through CAD analysis and finally during the installation.

Nevertheless, the components do not actuate separately, and each one influences the other. Therefore, a design procedure for a two-finger gripper may be proposed as a concurrent design of the components according to the scheme of Fig. 4.28, where the interconnecting lines are to be considered.

Each block in the flow diagram of Fig. 4.28 describes a design activity on one specific aspect of a gripper system. But even if a concurrent design approach cannot be considered, it is convenient to check the result of each block with the requirements for the next block, as indicated by the returning arrow between two blocks in Fig. 4.28.

The specific or general aim of a gripper is determined by the data of the objects that will be grasped.

In the first design block the mechanics of grasp is attached to evaluate the required grasping force as a function of the capability of the gripper. In this item the model and formulation for the equilibrium of the grasp is elaborated by using the fundamentals that are outlined in sections 4.4. and 4.5. Indeed, one important result of this design item can be the shape and size of the fingertips as a function of the shape and size of the objects but the relative position of the grasping force on the gripper fingers. In addition, this block will include the design of the control and sensor equipment as depending on the level of grasping force regulation that is required for the gripper application. Thus, the type of sensors will be properly chosen and sized for installation on the designed fingers and fingertips. The control algorithm will be designed for the required control type and will give requirements for the next design blocks for the actuator design and design of the gripper circuit. In the next section 4.8 the design of control and sensor equipment for grippers is outlined along with practical examples.

Once the grasping force has been determined, in the synthesis block for the gripping mechanism two main problems are addressed: the chain type and dimensional design of the gripping mechanism. Thus, a designer will choose a proper type of gripping mechanism by using an atlas, expert systems for searching, or even her/his own experience on certain mechanisms. Then, the dimensional design of the gripper mechanism can be approached by using a traditional technique for dimensional synthesis of mechanisms, but with the requirements for gripper application in terms of prescribed positions and force actions. Alternatively, a specific approach can be formulated for the design of the gripping mechanism by considering the peculiarities of the grasping purposes, as outlined in the next section 4.7.1.

Once the gripping mechanism is determined, it is possible to size the actuator action by looking at the mechanical efficiency of the gripping mechanism. Together with the motion characteristics in terms of input motion and operation velocity, it is possible to determine a suitable actuator that can be chosen conveniently from the market.

The last design block is concerned with the design of the components of the circuit that is needed for the controlled operation of the actuator. The circuit as well as the actuator can be pneumatic, hydraulic, or electric. The components are sized as a function of the performances for the actuator, gripping mechanism, and controlled grasping force, and generally they can be chosen from the available units in the market as the nearest solutions.

However, a concurrent design for a two-finger gripper can be expressed explicitly through suitable formulations for the design block of each component in Fig. 4.28 to give an analytical system of design equations. These design equations also express the relations between the components and their contemporaneous solution provide a well-proportioned gripper system.

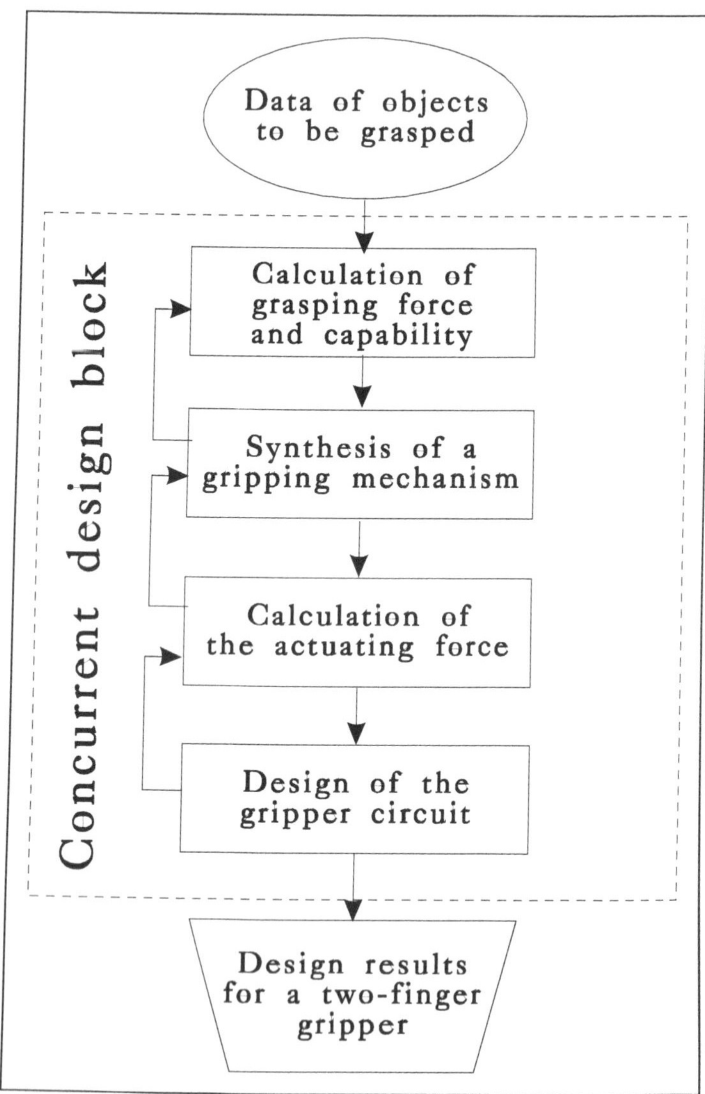

Fig. 4.28: A general scheme for a gripper design procedure.

Referring to the case of study shown in Fig. 4.13, the design parameters can be grouped in vectors as

$$l_{mec} = (l_1, l_2, l_3, l_4, l_f, r, d_t)$$

$$\alpha_{mec} = (\alpha_2, \alpha_3, \alpha_4, \alpha_f)$$

$$l_{act} = (S, V, d_A, d_s) \tag{4.7.1}$$

$$K_h = (k_1, k_2, \xi, \sigma_n, \beta)$$

$$K_c = (G_c, k_T, k_F)$$

where the geometrical parameters are shown in Fig. 4.13; k_1, k_2, ξ, and σ_n are the characteristics of the pneumatic/hydraulic circuit; β is a fluid bulk coefficient; and G_c, k_T, k_F, are the parameters of the control.
Both the grasping force F and grasping capability L are affected by the size set l_{mec} and the positional configuration set α_{mec} of the driving mechanism, the actuator size set l_{act}, the characteristics set K_h of the hydraulic circuit, and the parameters set K_c of the control. In fact, the grasping force F can be computed as a function of all those parameters by using the Laplace transformation in the form

$$\mathbf{F} = \frac{U - T_{mec}}{U} \mathbf{F}_{REF} - \frac{Z}{U} \dot{\mathbf{x}}_A \tag{4.7.2}$$

where

$$U = T_{mec} + \frac{k_1 \sigma_n^2 G_c}{\left(s^2 + 2\rho\sigma_n + \sigma_n^2\right)\left(\frac{V}{2\beta} s - k_2\right)} \tag{4.7.3}$$

$$Z = \frac{S}{\left(\frac{V}{2\beta} S - k_2\right)} - M_r s \tag{4.7.4}$$

in which M_r is the system inertia reduced to the actuator axis and the transmission coefficient T_{mec} of the driving mechanism can be expressed as

$$T_{mec} = 2\frac{l_2}{r}\left[\sin\alpha_2 + \left(\frac{l_f}{l_3}\sin(\alpha_3 - \alpha_f) - \sin\alpha_3\right)\left(\frac{\sin\alpha_2 - \cos\alpha_2 \, tg\alpha_4}{\sin\alpha_3 - \cos\alpha_3 \, tg\alpha_4}\right)\right] \quad (4.7.5)$$

Equations (4.7.2) to (4.7.5) express an example of the relations among the design parameters that can be used in a concurrent design according to the outlines of Fig. 4.28.

4.7.1 An optimum design procedure for gripping mechanisms

A proper formulation for the gripping mechanism design may help a designer in the optimum use of the gripping mechanism chains, since a suitable synthesis can also give optimum characteristics with respect to the grasping purpose.

The above-mentioned design and performance considerations address great importance to the gripping mechanism since it transmits the motion and force to the fingers for a suitable grasp. In addition, such considerations may strongly suggest using a design formulation for gripping mechanisms in the form of an optimization problem as

$$\max f \quad (4.7.6)$$

subject to

- design constraints
- material properties
- object characteristics
- peculiarities of the application

The critical point of such a design formulation is the choice of a suitable objective function f, which may include the peculiar aspects of the gripping mechanism together with its design parameters, and may give a solvable numerical formulation in the sense of obtaining practical solutions which are not trivial. The constraints can be expressed for so many different characteristics but they should be analytically formulated. This may limit the constraints to be considered in the design problem. Further complications may arise in formulating the constraints with the attempt to specialize the mechanism design to a gripping purpose, which nevertheless can be stated as general or as specific with respect to an object or a manipulation. However, the problem can be formulated in a general form, so that a number of constraints can be considered as or substituted by expressions for a specific application. The above-mentioned considerations and developments in previous sections may strongly suggest the use of a design formulation in the form of an optimization problem as

$$\min \frac{GI_{max} - GI_{min}}{GI_{med}} \quad (4.7.7)$$

Subject to $L_{min} \leq L \leq L_{max}$ (4.7.8)

in which min, max, and med indicate the minimum, maximum, and average of a quantity, respectively. The G.I. has been chosen as objective function since its formulation includes mechatronics aspects. L is the grasping capability of the gripping mechanism and it can be defined as the size of the largest object that the gripper can grasp. The global evaluation of the G.I., which has been expressed in Eq. (4.7.7) as the objective function, has been formulated to have a performance index that synthetically describes the grasping characteristics of a gripping mechanism within operation and design intervals. Thus, the optimization problem prescribes an optimum design in terms of maximum grasping performance, but in the form of a regular quasi-constant behavior within given limits for the grasping capability.

Furthermore, additional constraints on link sizes in the form of Eq. (4.7.8) can be formulated to give practical values for a practical mechanical design and to take into account specific limits and constraints of a specific application.

Indeed, this formulation can be easily extended to consider further mechatronic aspects by including all the characteristic aspects of the mechatronic layout. Nevertheless it can be unpractical due to the great computational efforts that are required to solve it.

A numerical procedure can be developed for the solution of the problem of Eqs (4.7.6) to (4.7.8) by using suitable analysis formulation for the gripping mechanism behavior so that commercial software can be properly used to solve the optimization problem.

Figure 4.29 illustrates basic items in order to solve the proposed optimization problem by means of commercial software and by using common kinematic analysis for gripping mechanisms. The main numerical features of the optimum design procedure in Fig. 4.29 can be recognized in the three blocks for formulating and computing the performance and requirements of gripping mechanisms in a way to fully exploit existing software for solving the optimization problem. In fact, a fourth block is indicated as a numerical procedure by commercial software that can be used by a gripper designer as a suitable computation means once the design problem has been properly formulated.

The first block in Fig. 4.29 is concerned with formulating design equations from the analysis of the gripping mechanism. This can be obtained by using the analysis of gripping mechanisms in terms of their kinematic behavior and performance indices, as outlined in section 4.6. The second block is related to the computation of the objective function. In Eq. (4.7.7) the G.I. has been used but many other expressions can be applied, even in a multi-objective function formulation. In the third block the computations are expressed for the constraints that are formulated as g functions.

The fact that planar 1 d.o.f. mechanisms are generally used as gripping mechanisms, can be very helpful in formulating the G.I. in efficient form. In fact, by assuming as negligible the friction at the joints of a gripping mechanism and applying the Principle of Virtual Work, it is possible to obtain a general expression for G.I. as

$$GI = \frac{\cos\psi}{\frac{\dot{x}}{v}\cos\psi + \frac{\dot{y}}{v}\sin\psi} \tag{4.7.9}$$

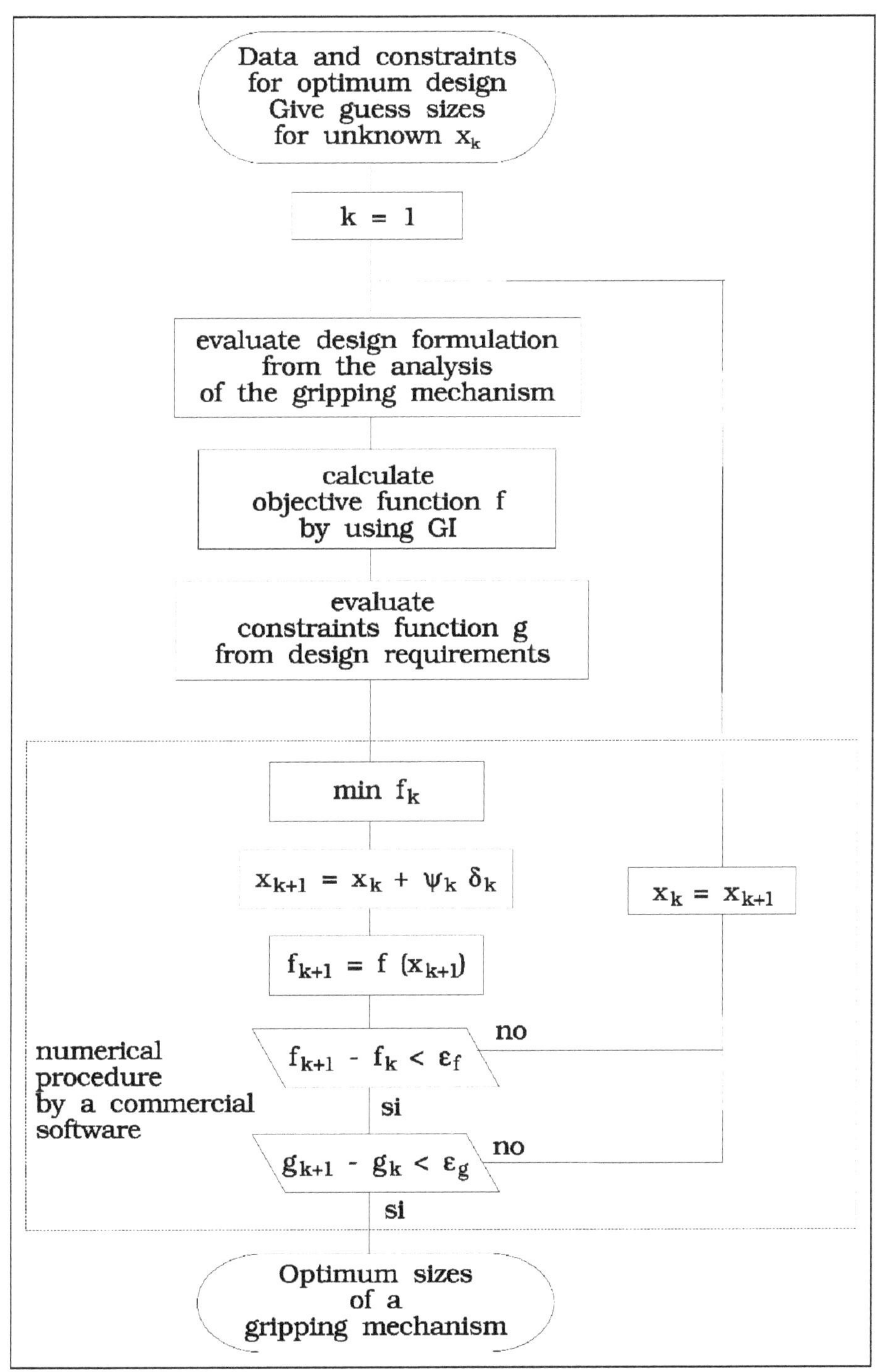

Fig. 4.29: A flow chart for an optimum design procedure of gripping mechanisms.

where v is the velocity of the input link; ψ is the orientation angle of the grasping force F with respect to the contact line, as shown in Fig. 4.17; dotted x and y are the Cartesian components of the velocity of the point where F is applied on the finger link. The velocity ratios are kinematic values, which do not depend on the input velocity but only on the mechanism configuration at the grasp. Therefore, Eq. (4.7.9) can be evaluated in an analytical form for a given gripping mechanism with evident computational advantages also for optimum design numerical procedures.
In addition, software can be used to perform the numerical solution with a Sequential Quadratic Programming procedure. This numerical procedure works as shown in the block of Fig. 4.29 for the numerical procedure in such a way that at each step k a solution is found along a search direction δ_k with a variable update Ψ_k. The iteration continues until the variables vector converges.
Indeed, the outlined numerical procedure can be developed so that the formulation for the gripping mechanism analysis can be easily included within the solving procedure for the optimization problem by using the facilities of a commercial software package. Once the numerical computations are convergent to a feasible solution, the velocity and the accuracy of the solution can be enhanced by a designer by updating the convergence parameters ε_f and ε_g, which refer to the objective function f and the constraint functions g, respectively.
Of course, the optimum solution is affected by the initial guess. But the algebraic formulation of the objective function f and constraint g can be useful to obtain optimum design also when the initial gripping mechanism is far away from the prescribed limits or it even violates some of the prescribed requirements.

4.7.1.1 A numerical example of optimum design

In Fig. 4.30 a widely-used finger gripping mechanism is shown with its design parameters. It is of the slider-crank type and a design problem can be formulated as concerned with the dimensions of the links and stroke range, but in order to obtain suitable behavior within a certain range of grasping capability.
Therefore, the optimization problem of Eqs (4.7.7) and (4.7.8) by using the G.I. can be conveniently applied with the numerical procedure of Fig. 4.29.
The specific kinematic chain of the gripping mechanism allows the opportunity to exploit the expression (4.7.9) for the G.I. since the closure equations for mechanism analysis give

$$\frac{\dot{x}}{v} = \frac{p \sin\alpha \cos(\gamma - \beta) - b \sin\alpha \cos\beta}{b \cos(\alpha + \beta)} \tag{4.7.9}$$

$$\frac{\dot{y}}{v} = \frac{b \cos\alpha \sin\beta - p \sin\alpha \sin(\gamma - \beta)}{b \cos(\alpha + \beta)} \tag{4.7.10}$$

and

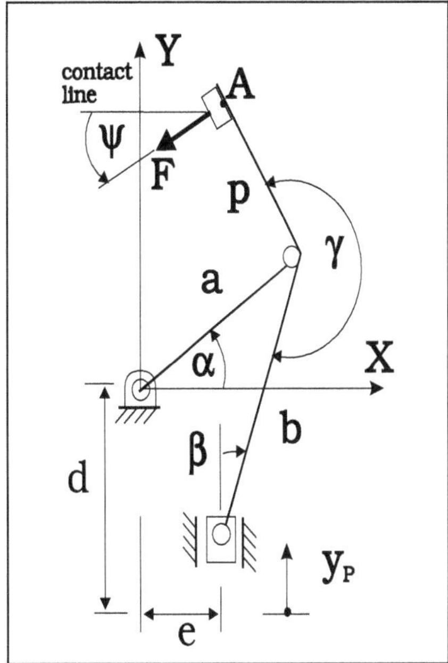

Fig. 4.30: The kinematic chain and its design parameters for a simple widely-used gripping mechanism.

$$\alpha = 2\tan^{-1}\frac{A - \sqrt{1 + A^2 - B^2}}{1 + B}$$

$$\beta = 2\tan^{-1}\frac{-1 - \sqrt{1 + A^2 - C^2}}{C - A} \tag{4.7.11}$$

$$\psi = \cos^{-1}(\gamma - \beta - \pi)$$

in which

$$A = \frac{(d - y_P)}{e}$$

$$B = \frac{b^2 - a^2 - e^2 - (d - y_P)^2}{2\,a\,e} \tag{4.7.12}$$

$$C = \frac{a^2 - b^2 - e^2 - (d - y_P)^2}{2\,b\,e}$$

The results of the optimum design procedure are shown in the plots of Fig. 4.31 in which two cases are illustrated as function of the type of a design constraint.
The sizes of the optimally designed gripping mechanism are shown in the simulation plots and they can be compared with the operation range in terms of grasping capability that is drawn by the asterisks for the fingertip points.

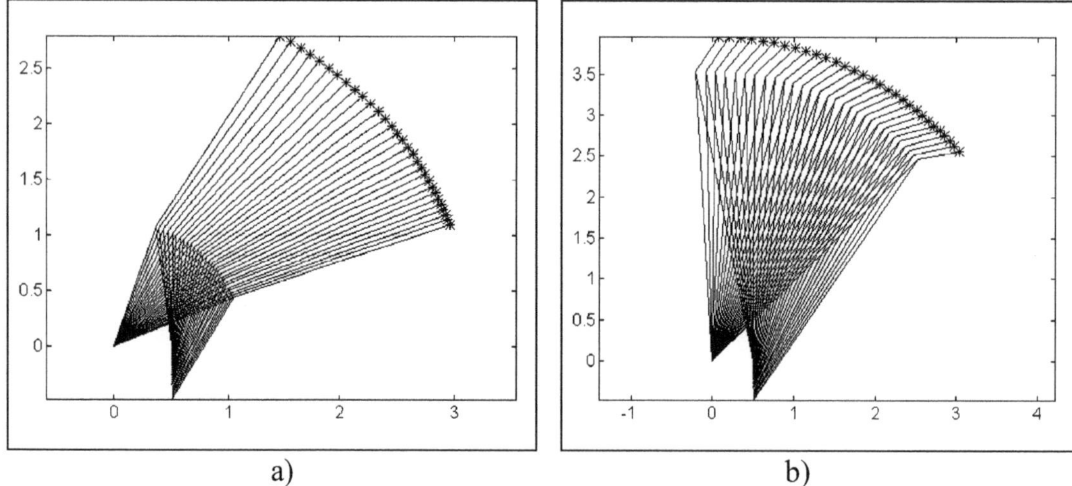

Fig. 4.31: Results of the optimum design procedure for the gripping mechanism of Fig. 4.30 when a design constraint is prescribed: a) for well-proportioned links; b) in a strong form as $d^2+e^2<(a+b)^2$.

4.8 Electropneumatic actuation and grasping force control

Industrial automatic systems are usually arranged so that any gripper may be installed as they are supplied with pipelines and given pneumatic or hydraulic pressure flow. Therefore, a design problem for gripper application can be stated as given a grasp force and an input displacement, to select an actuator with suitable size for an *a priori*-determined pressure flow. Further problems may arise in conceiving a hydraulic/pneumatic circuit and at most in selecting a suitable electrovalve to give a desirable response both from kinematic and force viewpoints. Thus, a hydraulic/pneumatic system for a gripper can be supplied through a component selection among available commercial means, which are designed for automation purposes.
Similarly, for electric actuation of grippers, the design problem for a gripper application can be approached in order to choose suitable motor, line size, and control equipment.
A control scheme for grippers can be designed as proposed in Fig. 4.27, which is specific for pneumatically/hydraulically actuated grippers. A similar control scheme can be designed for electrically actuated grippers by using suitable blocks for electrical and electronic components.
Figure 4.27 refers specifically to force control because the signal F regards the grasping force at the fingers. But the scheme can also be used for position control of the fingers when suitable position sensors are installed on the fingers to detect the relative position of the grasped object. Position control for gripper operation is usually obtained by

simply controlling the position of the actuating piston through sensors thereon, even in an open-loop scheme. Indeed in many industrial applications the gripper actuation is controlled only for opening and closing the fingers.

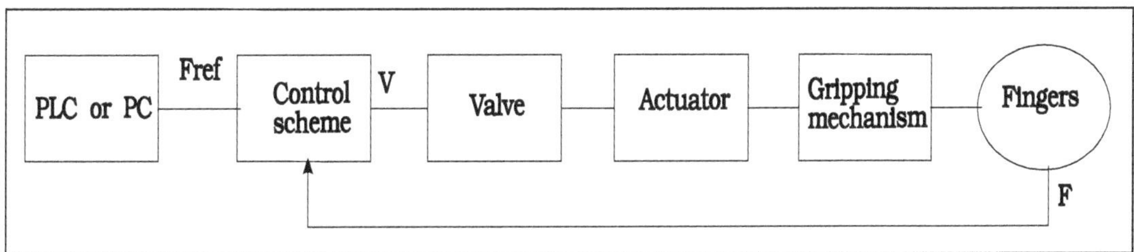

Fig. 4.32: A scheme for grippers with force controlled pneumatic/hydraulic actuation.

Each block in Fig. 4.32 represents both hardware and software of the design and operation for a force controlled gripper.
The PLC/PC block indicates systems that are used to store and run the control programs, but also to simulate grasping in order to compute the reference grasping force Fref for a static grasp through a suitable model and formulation. The value of Fref can be also given as a tension value V that is suitable for electrovalve and actuator commands.
The control block provides a proper tension V to the valve through suitable electric/electronic components in order to regulate the flow pressure in the actuator when the pneumatic/hydraulic circuit is designed, similar to the proposed schemes of Figs 4.33, 4.34, and 4.35. The control block can be obtained in the form of several architectures according to the Control Theory and required gripper behavior. Usually a PI (Proportional Derivative) algorithm is applied by using signals from position and force sensors.
The valve block indicates the regulation equipment for the operation of the actuator. The main component is the valve that, usually for electrically controlled grippers, is an electrovalve type. Other components are those of the circuit, which is needed for properly operating the actuator with a suitable flow.
The actuator block indicates the actuator with its sensors, which gives suitable power to the gripping mechanism in terms of force and movement, because of the signal commands and valve regulation.
The gripping mechanism block indicates the mechanical system that provides a suitable transformation of the actuator power to the grasping operation of the fingers. It can be identified as a gripping mechanism.
The fingers item indicates the functionality of the fingers during the grasp, whose characteristic grasping force F can be used to regulate the gripper operation.
Indeed, the scheme of Fig. 4.32 can be representative also of the position controlled grippers when the F signal is understood as a position signal. The same diagram can also be used for position and force controlled grippers when both position and force signals are used in a closed-loop control.
Nevertheless, new and enhanced applications require grippers that can be used by

regulating the grasping force. Grasping force control is required when a multi-task gripper grasps fragile objects or a large variety of objects.
When force control is adopted, usually a force sensor can be used as installed on one fingertip to mimic tactile sensing. The force signal is used through a suitable control system to regulate the pressure in the push chamber of the actuator, as illustrated in the schemes of Figs 4.33, 4.34, and 4.35.
Specific attention is addressed to sensors for fingers and several designs can be found in the literature using commercial means or innovative sensors and arrangements.
The scheme of Fig. 4.32 is general but useful to understand the influence of each component on the grasping action and avoid underestimation of any aspect or part of the control system. In particular, the sensed grasping force F in the feedback loop is compared with the *a priori*-calculated grasping force Fref that is usually evaluated by using suitable models for the mechanics of grasp, as in Fig. 4.17.
A simple and efficient force control can be obtained when a gripper is actuated by means of a pneumatic/hydraulic cylinder, as the cases shown in Figs 4.33, 4.34, and 4.35. In these cases designing a suitable pressure control system in order to vary the pressure in the push chamber of the cylinder can solve the force control problem. Several architectures can be used. Three practical solutions are reviewed in the form of the schemes shown in Figs 4.33, 4.34, and 4.35.
Referring to Figs 4.33, 4.34, and 4.35, a closing/opening movement of the gripper fingers is obtained in correspondence to a pulling/pushing movement of the cylinder when the chamber 1 is in pressure and the other chamber 2 is discharging the flow.
In Fig. 4.33 the pressure control in chamber 1 is obtained by directly using a pressure proportional electrovalve 3/2 (three ways/two positions), which can ensure the required pressure when a suitable electric analogue signal Vrif is imposed by the user or is sent by a control unit. Usually Vrif is computed as the V value in the Fig.4.32 in a preliminary step by considering the static equilibrium of the grasp, but even by suitable start-up calibration.

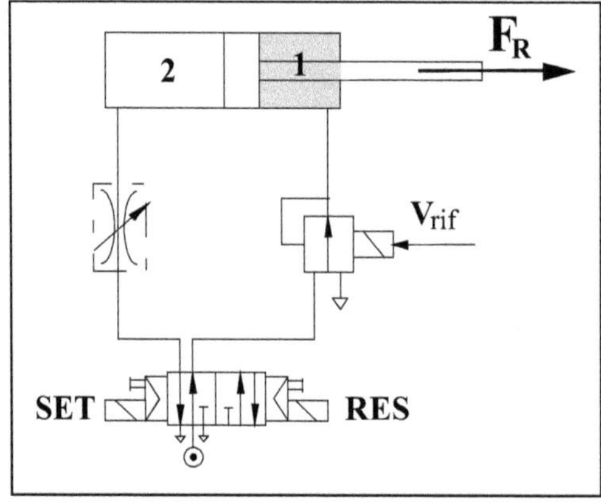

Fig. 4.33: A force control scheme for a pneumatic actuator of grippers by using a pressure proportional electrovalve.

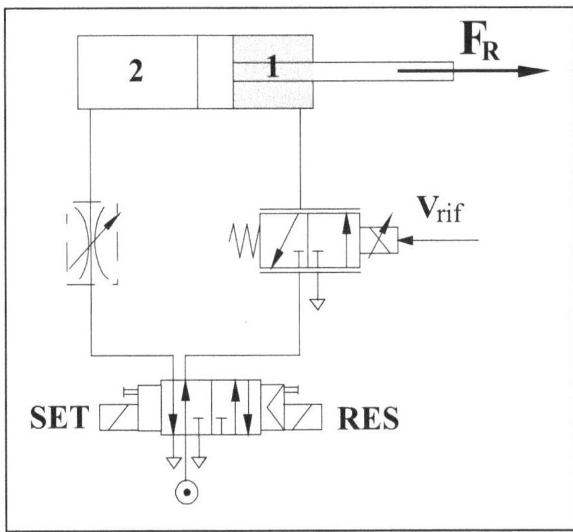

Fig. 4.34: A force control scheme for a pneumatic actuator of grippers by using a flow proportional electrovalve.

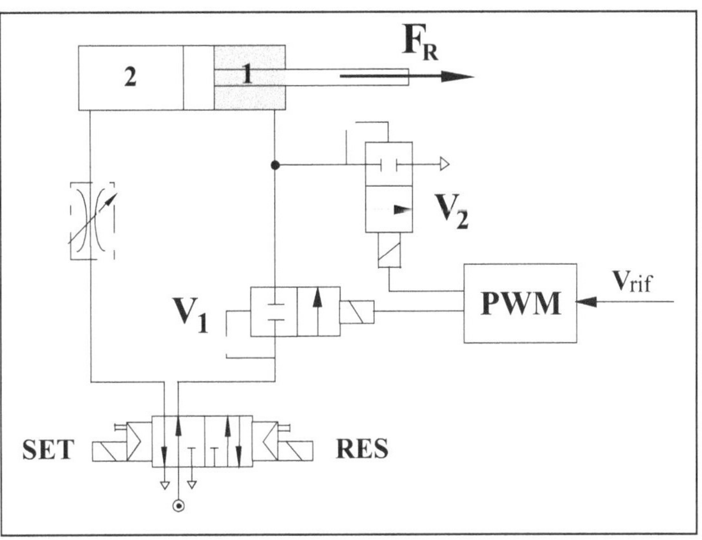

Fig. 4.35: A force control scheme for a pneumatic actuator of grippers by using digital electrovalves and PWM modulation.

In Fig. 4.34 the pressure control in chamber 1 is obtained indirectly by using a flow proportional electrovalve 3/2, which can vary the mass flow into chamber 1 when a suitable Vrif control signal is received by the electrovalve.

The electropneumatic circuit of Fig. 4.35 refers to a special solution, which has been proposed and analyzed at the Laboratory of Robotics and Mechatronics, in Cassino. It mainly consists of controlling the air pressure in chamber 1 by using two digital electrovalves 2/2, V_1 and V_2, which are connected to chamber 1 at the supply and discharge pipelines, respectively, in order to obtain an equivalent electrovalve 3/2. The mass flow to chamber 1 is controlled by means of a suitable PWM (Pulse Width Modulation) modulation of the Vrif control analogue signal.

In any case, in order to avoid damage to a grasped object a manual velocity regulation of the grasp closing movement can also be obtained through a suitable bi-directional flow regulator.

Each hardware control scheme corresponds to a control algorithm that can be designed for a suitable programming of PLC and operation of grippers.

For example Fig. 4.36 is an illustrative example of a control algorithm that has been developed specifically for the pneumatic actuation of Fig. 4.35.

The PI control scheme has been chosen to suitably regulate the grasping force at a stable *a priori*-determined level. The gain values Kp, Kv, Ki can be settled by trial and error technique depending also on the specific object that will be grasped in order to have a suitable practical response. Alternatively, the gain values can be given through a calibration procedure at the start-up of the gripper.

The proposed force control in Fig. 4.36 has been implemented in the test-bed two-finger gripper of Figs 4.37 and 4.38. The test-bed, as a typical industrial gripper, has been assembled by using commercial components and is composed of:

- a two-finger gripping mechanism;
- a pneumatic actuator, which is a double-acting pneumatic cylinder;
- a proportional pressure electrovalve;
- an adjustable choke-valve;
- a digital electrovalve;
- a load cell in one fingertip;
- a PLC, which is provided of a A/D–D/A (Analog/Digital–Digital/Analog) converter;
- a common PC, which has been connected to RS-232 serial port of the PLC.

A reference force Fref can be prescribed through the PC and consequently the PLC controls the mechatronic system by using its input/output device. An analog signal Vref controls the proportional electrovalve, and two digital signals A^+, A^- can be used to control gripper movements. The load cell sensor gives a feedback force signal Fout. The grasp force control has been obtained by controlling the pressure in the push chamber of the pneumatic cylinder by means of the proportional pressure electrovalve.

The adjustable choke-valve has been used to obtain slow movements of the piston through manual regulation. Thus, a small approaching velocity of two fingers to the object can be obtained without position/velocity control. The electrovalve has been used to obtain the closing and opening of the two-finger gripper by operating the double-acting cylinder in an on/off environment.

The load cell sensor has been installed on one fingertip only in order to have a feedback signal. The force sensor, whose gain coefficient has been settled with a value equal to 11.77 N/V, measures the force Fout. During experimental tests the load cell has also been used to measure the grasp force through a Data Acquisition System for monitoring purposes, in agreement with the scheme of Fig. 4.38.

The PLC controller has been connected to the proportional electrovalve and load cell by means of an A/D–D/A converter. The control of the digital electrovalve has been obtained by using two digital outputs of PLC. The PLC programming has been developed by means of the PC in Basic language.

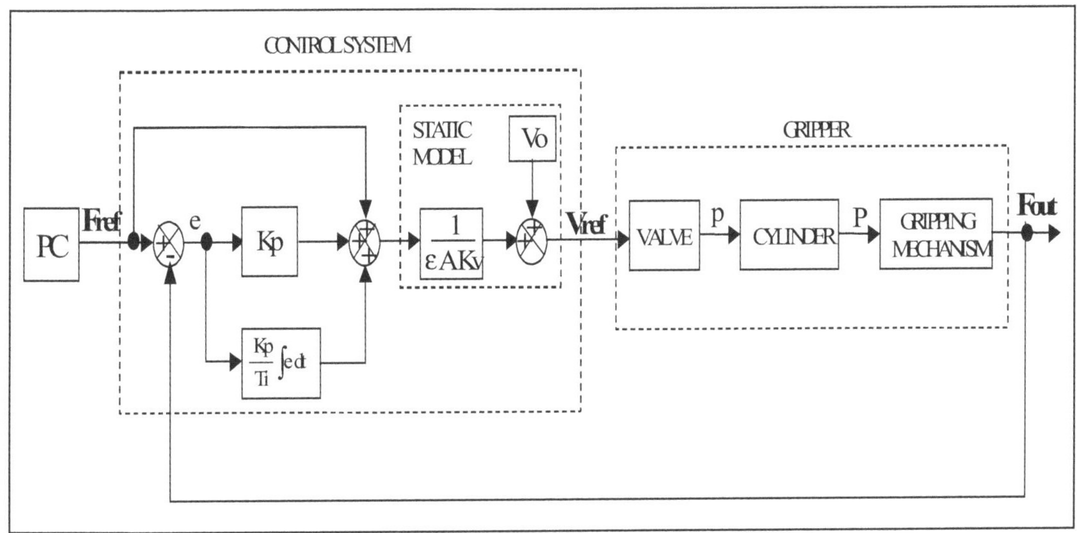

Fig. 4.36: A scheme for force control in two-finger grippers.

In Fig. 4.36 a closed-loop force control has been formulated through a suitable modeling of the proposed mechatronic system. The mechatronic system can be thought of as composed by three main blocks: PC, control system, and gripper.

A PC is needed for programming the control system and but it mainly computes the grasp action, which gives the reference value Fref of the grasping force at the fingers. It has also been used to monitor the grasping action.

The control system has been designed by using the PI scheme with gain coefficients Kp and Ti, and modeling the gripping behavior of the gripper through the mechanism efficiency ε, piston pushing area A, and static gain Kv of the electrovalve.

The gripper block is composed of the proportional pressure electrovalve, pneumatic cylinder, and gripping mechanisms, as indicated in Fig. 4.36.

Each block in Fig. 4.36 can be properly formulated and a suitable model can be obtained for the proposed mechatronic gripper system, and a practical example is illustrated in section 4.8.1.

A Fref is prescribed *a priori* for given application and object. Consequently, a suitable Vref can be sent to the proportional electrovalve because of the control system of Fig. 4.36. The pneumatic pressure p, which is obtained in the push chamber of the pneumatic cylinder, gives a suitable actuating force P for the gripping mechanism. The gripping mechanism transmits the actuating force P to each finger and a grasp force Fout is exerted on the object. The feedback signal by the force cell sensor has been used to correct any error by continuously using the closed-loop scheme.

In particular, the control system is composed of a PI algorithm and a static model for the grasping action of the gripper. The PI control algorithm is based on the proportional gain Kv and integral gain Kp/Ti by using the force error e as a difference between the measured grasp force Fout at a fingertip and desired value Fref that is evaluated *a priori*.

The statics model in the control system in Fig. 4.36 is based on the efficiency ε of the gripping mechanism and on the amplification factor Kv for the operation of an actuator with cross-section A. In addition a reference value Vo for its operation of the electric equipment is indicated. Thus, the electric signal Vref is produced to operate the electrovalve that can regulate the flow pressure p in the push chamber of the piston. The piston gives the actuation force P; the gripping mechanism transmits the force P to the fingertips at grasp, and the value Fout is measured by the sensor on the fingertip.

Summarizing, in Fig. 4.36 a mechatronic approach for grippers can be recognized in the blocks where mechanical efficiency, friction, and size coefficients are used together with the control gains to evaluate the force regulation.

4.8.1 An illustrative example for laboratory practice

A gripper test-bed has been built at the Laboratory of Robotics and Mechatronics, in Cassino both for teaching and investigating grippers and mechanics of grasp.

In fact, students do practice during courses to gain experience on the mechanics of grasp, gripping mechanisms, programming, and operating mechatronic systems. In addition, both design and teaching activities can be focused on the possibility to obtain and operate a gripper whose main characteristics are low-cost design and easy operation, in order to achieve the goals for a successful integration with robots.

The gripper test-bed of Fig. 4.37 has been built according to the mechatronic scheme of Figs 4.12 and 4.13, taking into account the design and operation considerations that have been discussed in previous sections.

The mechanical design of the gripper has been based on typical gripping mechanisms, such as those shown in Fig. 4.18, but it has been conceived and built with modular components so that several gripping mechanisms can easily be assembled in the gripper. A commercial robust force sensor has been used and installed on a fingertip, which has been properly designed for this force sensor but with general ideas to be easily removed and rearranged. Only one force sensor has been used because of the symmetry of the gripping mechanism and grasping configuration.

In Fig. 4.38 indicating the main sub-systems is shown the composition of the test-bed for two-finger grippers. In Fig. 4.38 an additional block is included to monitor the grasping force through the same force sensor for the force control. Suitable sensor equipment can be used as based on Labview virtual instruments. A PLC has been used to program the operation of the system by implementing the control scheme of Fig. 4.36, after suitable calibration start-up. The PC of the monitoring system can also be used for the programming of the PLC.

Figure 4.39 shows the design of a force sensored fingertip that can be easily changed.

The control block in Fig. 4.38 is obtained by using market pneumatic components, as shown in the scheme of Fig. 4.36.

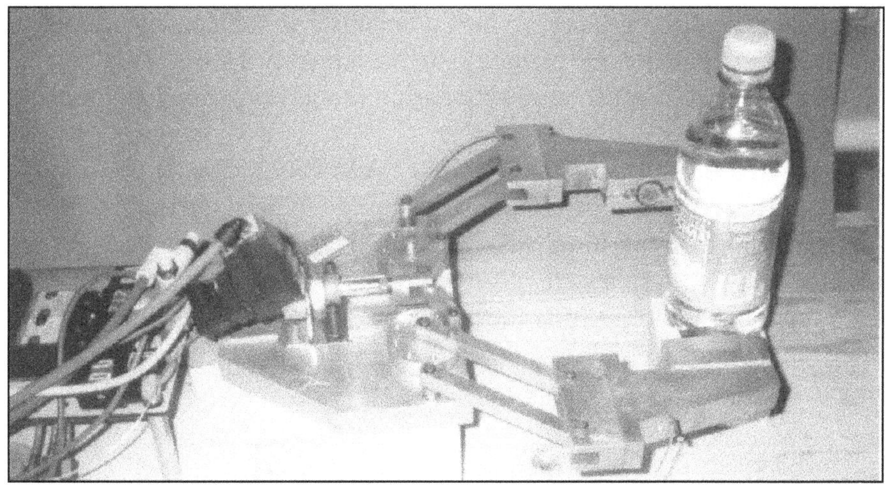

Fig. 4.37: A test-bed two-finger gripper at the Laboratory of Robotics and Mechatronics, in Cassino.

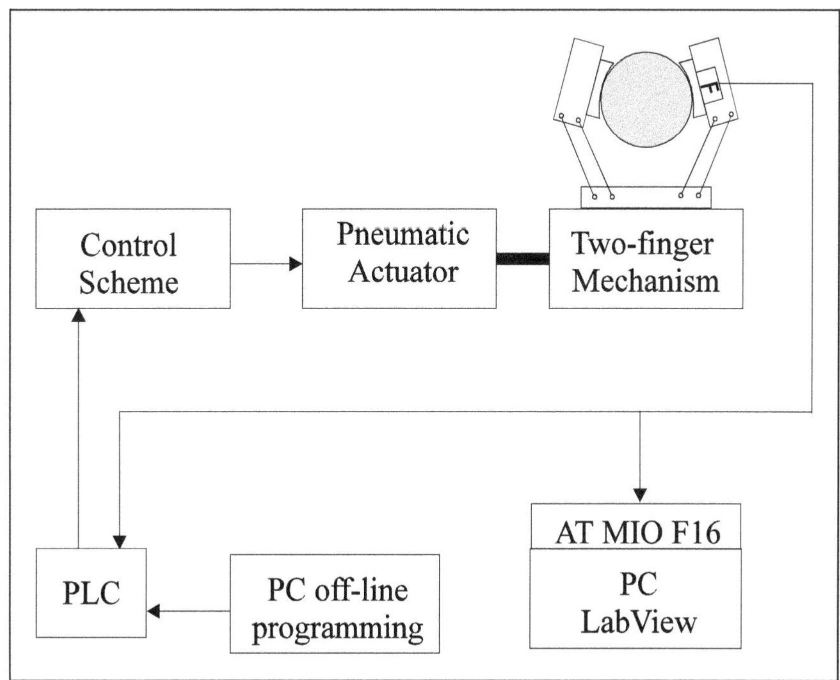

Fig. 4.38: A scheme for experimental set-up of a force controlled two-finger gripper.

Figure 4.40 shows some illustrative results that have been obtained in experimental tests when a small plastic bottle of water has been grasped, as also shown in Fig. 4.37. It is worthy of note in the plots of Fig. 4.40 that the grasping force increases slowly and with a suitably smooth path to the equilibrium value that has been calculated *a priori* and imposed through the PLC program. Using the choke-valve with a manual operation at a calibration set-up of the gripper, as depending also on the delicateness of the objects to be grasped, can regulate the velocity of the closure action. This regulation can be important in order to limit the impacts during the grasp phases, as described in section 4.4.

The stability of the static grasp in Fig. 4.40 a) can be verified by observing the response of the grasping force when external disturbance have been exerted in an impulsive form or for a short period of time, as shown in Fig. 4.40 b) and c), respectively. The disturbance action stimulates a change in the sensored grasping force in order to maintain the static equilibrium of the grasp, but the desired value of Fref is computed as the force acting on the object with a net value including the grasping component of the disturbance action.

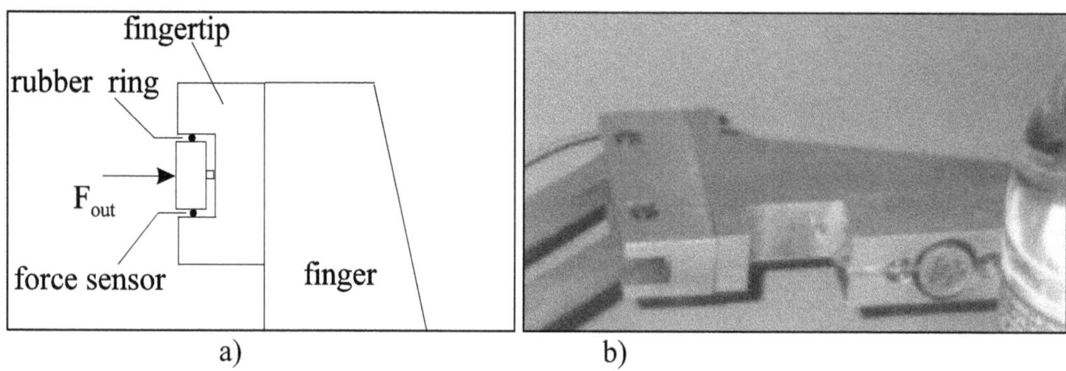

Fig. 4.39: Force sensored fingertip: a) a scheme; b) mechanical design of a prototype.

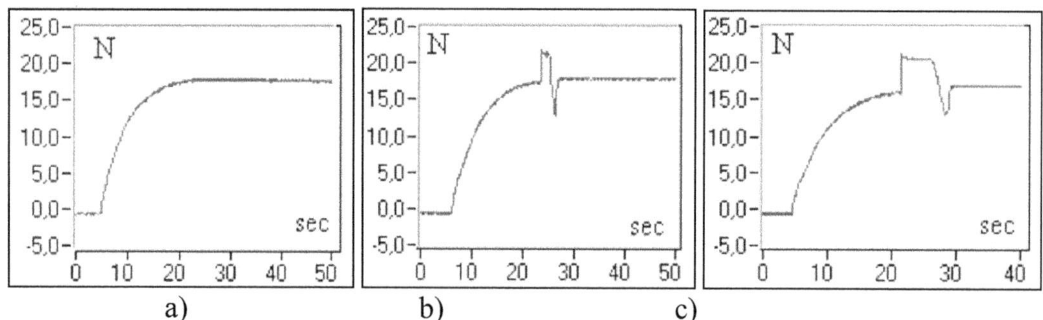

Fig. 4.40: Measured force at a fingertip of the gripper of Fig. 4.34 grasping a small bottle of water when acting with: a) no disturbance and static application; b) an impulsive external force; c) a continuous disturbance force due to motion of the gripper.

Thus, if the value of the measured grasping force at fingertips increases, the disturbance action is acting to decrease the grasping action on the object. Then, the value Fout decreases, and the disturbance action increases the grasping action on the object. It is of note from the plots of Fig. 4.40 b) and c) that the response of the force regulation is very quick and the waving of the measured force can be understood as the compliant behavior of the actuator because of the air flow, when the disturbance action disappears.

4.8.1.1 An acceleration sensored gripper

A two-finger gripper can be adjusted for advanced applications by using proper additional sensors, improved control scheme, and suitable mechanical design.

Figure 4.41 shows an illustrative example of how the test-bed gripper of Fig. 4.37 has been modified to obtain an acceleration sensored gripper. The corresponding block scheme is shown in Fig. 4.42 as a modification of the scheme of Fig. 4.38.

The aim of sensing the acceleration on the fingertips can be recognized in the possibility to regulate the grasping force as depending on the movement of the robotic arm, on which a gripper is installed. In addition it can also be used to control the movement of the robotic arm in order to limit the changes of the grasping force for ensuring a static grasp.

A control algorithm can be developed as an extension of the scheme of Fig. 4.36 and a resulting scheme is shown in Fig. 4.43. Results of an experimental test are shown in Fig. 4.44.

In particular, the mechanical design shown in Fig. 4.41 has been easily adjusted to install three accelerometers on one finger near the fingertip. The accelerometers have been installed to measure acceleration along the axis directions of a reference frame that is related to the contact line as X-axis, squeezing line as Y-axis, and slipping line as Z-axis, as shown in Fig. 4.42. The scheme of accelerometer positions in Fig. 4.42 shows the accelerometers Ax and Ay in the positions in Fig. 4.41. The accelerometer Az has been installed on the internal safe surface of the finger, as shown in Fig. 4.41, instead of in the external surface of the finger of the scheme in Fig. 4.42. The scheme of Fig. 4.42 shows how the signals from force sensors and accelerometers are used both for the control system through PLC, and monitoring of the grasp through LabView on a PC.

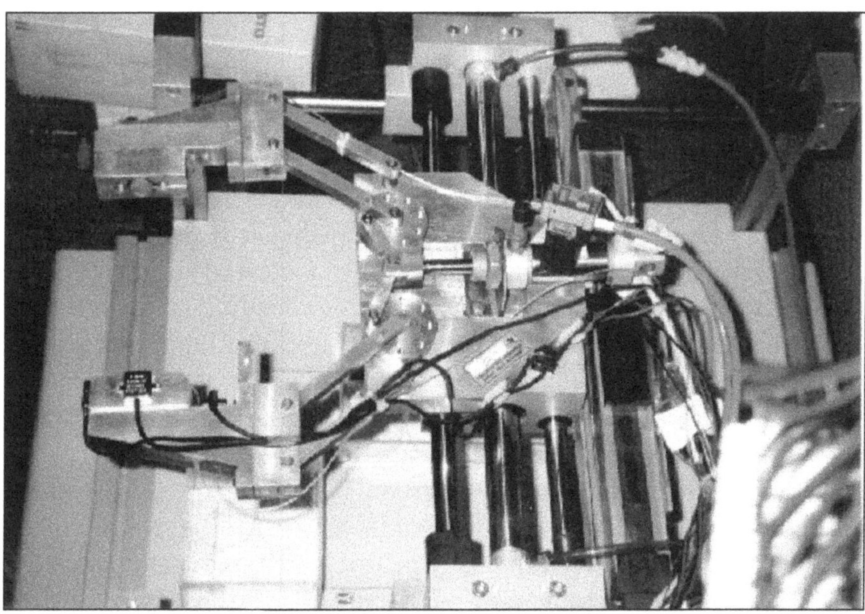

Fig. 4.41: The acceleration sensored two-finger test-bed gripper.

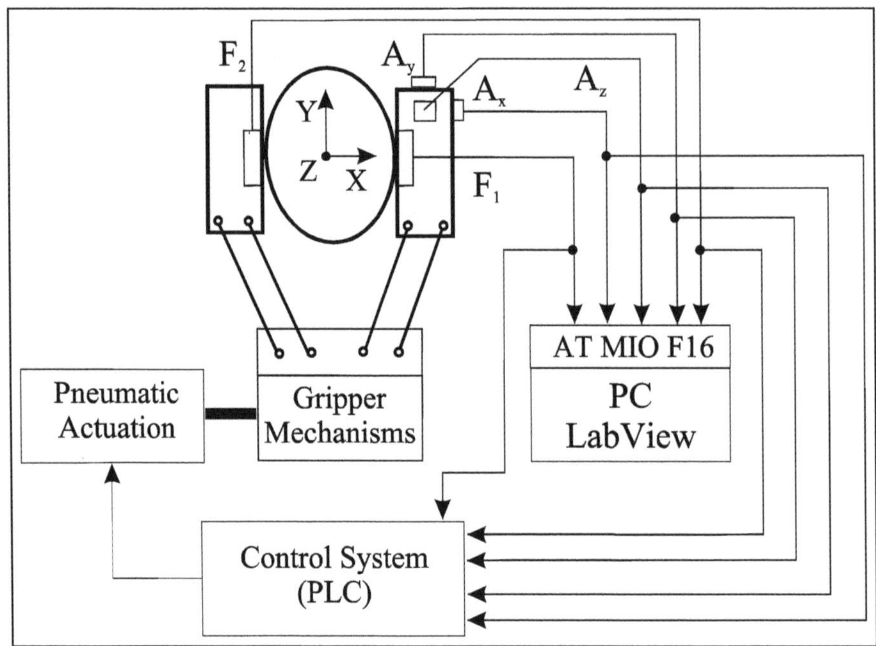

Fig. 4.42: A block scheme for the acceleration sensored two-finger gripper of Fig. 4.41.

The control system makes use of the sensor signals to run the control algorithm of Fig. 4.43. The control of grasping force error e is obtained through a PID (Proportional–Integrative–Derivative) algorithm that is characterized by the gains Kp, Ti, and Td. The scheme of the control algorithm has modified with respect to that in Fig. 4.36. The scheme with acceleration sensitivity takes into account the inertia effects of the motion by continuously computing the reference value Fref of the grasping force as a function of the acceleration of the motion, when a minimum value Fmin and a maximum value Fmax have been previously calculated. The minimum value Fmin can be obtained by considering that the inertia effects can reduce the action of the grasping force of the fingers and can even vanish it so that the static equilibrium of the grasp is not ensured. Therefore, Fmin can be calculated as the limit value for which the static equilibrium is still ensured. Thus the reference value Fref can be increased from Fmin. But the increase of the grasping force is limited by the object characteristics and mainly its resistance that gives the maximum value Fmax. These computations and limit evaluations for updating the grasping force are indicated in the block 'calculation of Fref' in Fig. 4.43. In some cases, it is not possible to adjust the grasping forces, and the only possible control consists of limiting the arm motion that is responsible for the inertia forces on the object. For example, this is the case when a person manipulates a fresh egg that requires a smooth arm motion in order to avoid a strengthening of the egg between the fingers.
In Fig. 4.43 the operation and efficiency of the advanced force controlled gripper with acceleration sensitivity are proved by the results of a test in which the gripper has been moved with high acceleration in a Cartesian planar manipulator, shown in Fig. 4.41.

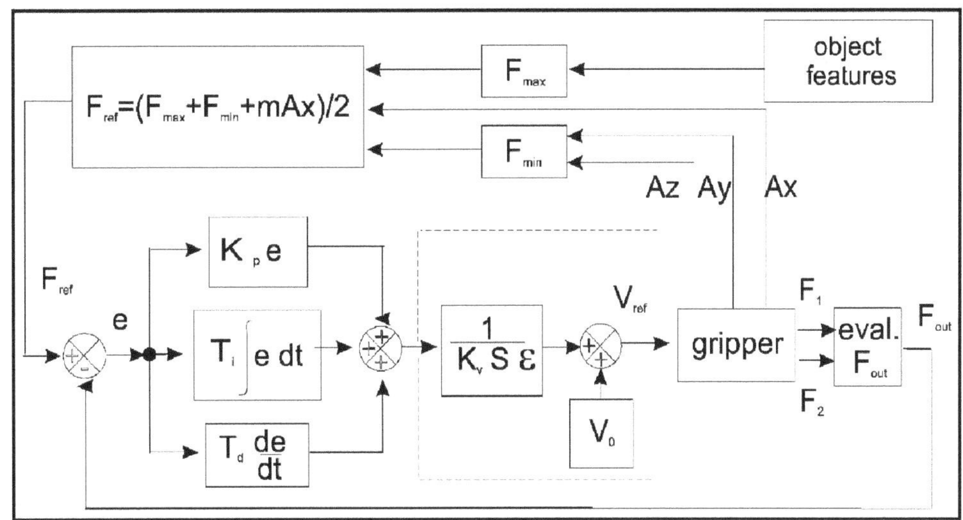

Fig. 4.43: A control scheme for the acceleration sensored two-finger gripper of Figs. 4.41 and 4.42.

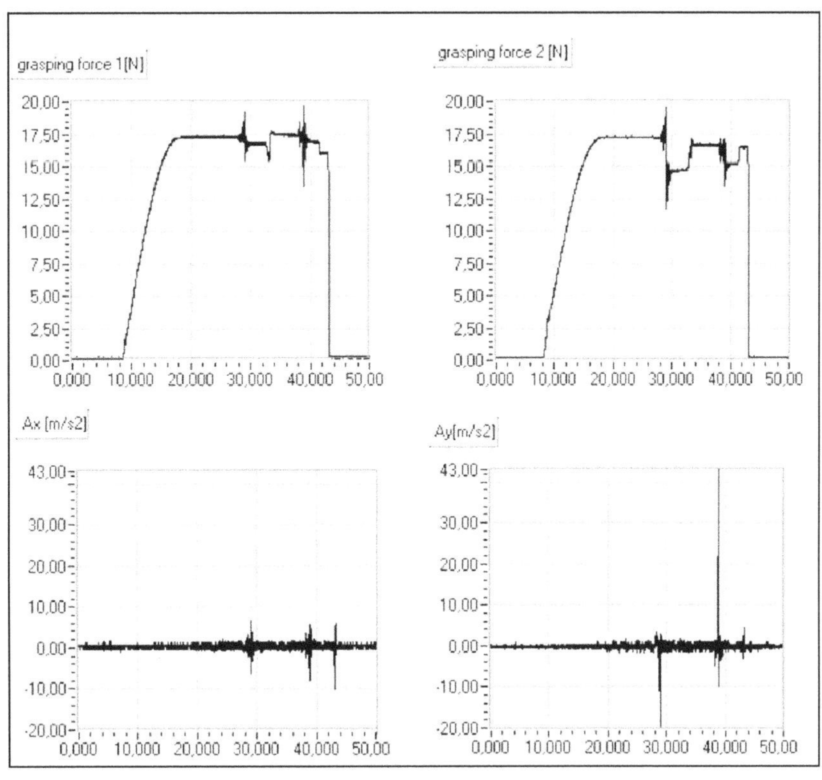

Fig. 4.44: Measured forces and accelerations at the fingertips of the force controlled two-finger gripper of Fig. 4.36 when it grasps a bottle of water of 5 N during an accelerated motion along the plane Y-Z.

The peaks of the measured accelerations in Fig. 4.44 correspond to a suitable regulation of the grasping forces by the two fingers as adjusted by the control algorithm of Fig. 4.43.
The application of advanced grippers with several sensors is limited in industrial applications because there are mainly practical problems for hardware and software installation.
The hardware problems are related to the installation of many sensors and their connections to a control unit that in addition should be dedicated specifically to the gripper operation.
The software problems are related to the implementation of suitable control programs that can require high performance in terms of signal elaboration and commands.
Nevertheless, as this example illustrates, there are not great complications in advancing the design and operation of two-finger grippers for severe grasps and manipulations.

4.9 Fundamentals on multi-finger grasp and articulated fingers

Multi-finger grasp for robotic devices has been developed as an extension of two-finger grasp but with the aim of mainly designing hands that have anthropomorphic architecture and capability. This requires attaching the design and operation problems for articulated fingers with human-like mechanical designs.
In this section fundamental concepts are overviewed to give outlines for design and operation of grasping devices with articulated fingers.
Several prototypes of human-like robotic hands have been designed and built in the last two decades and some are available on the market.
In Fig. 4.45 famous examples of human-like robotic hands are shown as referring to feasible designs with articulated fingers. The fingers are designed with anthropomorphic structures. The actuation is obtained usually by means of pulleys and ropes going to actuators that are installed far away from joints and the hand itself. Force sensors are used on the fingertips even with special solutions attempting tactile sensing that involves several features, such as for example the measure of force, displacement, slip, roughness, temperature, and pressure at the grasp area. The control equipment and hardware with high performances is aimed at regulating the action of several joints for the coordination of the fingers in advanced manipulative and grasping tasks.
Most of the robotic hands are usually used for research projects or specific applications. A special application can be advised in humanoid robots with the aim of replicating the human design and features.
Today, few robotic hands are applied in industrial environments and they are mainly of three-finger types.
In addition to human-like hands, grasping devices for three-dimensional applications are developed by extending the use of gripping mechanisms in multi-fingered non-anthropomorphic devices. In Fig. 4.46 two examples are shown in which one can recognize the basics of the design and features of gripping mechanisms as applied to a three-dimensional grasp in a three-dimensional gripper.

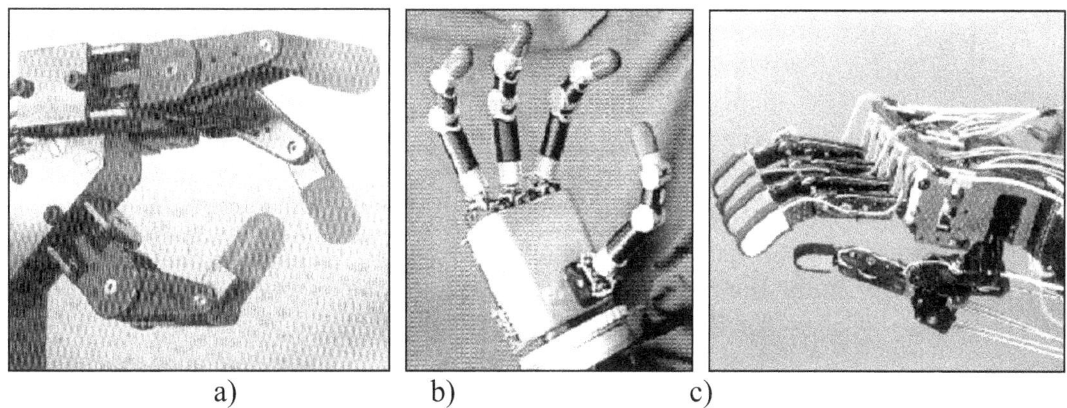

a) b) c)

Fig. 4.45: Examples of fingered anthropomorphic hands: a) Salisbury hand with three fingers; b) DRL Munich hand with four fingers; c) WAM-8R Waseda hand with five fingers. (Photos are taken from webpages).

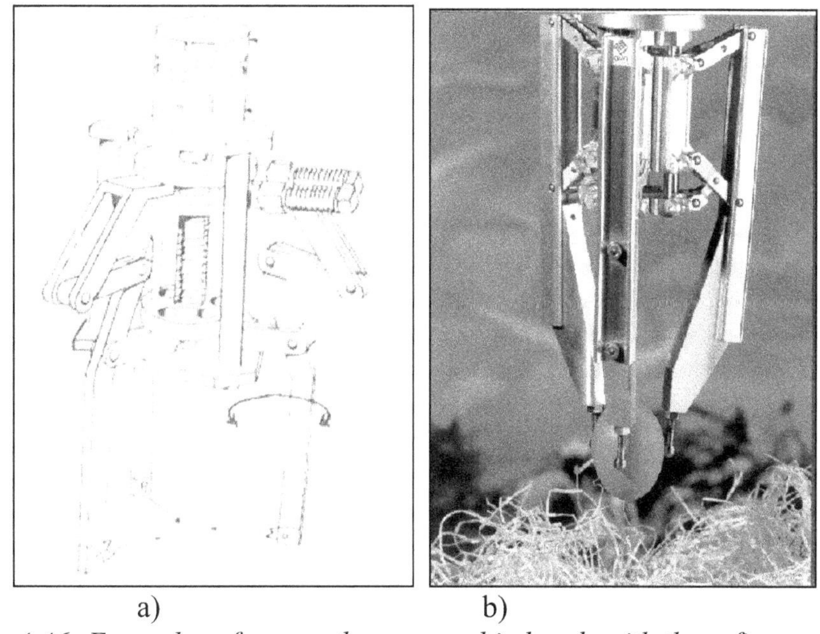

a) b)

Fig. 4.46: Examples of non-anthropomorphic hands with three fingers: a) a mechanical design; b) a prototype grasping en egg.

However, when attempting a three-dimensional grasp the aim can often be recognized in mimicking human hand operation. The basics of human grasp can be understood by looking at the mechanical design of a human hand by abstracting from the great complexity of its anatomy with bone structure, actuating muscles, sensing, and nervous systems.

In Fig. 4.47 the bone structures of a hand and a finger are shown in order to stress the variety of articulations that give a large grasping possibility so that the human hand can be considered as a universal gripper. In particular, in Fig. 4.47b) one can recognize a

scheme that permits the understanding of the 4 d.o.f.s for each finger and additional 1 d.o.f. as due to the palm connection. The scheme also shows the proportions of the phalanges that are usually also used in robotic designs.
Because of this architecture, a human hand has great versatility in grasping and manipulation. Basic movements of the fingers are shown in Fig. 4.48 as typical for many possibilities of grasp. Indeed, the study of the hand grasp is fundamental for developing suitable designs and also efficient grasping strategy.
The kinematic study of a human grasp can be attached as shown in Fig. 4.49 in which a kinematic model is determined by using suitable marks on finger articulations, Fig. 4.49a), to obtain the sketch of Fig. 4.49b). Then, in Fig. 4.49c) the motion sequence for approaching and grasping an object is recorded in order to evaluate both the grasping operation and finger motion. The scheme of Fig. 4.49b) refers to a planar grasp that is obtained by the thumb and index finger, whose motion is properly described through the corresponding joint angles θ_1, θ_2, and θ_3. Similar schemes can be determined for each of the many grasping configurations that a human hand can adapt to different objects.
The kinematics of a finger motion can also be studied as referring to a scheme of the finger as a planar 3R manipulator.

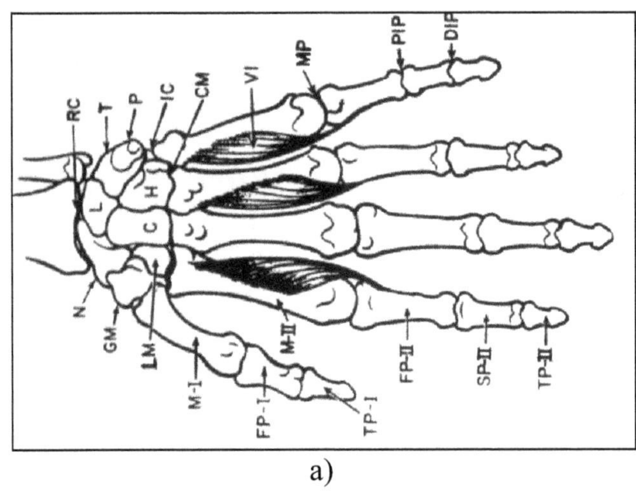

a)

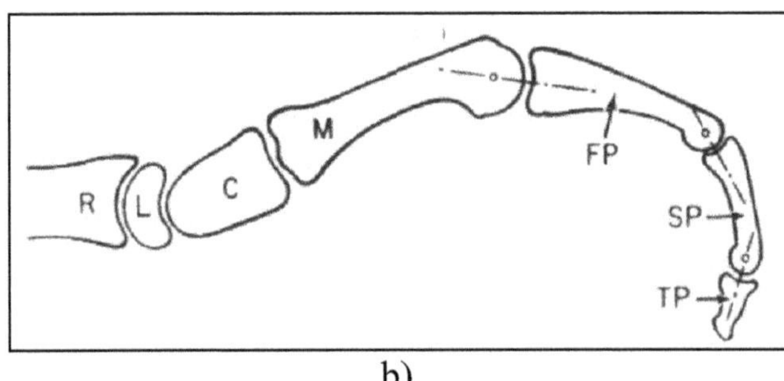

b)

Fig. 4.47: Bone architecture of human hands: a) the overall view; b) one finger with three phalanges FP, SP, TP, carp link B, and metacarpus link M.

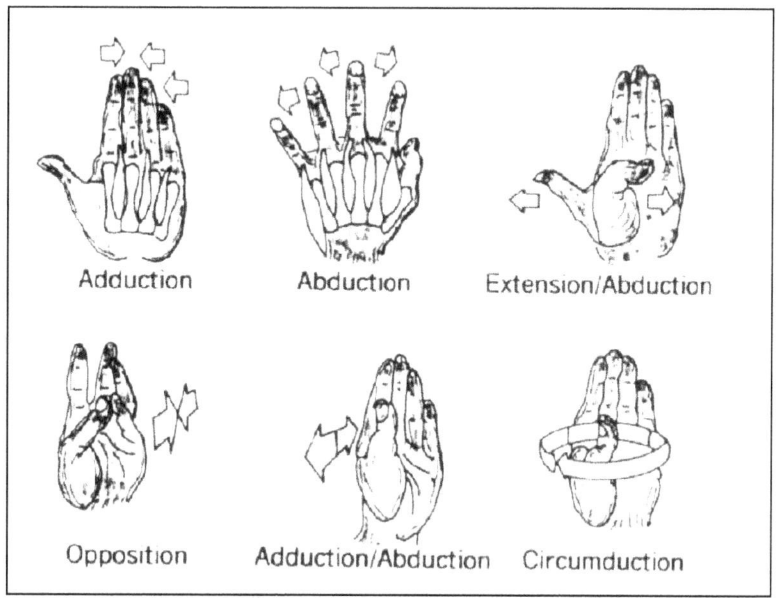

Fig. 4.48: Basic movements of the fingers in the human hand.

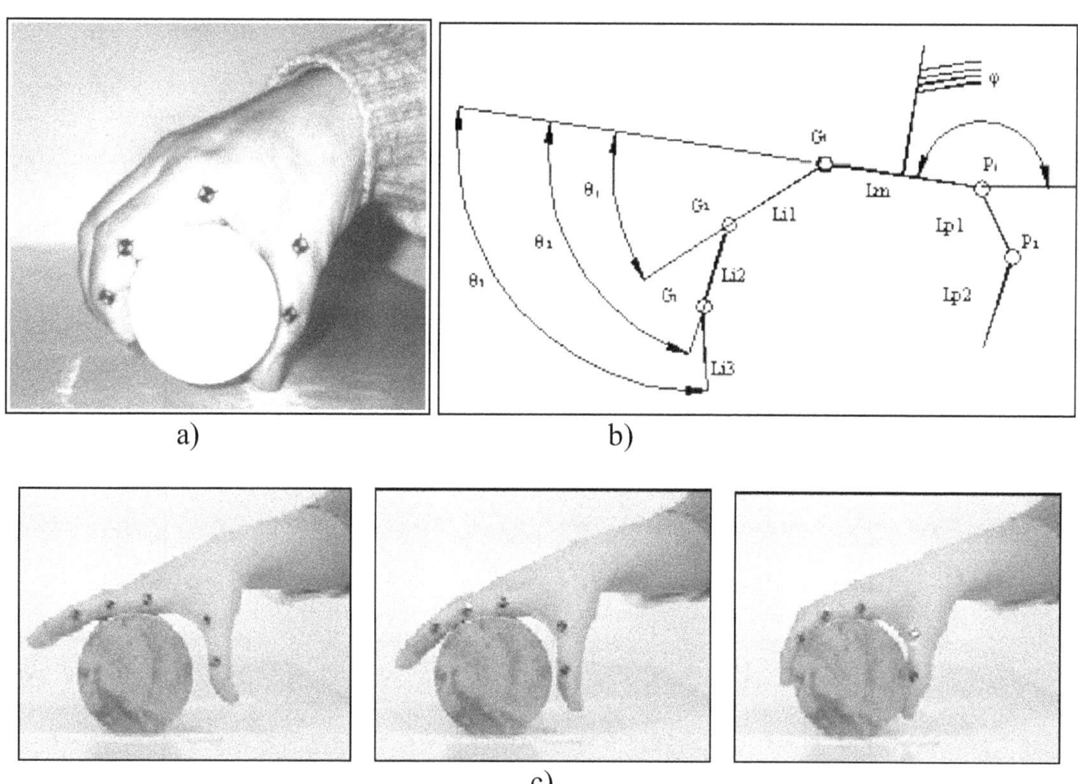

Fig. 4.49: An example of studying human grasp: a) at the grasp with marks on the finger articulations; b) a kinematic model; c) a sequence for approaching and grasping a cylindrical object.

Alternatively, planar mechanism can be used even to obtain 1 d.o.f. articulated fingers, as in the case of the diagram in Fig. 4.50. The 1 d.o.f. mechanism solution for fingers is of great interest since it permits the obtaining of simple low-cost designs, but it requires special attention for feasible operation. In particular the scheme of Fig. 4.50 can be used to obtain specific size of the linkages so that the phalanx bodies can move with a good approximation, as in a human finger. An example is shown in Fig. 4.51, in which the closing movement of the index finger is reproduced with a specific design of the linkage mechanism of Fig. 4.50.

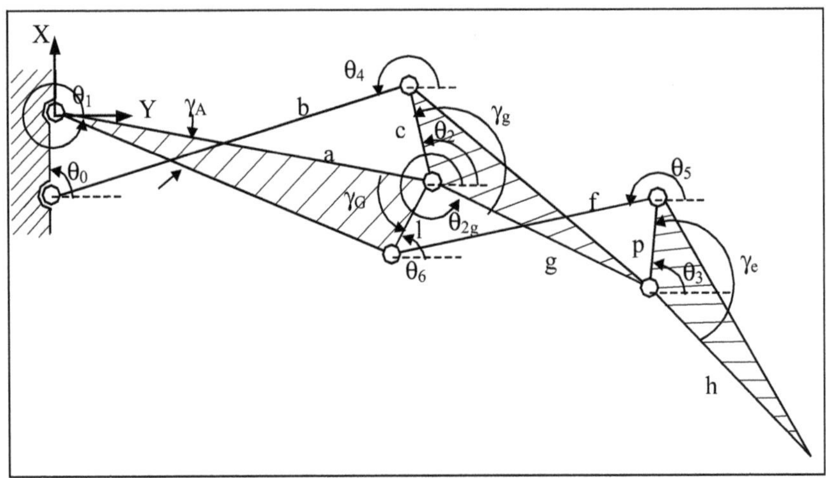

Fig. 4.50: A linkage mechanism for an articulated finger with 1 d.o.f.

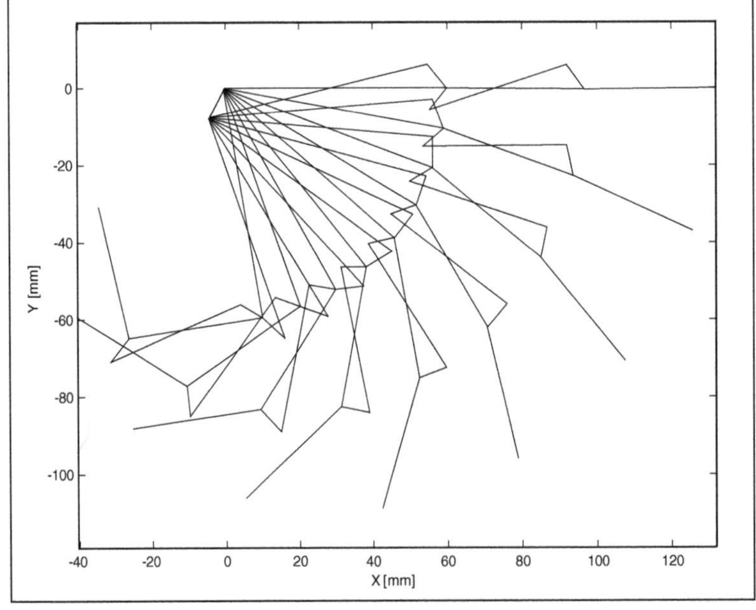

Fig. 4.51: A simulation of the finger mechanism of Fig. 4.50 in a closing motion.

Fig. 4.52: A low-cost prototype of an articulated finger with the 1 d.o.f. mechanism of Fig. 4.50, as built in Cassino.

Fig. 4.52 shows a prototype of a 1 d.o.f. articulated finger that has been built by using the mechanism layout of Fig. 4.50 with a specific dimensional synthesis to obtain a compact design in which the linkages are hidden in the finger body. The mechanism solution for an articulated finger of Fig. 4.50 can also be considered as convenient because it permits the application of features and requirements for design and operation performances that can be deduced similarly to those for grippers. An alternative design that is widely used, is shown in Fig. 4.53 in which the architecture of a planar 3R manipulator can be easily recognized. In the scheme the problem of multiple grasping contacts is also indicated by means of the determination of the contact points P_1, P_2, P_3. The action of a finger grasping an object through the forces F_1, F_2, F_3 at the contact points P_1, P_2, P_3 can be evaluated in a straightforward way by applying the Principle of Virtual work, when friction forces are assumed as negligible, in the form

$$\tau_1 \dot{\theta}_1 + \tau_2 \dot{\theta}_2 + \tau_3 \dot{\theta}_3 = F_1 v_1 + F_2 v_2 + F_3 v_3 \tag{4.9.1}$$

in which τ_i represents the actuating torque for each phalanx, dot θ_i is the angular velocity of each phalanx, and v_i is the velocity component of contact point P_i along $\mathbf{F}_i$ direction.

Looking at the static equilibrium of the grasped object can solve the problem of the distribution of contact forces. From a general viewpoint, at each contact point Pi one can mainly consider the action of:

- a grasping force $\mathbf{F}_i$ that is exerted by the finger link body;
- a friction force $\mathbf{F}_{fi}$ whose module can be evaluated as $\mu_i F_i$.

Thus, the static equilibrium of a grasped object with a multi-contact grasp can be formulated by the equilibrium of the forces in the form

$$\sum_1^N (\mathbf{F}_i + \mathbf{F}_{fi}) + \mathbf{W} + \mathbf{F}_e = 0 \tag{4.9.2}$$

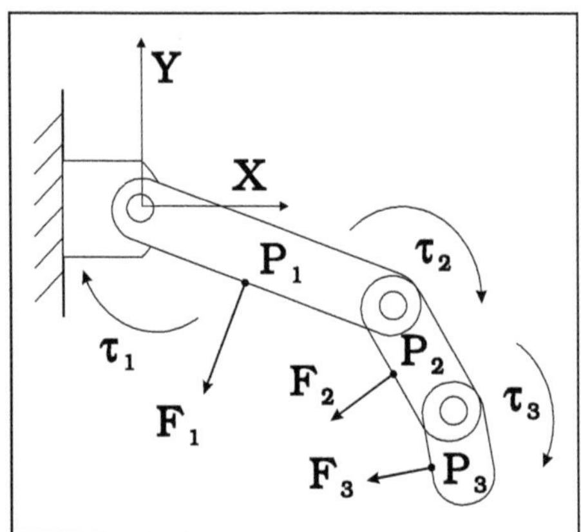

Fig. 4.53: A scheme for an articulated finger at the grasp.

and in terms of torque in the form

$$\sum_{1}^{N}(\mathbf{T}_i + \mathbf{T}_{fi}) + \mathbf{T}_W + \mathbf{T}_e = 0 \tag{4.9.3}$$

in which N is the number of grasp contacts; w is the weight of the object; Fe is the action of external forces that can also include inertia forces due to the hand motion; each force corresponds to a torque for the rotational equilibrium in Eq. (4.9.3).

In this general case, the determination of a frame of reference, such as the one for planar grasp is not obvious, although one can still identify contact line, squeezing line, and slipping line when the relative motion of the object among the fingers is analyzed.

It is of note that Eq. (4.9.2) and (4.9.3) cannot be solved in a straightforward way and several procedures are proposed in the specific literature depending on the assumed condition and application for the fingers and grasp task.

The statics problem of multiple contact for grasping can be formulated by considering the evaluation of both the contact positions and corresponding grasping forces.

In summary, similarly to the case of grippers, the design and operation of an articulated finger and multi-fingered hand can be approached through two alternative problems, namely a grasping force and size design problem, and a grasping configuration problem.

A grasping force and size design problem for an articulated finger can be defined as concerned with the determination of the grasping forces at given contact points, when the dimension of the finger is considered as a variable of the design problem.

A grasping configuration problem can be defined as concerned with the determination of contact points and their positions for given mechanical design and grasping force capability of a finger. Indeed, the first problem can be considered in the phase of designing an articulated finger for a robotic hand, while the latter problem is related to the practical operation of built fingers and hands.

Bibliography

Because of the general didactic character of the book, the bibliographic references are not cited in the presentation of arguments. A list of some significant works is proposed to allow a further reading of the matter from different points of view from that presented herein and for further information and methodology on Automation and Robotics. Few books are reported although a very rich literature of papers is published in Conference Proceedings and Journals, that are listed in Chapter 1.

Andeen G.B. (Ed.): "Robot Design Handbook-SRI International", McGraw-Hill, New York, 1988.
Angeles J.: "Fundamentals of Robotic Mechanical Systems", Springer-Verlag, NewYork, 1997.
Angulo J., Aviles R.: "Curso de Robotica", Paraninfo, Madrid, 1984.
Artobolevski I.I., Kobrinsi A.: "Les Robots", Mir Ed, Moscow, 1980.
Asada H., Slotine J.E.: "Robot Analysis and Control", Wiley, NewYork, 1986.
Balafoutis C.A., Patel R.V.: "Dynamic Analysis of Robot Manipulators: A Cartesian Tensor Approach", Kluwer, Dordrecht, 1991.
Beni G., Hackwood S. (Eds): "Recent Advances in Robotics", Wiley, New York, 1985.
Craig J.J.: "Introduction to Robotics: Mechanics and Control", Addison-Wesley, Reading, 1986 (2nd edition).
Crane C.D., Duffy J., "Kinematic Analysis of Robot Manipulators", Cambridge, 1998.
Dorf R.C., Nof S.Y.: "International Encyclopedia of Robotics", Wiley, 1988.
Duffy J.: "Analysis of Mechanisms and Robot Manipulators", Arnold, London, 1980.
Duffy J.: "Statics and Kinematics with Applications to Robotics", Cambridge University Press, Cambridge, 1996.
Featherstone R.: "Robot Dynamics Algorithms", Kluwer, Dordrecht, 1987.
Ferratè G.: "Robotica Industrial", Marcombo, Barcelona, 1986.
Fu K.S., Gonzales R.C., Lee C.S.G.: "Robotics", McGraw-Hill, NewYork,1987
Gupta K.C.: "Mechanics and Control of Robots", Springer-Verlag, NewYork, 1997.
Khalil W., Dombre E.: "Modeling, Identifications and Control of Robots", Hermes, London, 2002.
Klafter R.D., Chmielewski T.A., Negin M.: "Robotic Engineering: An Integrated Approach", Prentice-Hall, Englewood Cliffs, 1989.
Kobrinski A.A., Kobrinski A.E.: "Bras manipulateurs des robots", Mir Ed, Moscow, 1989. (French Edition).
Kozyrev Y.: "Industrial Robots Handbook", Mir Ed, Moscow, 1985.
Latombe J.-C.: "Robot Motion Planning", Kluwer, Dordrecht, 1991.
Lhote F., Kauffmann J., Taillard P.A., Taillard J.: "Robot Components and Systems", Kogan Page, London, 1984.
Legnani G.: "Robotrica Industriale", C.E.A., Milan, 2003.
Liegeois A.: "Les Robots: Analyse des Performances et C.A.O.", Hermes, Paris, 1985.

Mason M.T., Salisbury J.K.: "Robot Hands and the Mechanics of Manipulation", MIT Press, Cambridge, 1986.
Merlet J-P.: "Parallel Robots", Kluwer, Dordrecht, 2000.
Murray R.M., Li Z., Sastry S.S.: "A Mathematical Introduction to Robotic Manipulation", CRC Press, Boca Raton, 1994.
Nakamura Y.: "Advanced Robotics: Redundancy and Optimization", Addison-Wesley, Reading, 1991.
Nnaj B.O.: "Computer-Aided Design, Selection and Evaluation of Robots", Elsevier, Amsterdam, 1986.
Nof S.Y. (Ed.): "Handbook of Industrial Robotics", Wiley, New York, 1985.
Paul R.P.: "Robot Manipulators: Mathematics, Programming and Control", MIT Press, Cambridge,1981.
Pham D.T., Heginbotham W.B.: "Robot Grippers", IFS Publ., Bedford, 1986.
Popov E.P.: "Modern Robot Engineering", Mir Publ., Moscow, 1982.
Rivin E.I.: "Mechanical Design of Robots", McGraw-Hill, New York, 1985.
Rosheim M.E.: "Robot Evolution", Wiley, New York, 1994.
Ryan D.L., "Robotic Simulation", CRC Press, Boca Raton, 1994.
Schilling R.J.: "Fundamentals of Robotics: Analysis and Control", Prentice-Hall, Englewwood Cliffs, 1990.
Selig J.M.: "Geometrical Foundations of Robotics", Worls Scientific, London, 2000.
Song S.-M., Waldron K.J.: "Machines That Walk", MIT Press, Cambridge,1989.
Stadler W.: "Analytical Robotics and Mechatronics", McGraw-Hill, New York, 1995.
Todd D.J.: "Fundamentals of Robot Technology", Kogan Page, London, 1991.
Tsai L.W.: "Robot Analysis: The Mechanics of Serial and Parallel Manipulators", Wiley, New York, 1999.
Vukobratovic M.: "Introduction to Robotics", Springer-Verlag, Berlin, 1989
Yoshikawa T.: "Foundations of Robotics: Analysis and Control", MIT Press, 1990.
Xie M.: "Fundamentalks of Robotics", Worls Scientific, London, 2003.

Index

Biographical Notes

MARCO CECCARELLI

Marco Ceccarelli was born in Rome in 1958. He received the mechanical engineer degree cum laude in 1982 at the University "La Sapienza" of Rome. At the same University he completed a Ph.D. in Applied Mechanics in 1988. In 1987 he was visiting scholar at Stanford University, U.S.A., and in 1990 he received a CNR-NATO annual grant as visiting professor at the Technical University of Valencia, Spain. Since 1990 he teaches the courses "Mechanics of Machinery and Mechanisms" and "Mechanics of Robots" at the School of Engineering at the University of Cassino. Since 1996 he is Director of LARM, the Laboratory of Robotics and Mechatronics of DiMSAT, the Department of Mechanics, Structures, Environment and Territory at the University of Cassino. Since 2001 he has been appointed Full Professor of Mechanics of Machinery and Mechanisms at the University of Cassino.
He is member of ASME (The American Society of Mechanical Engineers), AIMETA (Italian Society for Theoretical and Applied Mechanics), SIRI (Italian Association of Robotics and Automation), IEEE (the Institute of Electrical and Electronics Engineers), FeIbIM (Iberoamerican Federation for Mechanical Engineering).
He is Chairman of the IFToMM Commission for History of Machine and Mechanism Science. He is also member of the IFToMM Commission for Robotics and member of the IFToMM Commission for Computational Kinematics. He is Chairman of the FeIbIM Commission for Mechatronics. He is Chairman of the Scientific Committee of MUSME, International Conference on Mechatronics and Multibody Systems. He is member of scientific Committees for several conferences; he is associate editor for Transactions of CSME (the Canadian Society of Mechanical Engineers), and he has served as reviewer for several international conferences and journals. He has given several invited lectures in many countries. He has been Scientific Editor for the Proceedings of HMM20000, International Symposium on History of Machines and

Mechanisms and Chairman for RAAD International Workshop on Robotics in Alpe-Adria Danube Region in 1997 and 2003.
In November 2003 he has received the Degree of Doctor Honoris Causa from UNI, National University of Lima, Perù, as recognizing his academic and scientific career, and his support to the activity of UNI in Mechatronic Engineering. He has been elected Secretary-General of IFToMM, International Federation for the Promotion of Mechanism and Machine Science, for the term 2004-2007.
His research interests cover aspects of Theory of Machines and Mechanisms (TMM) and Mechanics of Robots. Specific subjects of his interest are Analysis and Design of Workspace and Manipulation; Mechanical Design of Manipulators and Grippers; Mechanics of Grasp; History of TMM; and Mechanism Design.
He is author or co-author of more than two hundreds papers, which have been presented at Conferences or published in national and international journals.

More information at the web page
http://webuser.unicas.it/weblarm/larmindex.htm

prof. Marco Ceccarelli
Laboratory of Robotics and Mechatronics
DiMSAT – University of Cassino
Via Di Biasio 43, 03043 Cassino (Fr), Italy
Phone +39-0776-2993663/3652 Fax +39-0776-2993711
e-mail: ceccarelli@ing.unicas.it

International Series on

MICROPROCESSOR-BASED AND INTELLIGENT SYSTEMS ENGINEERING

Editor: Professor S. G. Tzafestas, *National Technical University, Athens, Greece*

---

1. S.G. Tzafestas (ed.): *Microprocessors in Signal Processing, Measurement and Control.* 1983 ISBN 90-277-1497-5
2. G. Conte and D. Del Corso (eds.): *Multi-Microprocessor Systems for Real-Time Applications.* 1985 ISBN 90-277-2054-1
3. C.J. Georgopoulos: *Interface Fundamentals in Microprocessor-Controlled Systems.* 1985 ISBN 90-277-2127-0
4. N.K. Sinha (ed.): *Microprocessor-Based Control Systems.* 1986 ISBN 90-277-2287-0
5. S.G. Tzafestas and J.K. Pal (eds.): *Real Time Microcomputer Control of Industrial Processes.* 1990 ISBN 0-7923-0779-8
6. S.G. Tzafestas (ed.): *Microprocessors in Robotic and Manufacturing Systems.* 1991 ISBN 0-7923-0780-1
7. N.K. Sinha and G.P. Rao (eds.): *Identification of Continuous-Time Systems.* Methodology and Computer Implementation. 1991 ISBN 0-7923-1336-4
8. G.A. Perdikaris: *Computer Controlled Systems.* Theory and Applications. 1991 ISBN 0-7923-1422-0
9. S.G. Tzafestas (ed.): *Engineering Systems with Intelligence.* Concepts, Tools and Applications. 1991 ISBN 0-7923-1500-6
10. S.G. Tzafestas (ed.): *Robotic Systems.* Advanced Techniques and Applications. 1992 ISBN 0-7923-1749-1
11. S.G. Tzafestas and A.N. Venetsanopoulos (eds.): *Fuzzy Reasoning in Information, Decision and Control Systems.* 1994 ISBN 0-7923-2643-1
12. A.D. Pouliezos and G.S. Stavrakakis: *Real Time Fault Monitoring of Industrial Processes.* 1994 ISBN 0-7923-2737-3
13. S.H. Kim: *Learning and Coordination.* Enhancing Agent Performance through Distributed Decision Making. 1994 ISBN 0-7923-3046-3
14. S.G. Tzafestas and H.B. Verbruggen (eds.): *Artificial Intelligence in Industrial Decision Making, Control and Automation.* 1995 ISBN 0-7923-3320-9
15. Y.-H. Song, A. Johns and R. Aggarwal: *Computational Intelligence Applications to Power Systems.* 1996 ISBN 0-7923-4075-2
16. S.G. Tzafestas (ed.): *Methods and Applications of Intelligent Control.* 1997 ISBN 0-7923-4624-6
17. L.I. Slutski: *Remote Manipulation Systems.* Quality Evaluation and Improvement. 1998 ISBN 0-7932-4822-2
18. S.G. Tzafestas (ed.): *Advances in Intelligent Autonomous Systems.* 1999 ISBN 0-7932-5580-6
19. M. Teshnehlab and K. Watanabe: *Intelligent Control Based on Flexible Neural Networks.* 1999 ISBN 0-7923-5683-7
20. Y.-H. Song (ed.): *Modern Optimisation Techniques in Power Systems.* 1999 ISBN 0-7923-5697-7
21. S.G. Tzafestas (ed.): *Advances in Intelligent Systems.* Concepts, Tools and Applications. 2000 ISBN 0-7923-5966-6
22. S.G. Tzafestas (ed.): *Computational Intelligence in Systems and Control Design and Applications.* 2000 ISBN 0-7923-5993-3
23. J. Harris: *An Introduction to Fuzzy Logic Applications.* 2000 ISBN 0-7923-6325-6

International Series on

MICROPROCESSOR-BASED AND INTELLIGENT SYSTEMS ENGINEERING

24. J.A. Fernández and J. González: *Multi-Hierarchical Representation of Large-Scale Space*. 2001 ISBN 1-4020-0105-3
25. D. Katic and M. Vukobratovic: *Intelligent Control of Robotic Systems*. 2003 ISBN 1-4020-1630-1
26. M. Vukobratovic, V. Potkonjak and V. Matijevic: *Dynamics of Robots with Contact Tasks*. 2003 ISBN 1-4020-1809-6
27. M. Ceccarelli: *Fundamentals of Mechanics of Robotic Manipulation*. 2004 ISBN 1-4020-1810-X

KLUWER ACADEMIC PUBLISHERS – DORDRECHT / BOSTON / LONDON